New Engineering Mathematics

Volume I

New Engineering Mathematics

Volume I

A Chandra Babu

C R Seshan

Alpha Science International Ltd.

Oxford, U.K.

A Chandra Babu
C R Seshan
Department of Mathematics
The American College
Madurai, India

Alpha Science International Ltd.
7200 The Quorum, Oxford Business Park North
Garsington Road, Oxford OX4 2JZ, U.K.

Printed from the camera-ready copy provided by the Authors

ISBN 1-84265-291-5

Printed in India

PREFACE

We have great pleasure in bringing out **New Engineering Mathematics** in three volumes.

Each Chapter begins with an introduction, the relevant definitions and concepts. A detailed discussion of the theory, inclusion of a large number of model problems and exercises with separate short and long answer questions are special features of these volumes.

Volume I contains 292 short answer questions and 726 long answer questions, of which 231 are solved as examples. Volume II contains 467 short answer questions and 832 long answer questions, of which 295 are solved as examples, to provide enough practice and to generate confidence in the subject. Volume III contains 206 short answer questions and 619 long answer questions, of which 184 are solved as examples, to provide enough practice and initiate further study.

A close reading of the theory portions is sufficient to answer all the questions in Part-A of the exercises. Further a student who works out the examples given under each topic will be able to solve all the problems in the exercises. The example problems are chosen with great care so that they cover necessary information and techniques for problem solving.

We hope this book will be useful to students of all levels. Critical evaluation and suggestions for improvement from students and teachers will be thankfully acknowledged and adapted in future editions.

We wish to express our sincere thanks to the post graduate students of American College Mr. Hasan Mohammed K. and Mr. Elango Valan S., who took great care in typing the manuscript. We are also thankful to Narosa Publishing House, who have brought out this book in a very short span of time. We have great pleasure to dedicate this book to our parents and teachers.

A Chandra Babu
C R Seshan

CONTENTS

Preface   *v*

VOLUME I

1. MATRICES

   1.0 Introduction   1.1
   1.1 Rank of Matrix   1.5
   1.2 The Characteristic Equation   1.18
   1.3 Cayley-Hamilton Theorem   1.33
   1.4 Reduction of a Quadratic Form to a Canonical Form   1.47

2. THREE DIMENSIONAL ANALYTICAL GEOMETRY

   2.0 Introduction   2.1
   2.1 Direction Cosines and Ratios of a Line   2.4
   2.2 The Plane   2.20
   2.3 The Straight Line   2.39
   2.4 The Sphere   2.62

3. GEOMETRICAL APPLICATIONS OF DIFFERENTIAL CALCULUS

   3.0 Introduction   3.1
   3.1 Curvature and Radius of Curvature   3.7
   3.2 Evolutes and Involutes   3.23
   3.3 Envelopes   3.33

4. FUNCTIONS OF SEVERAL VARIABLES

   4.0 Introduction   4.1
   4.1 Partial Derivatives   4.2
   4.2 Total Derivatives   4.11
   4.3 Taylor's Expansion of Functions of Two Variables   4.22
   4.4 Maxima and Minima of Functions of Two Variables   4.32

4.5 Jacobians   4.52

4.6 Differentiation Under the Integral Sign   4.61

5. **ORDINARY DIFFERENTIAL EQUATIONS**

5.0 Introduction   5.1

5.1 Linear Differential Equations with Constant Coefficients   5.5

5.2 Simultaneous Linear Differential Equations with Constant Coefficients   5.23

5.3 Euler's Homogeneous Linear Differential Equations   5.35

5.4 Linear Differential Equations with Variable Coefficients   5.46

5.5 Method of Variation of Parameters   5.66

Index   **I.1**

VOLUME II

1. **MULTIPLE INTEGRALS**

1.0 Introduction   1.1

1.1 Double Integrals   1.1

1.2 Change of Order of Integration   1.22

1.3 Triple Integrals   1.28

1.4 Change of Variables   1.38

2. **VECTOR CALCULUS**

2.0 Introduction   2.1

2.1 Gradient, Divergence and Curl   2.7

2.2 Vector Integration   2.29

2.3 Integral Theorems of Green, Gauss and Stoke   2.49

2.4 Evaluation of Integrals Using Integral Theorems   2.67

3. **ANALYTIC FUNCTIONS**

3.0 Introduction   3.1

3.1 Cauchy-Riemann Equations   3.8

3.2 Properties of Analytic Functions   3.23

3.3 Conformal Mappings   3.43

3.4 Bilinear Transformations   3.71

4. **COMPLEX INTEGRATION**

4.0 Introduction   4.1

4.1 Cauchy's Theorem and Applications   4.9

4.2 Taylor and Laurent Expansions   4.35
4.3 Singularities and Residues   4.57
4.4 Contour Integration   4.77

5. LAPLACE TRANSFORMS

5.0 Introduction   5.1
5.1 Laplace Transforms of Elementary Functions   5.1
5.2 Laplace Transforms of Derivatives and Integrals   5.18
5.3 Inverse Transforms   5.23
5.4 Convolution Theorem and Transforms of Periodic Functions   5.38
5.5 Application to Differential Equations   5.53

Index   II.1

VOLUME III

1. PARTIAL DIFFERENTIAL EQUATIONS

1.0 Introduction   1.1
1.1 Formation of Partial Differential Equations   1.2
1.2 Solutions of First Order Partial Differential Equations   1.11
1.3 Lagrange's Linear Equations   1.26
1.4 Linear Partial Differential Equations of Second and Higher Order   1.40

2. FOURIER SERIES

2.0 Introduction   2.1
2.1 General Fourier Series   2.2
2.2 Odd and Even Functions   2.21
2.3 Half Range Series   2.32
2.4 Complex Form of Fourier Series   2.43
2.5 Harmonic Analysis   2.51

3. BOUNDARY VALUE PROBLEMS

3.0 Introduction   3.1
3.1 Classification of Second Order Linear Partial Differential Equations 3.1
3.2 One Dimensional Wave Equation   3.4
3.3 One Dimensional Heat Equation   3.24
3.4 Two Dimensional Heat Equation   3.42

4. FOURIER TRANSFORMS

4.0 Introduction   4.1

4.1 Fourier Transforms   4.2

4.2 Properties of Fourier Transforms and Convolution Theorem   4.22

4.3 Finite Fourier Transforms   4.40

5. Z-TRANSFORMS AND DIFFERENCE EQUATIONS

5.0 Introduction   5.1

5.1 Z-Transforms and Elementary Properties   5.1

5.2 Inverse Z-Transform   5.17

5.3 Convolution Theorem   5.29

5.4 Solution of Difference Equations Using Z-Transforms   5.34

Index   **III.1**

CHAPTER 1

MATRICES

1.0   INTRODUCTION

We frequently come across matrices while representing data in real life. Matrices are widely used in many branches of applied mathematics and several other areas in science, humanities, sociology and management.

If there are four cities connected by rail or road, the distance or fare between any two cities can be represented by a 4 x 4 matrix. The prices of ten commodities during the last five years can be represented by a 10 x 5 matrix. Similarly while assigning three jobs to five persons, the cost or profit of doing any job by any person can be represented by a 3 x 5 matrix. Thus to represent data in two dimensions, always matrices are used.

Definition 1.0.1 An m x n matrix over real numbers is a set of mn real numbers arranged in a rectangular form along m rows and n columns and bounded by the brackets [ ] or ( ).

$$A = \begin{bmatrix} a_{11} & a_{12} & \cdots & a_{1j} & \cdots & a_{1n} \\ a_{21} & a_{22} & \cdots & a_{2j} & \cdots & a_{2n} \\ \cdots & \cdots & \cdots & \cdots & \cdots & \cdots \\ a_{i1} & a_{i2} & \cdots & a_{ij} & \cdots & a_{in} \\ \cdots & \cdots & \cdots & \cdots & \cdots & \cdots \\ a_{m1} & a_{m2} & \cdots & a_{mj} & \cdots & a_{mn} \end{bmatrix}$$

is an m x n matrix or a matrix of **order** m x n. The element in the i-th row and j-th column of A is a_{ij} and we write $A = [a_{ij}]$. When a_{ij} are complex numbers, A is said to be a matrix over complex numbers. In the following section we define different types of matrices.

1.0.1   TYPES OF MATRICES

Definition 1.0.2 An 1 x n matrix is called a **row matrix** or a **row vector** of dimension n. An m x 1 matrix is called a **column matrix** or a **column vector** of dimension m.

A row matrix has a single row. A column matrix has a single column.

Definition 1.0.3 An n x n matrix is called **a square matrix** of order n. It has n rows and n columns.

Definition 1.0.4 The elements a_{11}, a_{22} ... a_{nn} form the **leading** or the **principal diagonal** of the square matrix A = $[a_{ij}]$. The sum $a_{11} + a_{22} + ... + a_{nn}$ is called the **trace** of the square matrix A = $[a_{ij}]$.

Definition 1.0.5 The determinant having the same elements a_{ij} as in the square matrix A = $[a_{ij}]$ is called the **determinant of the matrix A** and is denoted by $|A|$ or $|a_{ij}|$ or det A.

Definition 1.0.6 A square matrix is said to be **singular** if its determinant is equal to zero. A square matrix is said to be **non-singular** if its determinant is not equal to zero.

Definition 1.0.7 A square matrix of order n is called a **diagonal matrix** of order n if all the elements except those in the leading diagonal are equal to zero. i.e., $a_{ij} = 0$ for $i \neq j$.

Definition 1.0.8 A diagonal matrix in which all the leading diagonal elements are equal is called a **scalar matrix**. A diagonal matrix of order n which has 1 for all its diagonal elements is called an **identity matrix** of order n and is denoted by I_n or simply I. If all the elements of an m x n matrix are equal to zero, it is called a **null matrix or zero matrix** and is denoted as O_{mn} or O. O_{m1} and O_{1n} are called **null vectors**.

Definition 1.0.9 A square matrix A = $[a_{ij}]$ all of whose elements below the leading diagonal are zero is called an **upper triangular matrix**. i.e., $a_{ij} = 0$ for $i > j$. A square matrix A = $[a_{ij}]$ all of whose elements above the leading diagonal are zero is called a **lower triangular matrix**. i.e., $a_{ij} = 0$ for $i < j$. A matrix that is either upper triangular or lower triangular is called a **triangular matrix.**

Definition 1.0.10 A square matrix A = $[a_{ij}]$ is said to be **symmetric** if $a_{ij} = a_{ji}$ for all i and j. A is said to be **skew-symmetric** if $a_{ij} = -a_{ji}$ for all i and j. In a skew-symmetric matrix all the leading diagonal elements are equal to zero.

Definition 1.0.11 Let A = $[a_{ij}]$ be an m x n matrix. If we select r rows and r columns from $|a_{ij}|$, deleting the remaining rows and columns, we get a determinant of order r called a **minor of A of order r.**

1.0.2 OPERATIONS ON MATRICES

1. Equality of Matrices

Definition 1.0.12 Two matrices A = $[a_{ij}]$ and B = $[b_{ij}]$ are said to be equal if
(i)   A and B have the same order m x n and

(ii) $a_{ij} = b_{ij}$ for all i and j.

When A and B are equal, we write A = B

2. Addition of Matrices

Definition 1.0.13 If A and B are two matrices of the same order, their sum A + B is obtained by adding the corresponding elements in A and B.

i.e., A + B = $[a_{ij} + b_{ij}]$. Similarly A - B = $[a_{ij} - b_{ij}]$.

Note:
(i)   Addition of matrices is commutative. i.e., A + B = B + A.
(ii)   Addition of matrices is associative. i.e., (A + B) +C = A + (B + C).
(iii)   A + 0 = A
(iv)   A + (-A) = 0

3. Scalar Multiplication

Definition 1.0.14 The scalar multiplication or the scalar product of a matrix A = $[a_{ij}]$ by a scalar k is the matrix kA obtained by multiplying the elements of A by k. i.e., kA = $[ka_{ij}]$

Note:
(i)   Scalar multiplication is distributive over addition. i.e., k(A + B) = kA + kB.
(ii)   0A = 0, the zero matrix.

4. Multiplication of Matrices

We have seen that two matrices can be added only if they have the same order. Similarly two matrices can be multiplied only if they satisfy certain conditions. If A and B are two matrices such that the number of columns in A is equal to the number of rows in B then A and B are said to be **conformable**. Two conformable matrices alone can be multiplied.

Definition 1.0.15 Let A = $[a_{ij}]$ be an m x n matrix and B = $[b_{ij}]$ be an n x p matrix. Then their product is the m x p matrix AB = $[C_{ij}]$ where

$$C_{ij} = a_{i1}b_{1j} + a_{i2}b_{2j} + a_{i3}b_{3j} + \ldots + a_{in}b_{nj} = \sum_{k=1}^{n} a_{ik}b_{kj} \quad 1 \leq i \leq m, \, 1 \leq j \leq p$$

C_{ij} is called the **inner product** of the ith row of A with the jth column of B.

Note:
(i)   Let A be an m x n matrix, then I_m A = A = A I_n
(ii)   AB = 0 does not imply that A = 0 or B = 0.
(iii)   Multiplication of matrices is not commutative. i.e., AB need not be equal to BA.
(iv)   Multiplication of matrices is associative. i.e., (AB)C = A(BC).
(v)   Multiplication of matrices is distributive over addition.
         i.e., A(B + C)=AB + AC

and   (A + B)C=AC + BC

1.0.3   POWERS OF A MATRIX

Definition 1.0.16 Let A be a square matrix. Then the powers of A are defined as
$$A^1 = A;$$
$$A^n = A^{n-1}A, \quad \text{for } n > 1$$

Definition 1.0.17 A square matrix A is called **idempotent** if $A^2 = A$.

Definition 1.0.18 A square matrix A is called **involutary** if $A^2 = I$.

Definition 1.0.19 A square matrix A is called **nilpotent** if $A^m = 0$ for some m.

1.0.4   TRANSPOSE OF A MATRIX

Definition 1.0.20 Let A be an m x n matrix. The **transpose** of A denoted by A^T or A'is the matrix obtained from A by interchanging its rows and columns. Thus, if $A = [a_{ij}]$, then $A^T = [b_{ij}]$ where $b_{ij} = a_{ji}$ for all i and j.

If A is an m x n matrix, then A^T is an n x m matrix.

Note:

(i) $(A^T)^T = A$

(ii) $(A + B)^T = A^T + B^T$

(iii) $(k\,A)^T = k\,A^T$

(iv) $(AB)^T = B^T A^T$

1.0.5   INVERSE OF A MATRIX

Definition 1.0.21 Let A be square matrices of order n. Then a square matrix B of order n is said to be the inverse of A if $AB = BA = I_n$.

Note:

(i)   Every square matrix need not have an inverse. If A has an inverse, A is said to be **invertible**

(ii)   The inverse of A is denoted by A^{-1}. i.e., $AA^{-1} = A^{-1}A = I_n$

(iii)   If A is an orthogonal matrix, then its inverse is A^T.

(iv)   If A has an inverse, there exists a matrix B such that $AB = BA = I_n$.
        i.e., $|AB| = |A|.|B| = |I_n| = 1$. Therefore $|A| \neq 0$ and A is non-singular.

(v) $(AB)^{-1} = B^{-1}A^{-1}$.

(vi) $(A^{-1})^{-1} = A$.

(vii) $(A^{-1})^T = (A^T)^{-1}$.

1.0.6   ORTHOGONAL MATRIX

Definition 1.0.22 A real square matrix A of order n is said to be **orthogonal** if $AA^T = A^T A = I_n$

Note:

(i)   If A is orthogonal, $A^{-1} = A^T$.

(ii)   Two column vectors $X = \begin{bmatrix} x_1 \\ x_2 \\ \vdots \\ x_n \end{bmatrix}$ and $Y = \begin{bmatrix} y_1 \\ y_2 \\ \vdots \\ y_n \end{bmatrix}$ are said to be **orthogonal**, if their

inner product $x_1y_1 + x_2y_2 + \ldots + x_ny_n = 0$ i.e., if $X^TY = 0$.

1.0.7   ADJOINT OF A SQUARE MATRIX

Definition 1.0.23 Let $A = [a_{ij}]$ be a square matrix of order n. The **minor** M_{ij} of an element a_{ij} is the determinant obtained by deleting the ith row and the jth column from $|a_{ij}|$. Thus the minor of a_{ij} is a determinant of order n-1. The **cofactor** of an element a_{ij} in the matrix A is $C_{ij} = (-1)^{i+j} M_{ij}$ where M_{ij} is the minor of a_{ij}. Let C be the matrix formed by the cofactors. i.e., $C = [C_{ij}]$. Then the transpose of C i.e., C^T is called the **adjoint of the matrix** A. Thus the adjoint of A is the transpose of the matrix of cofactors of A and is denoted by Adj A or adj A.

Note:

(i) $A.(Adj\ A) = (Adj\ A).A = |A|.I_n$

(ii) $A^{-1} = \dfrac{AdjA}{|A|}$ if $|A| \neq 0$

(iii)   A has an inverse if and only if A is non singular.

(iv)   Inverse of A is unique.

(v)   Inverse of the identity matrix I_n is the matrix I_n itself.

1.0.8   SIMILAR MATRICES

Definition 1.0.24 Two square matrices A and B are said to be **similar** if there exists a non-singular matrix P such that $B = P^{-1}AP$.

Definition 1.0.25 The transformation of a matrix A to a matrix B by a non-singular matrix P such that $B = P^{-1}AP$ is called a **similarity transformation**.

1.1   RANK OF A MATRIX

With each matrix over the set of real numbers we can associate a non-negative integer called the rank of the matrix.

Definition 1.1.1   A matrix A is said to be of **rank r** if

(i)   A has at least one nonzero minor of order r and

(ii)   Every minor of A of order higher than r is zero.

Rank of the matrix A is denoted by R(A) or $\rho(A)$.

Note:
(i)   If A is an m x n matrix, $R(A) \le \min\{m, n\}$
(ii)   If A has a nonzero minor of order r, then $R(A) \ge r$
(iii) If all minors of A of order r+1 are zero, then $R(A) \le r$
(iv)   Rank of the identity matrix I_n is n
(v)   Rank of a triangular matrix is the number of nonzero rows in it.
(vi)   Rank of a diagonal matrix is the number if nonzero entries on the diagonal.

1.1.1 ELEMENTARY TRANSFORMATIONS

The following three operations on the rows (columns) of a matrix are known as elementary row (column) operations or elementary row (column) transformations.
(i)   Interchange of any two rows (columns).
(ii)   Multiplication of any row (column) by a nonzero number.
(iii) Addition of a constant multiple of the elements of any row (column) to the corresponding elements of any other row (column)

The elementary row (column) transformations are denoted by the following symbols.

(i) R_{ij} (C_{ij}) for the interchange of ith and j^{th} rows (columns)
(ii) kR_i (kC_i) for multiplication of the ith row (column) by k.
(iii) $R_i + kR_j$ ($C_i + kC_j$) for addition to the ith row (column) k times the jth row (column).

1.1.2 EQUIVALENT MATRICES

Definition 1.1.2 Two matrices A and B are said to be **equivalent** if one can be obtained from the other by a sequence of elementary transformations. We write A ~ B when A and B are equivalent.

Note:
(i)   Elementary transformations do not change the order or the rank of a matrix.
(ii)   Two equivalent matrices have the same order and the same rank.
(iii) Using elementary row (column) transformations, a given matrix can be reduced to the upper triangular form.
(iv)   Rank of A is equal to the number of nonzero rows in the triangular form of A.

Example 1.1.1 Find the rank of $A = \begin{bmatrix} 1 & 2 & 3 \\ 2 & 4 & 7 \\ 3 & 6 & 10 \end{bmatrix}$

Solution: Rank of a matrix can be determined either by checking the nonzero minors or by using the elementary row (column transformations). Both these methods are explained in this problem.

Method 1.

Since A is a 3 x 3 matrix, A has minors of order 1, 2 and 3 only. The only minor of order 3 is

$$\begin{vmatrix} 1 & 2 & 3 \\ 2 & 4 & 7 \\ 3 & 6 & 10 \end{vmatrix} = 1\begin{vmatrix} 4 & 7 \\ 6 & 10 \end{vmatrix} - 2\begin{vmatrix} 2 & 7 \\ 3 & 10 \end{vmatrix} + 3\begin{vmatrix} 2 & 4 \\ 3 & 6 \end{vmatrix}$$

$$= 1(40 - 42) -2(20 - 21) + 3(12 - 12)$$
$$= 1(-2) - 2(-1) + 3(0)$$
$$= 0$$

A minor of order 2 is $\begin{vmatrix} 2 & 3 \\ 4 & 7 \end{vmatrix} = 14 - 12 = 2 \neq 0.$

Hence there is one nonzero minor of order 2, whereas all minors of order 3 are zero. Hence R(A) = 2.

Method 2.

Rank of a given matrix can be determined by using elementary row (column) transformations. First reduce the matrix A to the triangular form using elementary row transformations. Then the rank of A is the number of nonzero rows in the triangular form of A.

$$A = \begin{bmatrix} 1 & 2 & 3 \\ 2 & 4 & 7 \\ 3 & 6 & 10 \end{bmatrix} \sim \begin{bmatrix} 1 & 2 & 3 \\ 0 & 0 & 1 \\ 0 & 0 & 1 \end{bmatrix} \qquad \begin{array}{l} R_2 \rightarrow R_2 - 2R_1 \\ R_3 \rightarrow R_3 - 3R_1 \end{array}$$

$$\sim \begin{bmatrix} 1 & 2 & 3 \\ 0 & 0 & 1 \\ 0 & 0 & 0 \end{bmatrix} \qquad R_3 \rightarrow R_3 - R_2$$

Since the number of nonzero rows in the triangular form = 2, we have R(A) = 2.

Note:

Note that in the above problem we have carried out two elementary row operations simultaneously. This way one can obtain the triangular form of a given matrix in less number of steps. Two or more elementary row (column) transformations of the type $R_i \rightarrow R_i + kR_j$ $(C_i \rightarrow C_i + kC_j)$ can be carried out simultaneously if R_i (C_i) is kept fixed or R_j (C_j) is kept fixed.

Example 1.1.2 Find the rank of $A = \begin{bmatrix} 4 & 3 & 0 & -2 \\ 3 & 4 & -1 & -3 \\ -7 & -7 & 1 & 5 \end{bmatrix}$

Solution: We reduce A to the triangular form using elementary row transformations.

$$A = \begin{bmatrix} 4 & 3 & 0 & -2 \\ 3 & 4 & -1 & -3 \\ -7 & -7 & 1 & 5 \end{bmatrix} \sim \begin{bmatrix} 4 & 3 & 0 & -2 \\ 12 & 16 & -4 & -12 \\ -28 & -28 & 4 & 20 \end{bmatrix} \quad \begin{matrix} R_2 \rightarrow 4R_2 \\ R_3 \rightarrow 4R_3 \end{matrix}$$

$$\sim \begin{bmatrix} 4 & 3 & 0 & -2 \\ 0 & 7 & -4 & -6 \\ 0 & -7 & 4 & 6 \end{bmatrix} \quad \begin{matrix} R_2 \rightarrow R_2 - 3R_1 \\ R_3 \rightarrow R_3 + 7R_1 \end{matrix}$$

$$\sim \begin{bmatrix} 4 & 3 & 0 & -2 \\ 0 & 7 & -4 & -6 \\ 0 & 0 & 0 & 0 \end{bmatrix} \quad R_3 \rightarrow R_3 + R_2$$

R(A) = number of nonzero rows in the triangular form = 2.

Example 1.1.3 Find the rank of $A = \begin{bmatrix} 1 & 2 & -2 & 3 \\ 2 & 5 & -4 & 6 \\ -1 & -3 & 2 & -2 \\ 2 & 4 & -1 & 6 \end{bmatrix}$

Solution: We reduce the matrix A to the triangular form.

$$A = \begin{bmatrix} 1 & 2 & -2 & 3 \\ 2 & 5 & -4 & 6 \\ -1 & -3 & 2 & -2 \\ 2 & 4 & -1 & 6 \end{bmatrix} \sim \begin{bmatrix} 1 & 2 & -2 & 3 \\ 0 & 1 & 0 & 0 \\ 0 & -1 & 0 & 1 \\ 0 & 0 & 3 & 0 \end{bmatrix} \quad \begin{matrix} R_2 \rightarrow R_2 - 2R_1 \\ R_3 \rightarrow R_3 + R_1 \\ R_4 \rightarrow R_4 - 2R_1 \end{matrix}$$

$$\sim \begin{bmatrix} 1 & 2 & -2 & 3 \\ 0 & 1 & 0 & 0 \\ 0 & 0 & 0 & 1 \\ 0 & 0 & 3 & 0 \end{bmatrix} \quad R_3 \rightarrow R_3 + R_2$$

$$\sim \begin{bmatrix} 1 & 2 & -2 & 3 \\ 0 & 1 & 0 & 0 \\ 0 & 0 & 3 & 0 \\ 0 & 0 & 0 & 1 \end{bmatrix} \quad R_{34}$$

R(A) = number of nonzero rows in the triangular form of A.= 4.

1.1.3 LINEAR DEPENDENCE OF VECTORS

Definition 1.1.3 The vectors (row vectors or column vectors) $X_1, X_2, ..., X_m$ are said to be **linearly dependent** if scalars $\lambda_1, \lambda_2, ..., \lambda_m$ (not all zero simultaneously) can be found such that $\lambda_1 X_1 + \lambda_2 X_2 + ... + \lambda_m X_m = 0$. If no such scalars, other than zero exist, then the vectors are said to be **linearly independent**.

Note:

(i)   If a vector $X = \lambda_1 X_1 + \lambda_2 X_2 + ... + \lambda_m X_m$, then X is said to be a linear combination of the vectors $X_1, X_2... X_m$.

(ii)   If m vectors are linearly dependent, then atleast one of them can be expressed as a linear combination of the remaining vectors.

(iii)   Let $X_1, X_2... X_m$ be vectors of dimension n. With $X_1, X_2... X_m$ as m rows, a matrix A of dimension mxn can be formed. If $R(A) = m$, then the m vectors $X_1, X_2 ... X_m$ are linearly independent. Otherwise (i.e., $R(A) < m$), then the m vectors are linearly dependent.

1.1.4 LINEAR SYSTEMS OF EQUATIONS

In this section we formulate a general method for solving linear systems of equations. Finding the rank of a certain matrix one can conclude whether the system has a solution or not and if it has solutions they can be found explicitly. This method is usually known as the matrix method of solving linear systems of equations.

$$a_{11}x_1 + a_{12}x_2 + ... + a_{1n}x_n = b_1$$
$$a_{21}x_1 + a_{22}x_2 + ... + a_{2n}x_n = b_2$$
$$.................................$$
$$a_{m1}x_1 + a_{m2}x_2 + ... + a_{mn}x_n = b_m$$

$$(1)$$

where a_{ij} and b_j are real number is called a system of m linear equations in n variables $x_1, x_2, ..., x_n$.

The system (1) can be expressed in matrix form as $AX = B$, where

$$A = \begin{bmatrix} a_{11} & a_{12} & \cdots & a_{1n} \\ a_{21} & a_{22} & \cdots & a_{2n} \\ \cdots & \cdots & \cdots & \cdots \\ a_{m1} & a_{m2} & \cdots & a_{mn} \end{bmatrix}$$

$$X = \begin{bmatrix} x_1 \\ x_2 \\ \vdots \\ x_n \end{bmatrix} \qquad B = \begin{bmatrix} b_1 \\ b_2 \\ \vdots \\ b_m \end{bmatrix}$$

The matrix $A = [a_{ij}]$ is called the **coefficient matrix**. The matrix

$$[A, B] = \begin{bmatrix} a_{11} & a_{12} & \cdots & a_{1n} & b_1 \\ a_{21} & a_{22} & \cdots & a_{2n} & b_2 \\ \cdots & \cdots & \cdots & \cdots & \\ a_{m1} & a_{m2} & \cdots & a_{mn} & b_m \end{bmatrix} \text{ is called the } \textbf{augmented}$$

matrix of the system(1).

If B=0, a zero matrix, then the system (1) is called a system of **linear homogeneous** equations; otherwise it is called a system of **linear non-homogeneous** equations.

A set of numbers x_1, x_2, ...,x_n satisfying the m equations in the system is called a **solution** of the system of equations. A system of equations is said to be **consistent** if it has a solution. Otherwise it is said to be **inconsistent**. A given system of equations may have no solution, one solution (i.e., unique solution) or infinite number of solutions.

Note:

(i) **Rouche's theorem:** The system of equations AX = B is consistent if and only if the coefficient matrix A and the augmented matrix [A, B] have the same rank. Otherwise the system is inconsistent.

(ii) Let the rank of A be r and that of [A, B] be r'. If r ≠ r', then the system of equations is inconsistent. If r = r' = n, the number of variables, then the system is consistent and has a unique solution. If r = r' < n, then the system is consistent and there are infinite number of solutions.

(iii) Consider the system AX = 0. If r = n, then the system has only a trivial solution $x_1 = x_2 = ...= x_n = 0$. If r< n, then the system has infinite number of solutions. Arbitrary values can be assigned to n-r variables and the values of the remaining variables can be uniquely determined.

Example 1.1.4 Show that the vectors $X_1 = (1, -1, 2, 3)$; $X_2 = (1, 0, -1, 2)$ and $X_3 = (1, 1, -4, 0)$ are linearly independent.

Solution:

Method 1. The given vectors are linearly independent if the matrix with these vectors as row vectors has rank 3

$$A = \begin{bmatrix} X_1 \\ X_2 \\ X_3 \end{bmatrix} = \begin{bmatrix} 1 & -1 & 2 & 3 \\ 1 & 0 & -1 & 2 \\ 1 & 1 & -4 & 0 \end{bmatrix} \sim \begin{bmatrix} 1 & -1 & 2 & 3 \\ 0 & 1 & -3 & -1 \\ 0 & 2 & -6 & -3 \end{bmatrix} \quad \begin{matrix} R_2 \rightarrow R_2 - R_1 \\ R_3 \rightarrow R_3 - R_1 \end{matrix}$$

$$\sim \begin{bmatrix} 1 & -1 & 2 & 3 \\ 0 & 1 & -3 & -1 \\ 0 & 0 & 0 & -1 \end{bmatrix} \quad R_3 \rightarrow R_3 - 2R_2$$

R(A) = number of nonzero vectors = 3.

∴ X_1, X_2, X_3 are linearly independent.

Method 2.

Let $\lambda_1 X_1 + \lambda_2 X_2 + \lambda_3 X_3 = 0$.

i.e, $\lambda_1(1, -1, 2, 3) + \lambda_2(1, 0, -1, 2) + \lambda_3(1, 1, -4, 0) = (0, 0, 0, 0)$

Equating the corresponding elements on both sides we get,

$$\lambda_1 + \lambda_2 + \lambda_3 = 0 \qquad (1)$$
$$-\lambda_1 \qquad + \lambda_3 = 0 \qquad (2)$$
$$2\lambda_1 - \lambda_2 - 4\lambda_3 = 0 \qquad (3)$$
$$3\lambda_1 + 2\lambda_2 \qquad = 0 \qquad (4)$$

From (2) $\qquad\qquad \lambda_1 = \lambda_3$

∴ From (1) $\qquad 2\lambda_1 + \lambda_2 = 0 \qquad\qquad\qquad (5)$

From (4) and (5) we get, $\lambda_1 = \lambda_2 = 0$

∴ $\lambda_1 = \lambda_2 = \lambda_3 = 0$

∴ The vectors X_1, X_2, X_3 are linearly independent.

Example 1.1.5 Show that the three vectors $X_1 = (1, 2, -1, 3)$; $X_2 = (2, -1, 3, 2)$; and $X_3 = (-1, 8, -9, 5)$ are linearly dependent. Find the linear relationship among them.

Solution:

Method 1.

$$A = \begin{bmatrix} X_1 \\ X_2 \\ X_3 \end{bmatrix} = \begin{bmatrix} 1 & 2 & -1 & 3 \\ 2 & -1 & 3 & 2 \\ -1 & 8 & -9 & 5 \end{bmatrix}$$

$$\sim \begin{bmatrix} 1 & -2 & -1 & 3 \\ 0 & -5 & 5 & -4 \\ 0 & 10 & -10 & 8 \end{bmatrix} \qquad \begin{aligned} R_2 &\to R_2 - 2R_1 \\ R_3 &\to R_3 + R_1 \end{aligned}$$

$$\sim \begin{bmatrix} 1 & -2 & -1 & 3 \\ 0 & -5 & 5 & -4 \\ 0 & 0 & 0 & 0 \end{bmatrix} \qquad R_3 \to R_3 + 2R_2$$

∴ R(A) = 2 < number of given vectors.

∴ X_1, X_2, X_3 are linearly dependent.

Also the last row $\quad 0 = R_3$

$$= R_3 + 2R_2$$
$$= (R_3 + R_1) + 2(R_2 - 2R_1)$$
$$= -3R_1 + 2R_2 + R_3$$

∴ $\quad -3R_1 + 2R_2 + R_3 = 0$

i.e., $-3X_1 + 2X_2 + X_3 = 0$ is the linear relationship among X_1, X_2, and X_3.

Method 2.
Let $\lambda_1 X_1 + \lambda_2 X_2 + \lambda_3 X_3 = 0$.
i.e, $\lambda_1(1, 2, -1, 3) + \lambda_2(2, -1, 3, 2) + \lambda_3(-1, 8, -9, 5) = (0, 0, 0, 0)$
Equating the corresponding elements on both sides we get,

$$\therefore \lambda_1 + 2\lambda_2 + \lambda_3 = 0 \tag{1}$$
$$2\lambda_1 - \lambda_2 + \lambda_3 = 0 \tag{2}$$
$$-\lambda_1 + 3\lambda_2 - 9\lambda_3 = 0 \tag{3}$$
$$3\lambda_1 + 2\lambda_2 + 5\lambda_3 = 0 \tag{4}$$

From (1) and (2) $10\lambda_1 + 15\lambda_2 = 0$,
i.e. $\lambda_1 = (-3/2)\lambda_2$
From (2) and (3), we get, $5\lambda_2 - 10\lambda_3 = 0$,
i.e. $\lambda_3 = (1/2)\lambda_2$
Taking $\lambda_2 = 2$ we get, $\lambda_1 = -3$ and $\lambda_3 = 1$.
These values satisfy equation (4) also.
$\therefore -3X_1 + 2X_2 + X_3 = 0$
and this is the linear relationship among the given vectors X_1, X_2, and X_3.
Hence the vectors X_1, X_2, X_3 are linearly dependent.

Example 1.1.6 Solve the following equations if consistent.
$$\begin{aligned} x + 2y - z &= 3 \\ 3x - y + 2z &= 1 \\ 2x - 2y + 3z &= 2 \\ x - y + z &= -1 \end{aligned}$$

Solution:
First we compute the ranks of the augmented matrix and the coefficient matrix. If they are same, the system of equations will be consistent. If consistent, the system will have solutions. The augmented matrix is

$$[A, B] = \begin{bmatrix} 1 & 2 & -1 & 3 \\ 3 & -1 & 2 & 1 \\ 2 & -2 & 3 & 2 \\ 1 & -1 & 1 & -1 \end{bmatrix}$$

$$\sim \begin{bmatrix} 1 & 2 & -1 & 3 \\ 0 & -7 & 5 & -8 \\ 0 & -6 & 5 & -4 \\ 0 & -3 & 2 & -4 \end{bmatrix} \quad \begin{aligned} R_2 &\to R_2 - 3R_1 \\ R_3 &\to R_3 - 2R_1 \\ R_4 &\to R_4 - R_1 \end{aligned}$$

$$\sim \begin{bmatrix} 1 & 2 & -1 & 3 \\ 0 & -7 & 5 & -8 \\ 0 & 42 & -35 & 28 \\ 0 & 21 & -14 & 28 \end{bmatrix} \quad \begin{aligned} R_3 &\rightarrow -7R_3 \\ R_4 &\rightarrow -7R_4 \end{aligned}$$

$$\sim \begin{bmatrix} 1 & 2 & -1 & 3 \\ 0 & -7 & 5 & -8 \\ 0 & 0 & -5 & -20 \\ 0 & 0 & 1 & 4 \end{bmatrix} \quad \begin{aligned} R_3 &\rightarrow R_3 + 6R_2 \\ R_4 &\rightarrow R_4 + 3R_2 \end{aligned}$$

$$\sim \begin{bmatrix} 1 & 2 & -1 & 3 \\ 0 & -7 & 5 & -8 \\ 0 & 0 & 1 & 4 \\ 0 & 0 & 1 & 4 \end{bmatrix} \quad R_3 \rightarrow -\frac{1}{5}R_3$$

$$\sim \begin{bmatrix} 1 & 2 & -1 & 3 \\ 0 & -7 & 5 & -8 \\ 0 & 0 & 1 & 4 \\ 0 & 0 & 0 & 0 \end{bmatrix} \quad R_4 \rightarrow R_4 - R_3$$

R(A,B) = Number of nonzero rows in the triangular form = 3.
Also the coefficient matrix

$$A \sim \begin{bmatrix} 1 & 2 & -1 \\ 0 & -7 & 5 \\ 0 & 0 & 1 \\ 0 & 0 & 0 \end{bmatrix}$$

∴   R(A) = 3.

Since R(A, B) = R(A) = 3, the number of variables, the system is consistent and has a unique solution.

The given system of equations is equivalent to

$$\begin{aligned} x + 2y - z &= 3 & (1) \\ -7y + 5z &= -8 & (2) \\ z &= 4 & (3) \end{aligned}$$

From (3) and (2), y = 4.   From (1), x = -1

∴ The unique solution is x = -1, y = 4, z = 4.

Note: While solving a system of equations, one should not use elementary column operations, since it changes the variables.

Example 1.1.7 Solve the system of equations

$$4x + 2y + z + 3w = 0$$
$$6x + 3y + 4z + 7w = 0$$
$$2x + y \qquad + w = 0$$

Solution:

$$A = \begin{bmatrix} 4 & 2 & 1 & 3 \\ 6 & 3 & 4 & 7 \\ 2 & 1 & 0 & 1 \end{bmatrix}$$

$$\sim \begin{bmatrix} 4 & 2 & 1 & 3 \\ 24 & 12 & 16 & 28 \\ 8 & 4 & 0 & 4 \end{bmatrix} \quad \begin{array}{l} R_2 \to 4R_2 \\ R_3 \to 4R_3 \end{array}$$

$$\sim \begin{bmatrix} 4 & 2 & 1 & 3 \\ 0 & 0 & 10 & 10 \\ 0 & 0 & -2 & -2 \end{bmatrix} \quad \begin{array}{l} R_2 \to R_2 - 6R_1 \\ R_3 \to R_3 - 2R_1 \end{array}$$

$$\sim \begin{bmatrix} 4 & 2 & 1 & 3 \\ 0 & 0 & 1 & 1 \\ 0 & 0 & 1 & 1 \end{bmatrix} \quad \begin{array}{l} R_2 \to \dfrac{1}{10}R_2 \\ R_3 \to -\dfrac{1}{2}R_3 \end{array}$$

$$\sim \begin{bmatrix} 4 & 2 & 1 & 3 \\ 0 & 0 & 1 & 1 \\ 0 & 0 & 0 & 0 \end{bmatrix} \quad R_3 \to R_3 - R_2$$

$R(A) = 2 < 4$, the number of variables.

∴ The system has infinite number of solutions.

Arbitrary values can be assigned to 4 - 2 = 2 variables (any two)

The system is equivalent to

$$4x + 2y + z + 3w = 0 \qquad (1)$$
$$z + w = 0 \qquad (2)$$

Choose x = s, z = t (or y = s, w = t)

From (2)   w = -t

From (1)   2y = -4s - t +3t

           y = -2s + t

Thus the system of equations has infinite number of solutions. They are given by $x = s$, $y = -2s + t$, $z = t$, $w = -t$, where s and t are real numbers (called parameters).

Example 1.1.8 Investigate for what values of λ, μ the following system has
(a) no solution (b) a unique solution (c) an infinite number of solutions.

$$x + y + z = 6$$
$$x + 2y + 3z = 10$$
$$x + 2y + \lambda z = \mu$$

Solution:

The augmented matrix $[A, B] = \begin{bmatrix} 1 & 1 & 1 & 6 \\ 1 & 2 & 3 & 10 \\ 1 & 2 & \lambda & \mu \end{bmatrix}$

$$\sim \begin{bmatrix} 1 & 1 & 1 & 6 \\ 0 & 1 & 2 & 4 \\ 0 & 1 & \lambda-1 & \mu-6 \end{bmatrix} \quad \begin{matrix} R_2 \rightarrow R_2 - R_1 \\ R_3 \rightarrow R_3 - R_1 \end{matrix}$$

$$\sim \begin{bmatrix} 1 & 1 & 1 & 6 \\ 0 & 1 & 2 & 4 \\ 0 & 0 & \lambda-3 & \mu-10 \end{bmatrix} \quad R_3 \rightarrow R_3 - R_2$$

(a) **Case 1.** $\lambda = 3$, $\mu \neq 10$
   In this case $R(A) = 2$, $R(A, B) = 3$.
   $\therefore R(A) \neq R(A, B)$
   $\therefore$ The system of equations has no solutions.

(b) **Case 2.** $\lambda \neq 3$.
   In this case $R(A) = R(A, B) = 3 =$ number of variables.
   $\therefore$ The system of equations has a unique solution.

(c) **Case 3.** $\lambda = 3$, $\mu = 10$
   In this case $R(A) = R(A, B) = 2 <$ number of variables.
   $\therefore$ The system of equations has infinite number of solutions.
   The given system is equivalent to
   $$x + y + z = 6 \qquad\qquad (1)$$
   $$y + 2z = 4 \qquad\qquad (2)$$
   Choosing $z = s$, we have from (2), $y = 4 - 2s$.
   Again from (1) $x = 6 - (4 - 2s) - s$
   $$x = 2 + s$$
   $\therefore$ The infinite number of solutions are given by $x = 2 + s$, $y = 4 - 2s$, $z = s$, where s is any real number.

EXERCISE 1.1

PART – A

1. Define the rank of a matrix.
2. What is the rank of identity matrix of order n?
3. What is the rank of $\begin{bmatrix} 4 & 2 & 2 & 3 \\ 0 & 4 & 2 & 3 \\ 0 & 0 & 0 & 5 \\ 0 & 0 & 4 & 5 \end{bmatrix}$?
4. What is the condition for which the equations $a_1x + b_1y + c_1z = 0$, $a_2x + b_2y + c_2z = 0$, $a_3x + b_3y + c_3z = 0$ have nontrivial solutions?
5. Define the linear dependence of a set of vectors.
6. Define the linear independence of a set of vectors.
7. If a set of vectors are linearly dependent, show that at least one of the vectors can be expressed as a linear combination of the other vectors.
8. Prove that the vectors $(1, 0, 0)$, $(0, 1, 0)$ and $(0, 0, 1)$ are linearly independent.
9. Prove that the vectors $(1, 0, 0)$, $(2, 5, 7)$ and $(0, 5, 7)$ are linearly dependent.
10. Find the value of a for which the vectors $(2, -1, 0)$, $(4, 1, 1)$ and $(a, -1, 1)$ are linearly dependent
11. What do you mean by consistent and inconsistent systems of equations?
12. State Rouche's theorem.
13. State the condition for a system of m equations in n unknowns to have (i) one solution (ii) many solutions and (iii) no solution.
14. Find the values of a and b if the equations $2x - 3y = 5$ and $ax + by = -10$ have many solutions.
15. If the augmented matrix of a system of equations is equivalent to $\begin{bmatrix} 1 & 2 & 3 & 4 \\ 0 & -4 & -2 & -5 \\ 0 & 0 & 0 & \lambda \end{bmatrix}$, find the values of λ for which the system has a unique solution.
16. If the augmented matrix of a system of equations is equivalent to $\begin{bmatrix} 1 & 2 & 4 & 5 \\ 0 & 3 & -2 & 6 \\ 0 & 0 & \lambda-1 & \mu+3 \end{bmatrix}$, find the values of λ and μ for which the system has a unique solution.
17. If the augmented matrix of a system of equations is equivalent to $\begin{bmatrix} 1 & 4 & -2 & 3 \\ 0 & 3 & 4 & -5 \\ 0 & 0 & \lambda-2 & \mu-3 \end{bmatrix}$, find the values of λ and μ for which the system has many solutions.

18. If the equations $x + 2y + z = 0$, $5x + y - z = 0$ and $x + 5y + \lambda z = 0$ have a non-trivial solution, find the value of λ.

PART – B
Find the rank of the following matrices.

19. $\begin{bmatrix} 4 & 3 & 0 & -2 \\ 3 & 4 & -1 & -3 \\ -7 & -7 & 1 & 5 \end{bmatrix}$

20. $\begin{bmatrix} 1 & -2 & 3 & 4 \\ -2 & 4 & -1 & -3 \\ -1 & 2 & 7 & 6 \end{bmatrix}$

21. $\begin{bmatrix} 2 & 3 & -1 & -1 \\ 1 & -1 & -2 & -4 \\ 3 & 1 & 3 & -2 \\ 6 & 3 & 0 & -7 \end{bmatrix}$

Show that the following sets of vectors are linearly dependent. Find their relationship in each case.

22. $X_1 = (1, 1, -1, 1)$, $X_2 = (1, -1, 2, -1)$, $X_3 = (3, 1, 0, 1)$.
23. $X_1 = (1, 2, -1, 4)$, $X_2 = (2, 4, 3, 5)$, $X_3 = (-1, -2, 6, -7)$.
24. $X_1 = (1, -2, 4, 1)$, $X_2 = (1, 0, 6, -5)$, $X_3 = (2, -3, 9, -1)$, $X_4 = (2, -5, 7, 5)$.

Show that the following sets of vectors are linearly independent.

25. $X_1 = (1, 1, 1)$, $X_2 = (1, 2, 3)$, $X_3 = (2, 3, 8)$.
26. $X_1 = (1, 4, 2)$, $X_2 = (2, 1, 3)$, $X_3 = (-4, 1, 2)$.
27. $X_1 = (-1, 2, 2, 1)$, $X_2 = (2, 0, 1, -3)$, $X_3 = (3, 5, -7, 2)$.

Test for consistency of the following systems of equations and if consistent solve them.

28. $2x + y + 5z = 4$. $3x - 2y + 2z = 2$, $5x - 8y - 4z = 1$
29. $x + 2y + z = 3$. $2x + 3y + 2z = 5$, $3x - 5y + 5z = 2$, $3x + 9y - z = 4$.
30. $x - 3y - 8z = -10$, $3x + y - 4z = 0$, $2x + 5y + 6z = 13$.

31. For what values of a and b do the equations $x + 2y + 3z = 6$, $x + 3y + 5z = 9$ and $2x + 5y + az = b$ have (i) no solution (ii) unique solution (iii) infinite number of solutions.
32. Find the condition on a, b, c, so that the equations $x + y + z = a$, $x + 2y + 3z = b$, $3x + 5y + 7z = c$ may have a one-parameter family of solutions.

1.2   THE CHARACTERISTIC EQUATION

Definition 1.2.1 Let A be a square matrix of order n. I be the identity matrix of order n and λ be any scalar. Then the matrix $A - \lambda I$ is called the **characteristic matrix of A**. The determinant $|A - \lambda I|$ is a polynomial of degree n in λ called **characteristic polynomial of A**. The polynomial equation $|A - \lambda I| = 0$ is called the **characteristic equation of A**.

Definition 1.2.2 The roots of the characteristic equation are called the **characteristic roots** or **latent roots** or **eigen values** of the matrix A.

Note:
(i)   Eigen values of a matrix may repeat.
(ii)   Eigen values of matrix may be complex numbers.
Let λ be an eigen value of the square matrix of order n. Then $|A - \lambda I| = 0$. Let X be any column vector of dimension n. Then $(A - \lambda I)X = 0$ is a system of n linear homogeneous equations. Since the determinant $|A - \lambda I|$ of the coefficient matrix $A - \lambda I$ is equal to zero, the system of equations has non-trivial solutions.

Definition 1.2.3 Let λ be an eigen value of the matrix A. The non-trivial solutions of the system of equations $(A - \lambda I)X = 0$ are called the **characteristic vectors** or **latent vectors** or **eigen vectors** of A corresponding to the eigen value λ.

The transformation $Y = AX$ carries a column vector X into another column vector Y. If the vector X is transformed into the vector λX under the above transformation, then $AX = \lambda X$. Such a vector X is called an invariant vector under the transformation. Also $AX = \lambda X$ means $(A - \lambda I) X = 0$. Therefore the characteristic vectors corresponding to λ are the invariant vectors under the transformation $Y = AX$.

Note:
(i)   The equation $|A - \lambda I| = 0$ can be expanded as a polynomial in λ in the form,
$$\lambda^n - D_1 \lambda^{n-1} + D_2 \lambda^{n-2} - \ldots + (-1)^n D_n = 0$$
where D_i is the sum of all the minors of A of order i, whose principal diagonals lie along the principal diagonal of A.

Example 1.2.1 Find the characteristic equation and characteristic roots of the following

matrix. $\begin{bmatrix} 1 & 1 & 3 \\ 1 & 5 & 1 \\ 3 & 1 & 1 \end{bmatrix}$.

Solution: The characteristic equation is given by $\begin{vmatrix} 1-\lambda & 1 & 3 \\ 1 & 5-\lambda & 1 \\ 3 & 1 & 1-\lambda \end{vmatrix} = 0$

i.e., $(1 - \lambda) [(5 - \lambda).(1 - \lambda) -1] -1[1.(1 - \lambda) - 3] - 3[1 - 3.(5 - \lambda)] = 0$

i.e., $(1 - \lambda) [\lambda^2 - 6\lambda + 4] - 1[-2 - \lambda] -3.[-14 + 3\lambda] = 0$

i.e., $-\lambda^3 + 7\lambda^2 - 36 = 0$

i.e., $\lambda^3 - 7\lambda^2 + 36 = 0$

Since $(-2)^3 - 7(-2)^2 + 36 = 0$, -2 is a root.

Divide $\lambda^3 - 7\lambda^2 + 36$ by the factor $\lambda + 2$

i.e.,

$$\begin{array}{r|rrrr} -2 & 1 & -7 & 0 & 36 \\ & 0 & -2 & 18 & -36 \\ \hline & 1 & -9 & 18 & 0 \end{array}$$

The remaining roots are given by $\lambda^2 - 9\lambda + 18 = 0$

i.e., $(\lambda - 3) (\lambda - 6) = 0$

i.e., $\lambda = 3, 6$.

∴ The characteristic roots are $-2, 3$ and 6.

Example 1.2.2 Find the eigen values and eigen vectors of $A = \begin{bmatrix} 2 & 1 \\ -1 & 2 \end{bmatrix}$

Solution:

The characteristic equation is $| A - \lambda I | = 0$

i.e., $\begin{vmatrix} 2-\lambda & 1 \\ -1 & 2-\lambda \end{vmatrix} = 0$

i.e., $(2 - \lambda)^2 + 1 = 0$

i.e., $\lambda^2 - 4\lambda + 5 = 0$

∴ $\lambda = \dfrac{4 \pm \sqrt{16-20}}{2} = 2 \pm i$

$\lambda = 2 + i, 2 - i$

∴ The eigen values are $2 + i$ and $2 - i$.

The eigen vectors are given by $(A - \lambda I) X = 0$

i.e., $\begin{bmatrix} 2-\lambda & 1 \\ -1 & 2-\lambda \end{bmatrix} \begin{bmatrix} x_1 \\ x_2 \end{bmatrix} = \begin{bmatrix} 0 \\ 0 \end{bmatrix}$

Case 1. $\lambda = 2 + i$

$\begin{bmatrix} -i & 1 \\ -1 & -i \end{bmatrix} \begin{bmatrix} x_1 \\ x_2 \end{bmatrix} = \begin{bmatrix} 0 \\ 0 \end{bmatrix}$

i.e., $-ix_1 + x_2 = 0$ (1)

$-x_1 - ix_2 = 0$ (2)

$x_1 = 1$, $x_2 = i$ is a solution of equations (1) and (2).

∴ Eigen vector corresponding to $\lambda = 2 + i$ is $\begin{bmatrix} 1 \\ i \end{bmatrix}$

Case 2. $\lambda = 2 - i$

$$\begin{bmatrix} i & 1 \\ -1 & i \end{bmatrix} \begin{bmatrix} x_1 \\ x_2 \end{bmatrix} = \begin{bmatrix} 0 \\ 0 \end{bmatrix}$$

i.e., $ix_1 + x_2 = 0$ (3)

$-x_1 + ix_2 = 0$ (4)

$x_1 = 1$, $x_2 = -i$ is a solution of equations (3) and (4).

$\therefore$ Eigen vector corresponding to $\lambda = 2 - i$ is $\begin{bmatrix} 1 \\ -i \end{bmatrix}$

Example 1.2.3 Find the eigen values and eigen vectors of the matrix

$$A = \begin{bmatrix} -2 & 2 & -3 \\ 2 & 1 & -6 \\ -1 & -2 & 0 \end{bmatrix}$$

Solution:

The characteristic equation is $|A - \lambda I| = 0$.

i.e., $$\begin{vmatrix} -2-\lambda & 2 & -3 \\ 2 & 1-\lambda & -6 \\ -1 & -2 & 0-\lambda \end{vmatrix} = 0$$

i.e., $(-2 - \lambda) [-\lambda(1 - \lambda) -12] - 2[-2\lambda - 6] -3[-4 + 1 - \lambda] = 0$

i.e., $(-2 - \lambda) [\lambda^2 - \lambda -12] + 4\lambda + 12 + 9 + 3\lambda = 0$

i.e., $\lambda^3 + \lambda^2 - 21\lambda - 45 = 0$ (1)

Now, $(-3)^3 + (-3)^2 - 21(-3) - 45 = -27 + 9 + 63 - 45$

$= -72 + 72 = 0$

$\therefore$ -3 is a root of equation (1).

Dividing $\lambda^3 + \lambda^2 - 21\lambda - 45$ by $\lambda + 3$

$$\begin{array}{r|rrrr} -3 & 1 & 1 & -21 & -45 \\ & 0 & -3 & 6 & 45 \\ \hline & 1 & -2 & -15 & 0 \end{array}$$

Remaining roots are given by

$\lambda^2 - 2\lambda - 15 = 0$

i.e., $(\lambda + 3)(\lambda - 5) = 0$ i.e., $\lambda = -3, 5$.

$\therefore$ The eigen values are -3, -3, 5

The eigen vectors of A are given by $\begin{bmatrix} -2-\lambda & 2 & -3 \\ 2 & 1-\lambda & -6 \\ -1 & -2 & -\lambda \end{bmatrix} \begin{bmatrix} x_1 \\ x_2 \\ x_3 \end{bmatrix} = \begin{bmatrix} 0 \\ 0 \\ 0 \end{bmatrix}$

Case 1. $\lambda = -3$

Now $\begin{bmatrix} -2+3 & 2 & -3 \\ 2 & 1+3 & -6 \\ -1 & -2 & 3 \end{bmatrix} \sim \begin{bmatrix} 1 & 2 & -3 \\ 2 & 4 & -6 \\ -1 & -2 & 3 \end{bmatrix}$

$\sim \begin{bmatrix} 1 & 2 & -3 \\ 0 & 0 & 0 \\ 0 & 0 & 0 \end{bmatrix}$

$\therefore$ $x_1 + 2x_2 - 3x_3 = 0$

Put $x_2 = k_1$, $x_3 = k_2$

Then $x_1 = 3k_2 - 2k_1$

$\therefore$ The general eigen vectors corresponding to $\lambda = -3$ is $\begin{bmatrix} 3k_2 - 2k_1 \\ k_1 \\ k_2 \end{bmatrix}$

When $k_1 = 0$, $k_2 = 1$, we get the eigen vector $\begin{bmatrix} 3 \\ 0 \\ 1 \end{bmatrix}$

When $k_1 = 1$, $k_2 = 0$, we get the eigen vector $\begin{bmatrix} -2 \\ 1 \\ 0 \end{bmatrix}$

Hence the two eigen vectors corresponding to $\lambda = -3$ are $\begin{bmatrix} 3 \\ 0 \\ 1 \end{bmatrix}$ and $\begin{bmatrix} -2 \\ 1 \\ 0 \end{bmatrix}$.

These two eigen vectors corresponding to $\lambda = -3$ are linearly independent.

Case 2. $\lambda = 5$.

$\begin{bmatrix} -2-5 & 2 & -3 \\ 2 & 1-5 & -6 \\ -1 & -2 & -5 \end{bmatrix} \sim \begin{bmatrix} -7 & 2 & -3 \\ 2 & -4 & -6 \\ -1 & -2 & -5 \end{bmatrix}$

$\sim \begin{bmatrix} -1 & -2 & -5 \\ 0 & -8 & -16 \\ 0 & 0 & 0 \end{bmatrix}$

$\therefore$ $-x_1 - 2x_2 - 5x_3 = 0$

$-8x_2 - 16x_3 = 0$

A solution is $x_3 = 1$, $x_2 = -2$, $x_1 = -1$

$\therefore$ Eigen vector corresponding to $\lambda = 5$ is $\begin{bmatrix} -1 \\ -2 \\ 1 \end{bmatrix}$.

Example 1.2.4 Find the eigen values and the corresponding eigen vectors of the matrix
$\begin{bmatrix} 8 & -6 & 2 \\ -6 & 7 & -4 \\ 2 & -4 & 3 \end{bmatrix}$

Solution:
The eigen values are given by the equation $|A - \lambda I| = 0$

$$\text{i.e.,} \quad \begin{vmatrix} 8-\lambda & -6 & 2 \\ -6 & 7-\lambda & -4 \\ 2 & -4 & 3-\lambda \end{vmatrix} = 0$$

i.e., $(8 - \lambda) [(7 - \lambda)(3 - \lambda) -16] + 6[-6(3 - \lambda) + 8] + 2[24 - 2(7 - \lambda)] = 0$

$$\text{i.e., } \lambda^3 - 18\lambda^2 + 45\lambda = 0$$
$$\text{i.e., } \lambda(\lambda - 3)(\lambda - 15) = 0$$
$$\text{i.e., } \lambda = 0, 3, 15$$

The eigen values are 0, 3 and 15.
The eigen vectors of A are given by the system $(A - \lambda I) X = 0$.

$$\text{i.e.,} \quad \begin{bmatrix} 8-\lambda & -6 & 2 \\ -6 & 7-\lambda & -4 \\ 2 & -4 & 3-\lambda \end{bmatrix} \begin{bmatrix} x_1 \\ x_2 \\ x_3 \end{bmatrix} = \begin{bmatrix} 0 \\ 0 \\ 0 \end{bmatrix} \quad (1)$$

Case 1. Put $\lambda = 0$ in (1) and find a non-trivial solution, by reducing the coefficient matrix to triangular form

$$\begin{bmatrix} 8 & -6 & 2 \\ -6 & 7 & -4 \\ 2 & -4 & 3 \end{bmatrix} \sim \begin{bmatrix} 4 & -3 & 1 \\ 0 & 1 & -1 \\ 0 & 0 & 0 \end{bmatrix}$$

$$\therefore \ 4x_1 - 3x_2 + x_3 = 0$$
$$x_2 - x_3 = 0$$

A non-trivial solution is $x_2 = x_3 = 2$, $x_1 = 1$.

$\therefore$ An eigen vector corresponding to $\lambda = 0$ is $\begin{bmatrix} 1 \\ 2 \\ 2 \end{bmatrix}$

Case 2. Put $\lambda = 3$ in (1) and find a non-trivial solution.

$$\begin{bmatrix} 5 & -6 & 2 \\ -6 & 4 & -4 \\ 2 & -4 & 0 \end{bmatrix} \sim \begin{bmatrix} 2 & -4 & 0 \\ 0 & 2 & 1 \\ 0 & 0 & 0 \end{bmatrix}$$

$\therefore \qquad \qquad 2x_1 - 4x_2 = 0$

$\qquad \qquad \qquad 2x_2 + x_3 = 0$

A non-trivial solution is $x_3 = -2$, $x_2 = 1$, $x_1 = 2$.

$\therefore$ An eigen vector corresponding to $\lambda = 3$ is $\begin{bmatrix} 2 \\ 1 \\ -2 \end{bmatrix}$

Case 3. Put $\lambda = 15$ in (1) and find a non-trivial solution.

$$\begin{bmatrix} 7 & -6 & 2 \\ -6 & -8 & -4 \\ 2 & -4 & -12 \end{bmatrix} \sim \begin{bmatrix} 2 & -4 & -12 \\ 0 & 1 & 2 \\ 0 & 0 & 0 \end{bmatrix}$$

$\therefore \qquad \qquad 2x_1 - 4x_2 - 12x_3 = 0$

$\qquad \qquad \qquad x_2 + 2x_3 = 0$

A non-trivial solution is $x_3 = 1$, $x_2 = -2$, $x_1 = 2$

$\therefore$ An eigen vector corresponding to $\lambda = 15$ is $\begin{bmatrix} 2 \\ -2 \\ 1 \end{bmatrix}$

1.2.1 SOME PROPERTIES OF EIGEN VALUES AND EIGEN VECTORS

(1) The eigen vector corresponding to an eigen value is not unique.

Proof: If X_i is an eigen vector corresponding to an eigen value λ_i, then we have,

$AX_i = \lambda_i X_i$.

If c is any non-zero scalar, then $cAX_i = c\lambda_i X_i$

i.e., $\qquad \qquad \qquad A(cX_i) = \lambda_i(cX_i)$

$\therefore$ cX_i is also an eigen vector of A corresponding to λ_i.

Thus any scalar multiple of an eigen vector is also an eigen vector.

Note:

There can be two independent eigen vectors corresponding to a given eigen value. (Refer Example 1.2.3)

(2) An eigen vector cannot correspond to two different eigen values.

Proof: Let λ_1, λ_2 be two distinct eigen values of A and suppose X be an eigen vector corresponding to λ_1 and λ_2.

Then $AX = \lambda_1 X$ and $AX = \lambda_2 X$.

$$\therefore \qquad \lambda_1 X = \lambda_2 X.$$

i.e., $(\lambda_1 - \lambda_2) X = 0$

Since $\lambda_1 \neq \lambda_2, \quad X = 0.$

$\therefore$ X cannot be an eigen vector.

$\therefore$ An eigen vector can not correspond to two different eigen values.

(3) The eigen values of an orthogonal matrix have the absolute value 1.

Proof: Let A be an orthogonal matrix. Then $AA^T = A^TA = I$. Let λ be an eigen value of A and X be an eigen vector corresponding to λ. Then $AX = \lambda X$.

$$\therefore \qquad (AX)^T = (\lambda X)^T$$

i.e., $\qquad\qquad X^T A^T = \lambda X^T$

$$\therefore \qquad (X^T A^T)(AX) = \lambda X^T . \lambda X$$

i.e., $\qquad\quad X^T (A^T A) X = \lambda^2 X^T X$

i.e., $\qquad\qquad X^T I X = \lambda^2 X^T X$

i.e., $\qquad\qquad X^T X = \lambda^2 X^T X$

i.e., $\qquad\quad (\lambda^2 - 1) X^T X = 0$

Since $X^T X \neq 0, \lambda^2 - 1 = 0$

i.e., $\qquad |\lambda| = \pm 1$

(4) If the eigen values of a matrix are all distinct, then the corresponding eigen vectors are linearly independent.

Proof: Let $\lambda_1, \lambda_2, ..., \lambda_n$ be the eigen values of a square matrix of order n and $\lambda_i \neq \lambda_j$ for $i \neq j$

Let $X_1, X_2, ..., X_n$ be the corresponding eigen vectors.

Then $\qquad AX_i = \lambda_i X_i \quad i = 1, 2, ..., n \qquad\qquad\qquad\qquad (1)$

To prove that $X_1, X_2, X_3, ..., X_n$ are linearly independent.

We apply induction on n.

$\alpha_1 X_1 = 0$ implies $\alpha_1 = 0 \ (\because X_1 \neq 0)$

$\therefore$ Any single eigen vector is linearly independent.

Assume that any set of n-1 or lesser number of eigen vector are linearly independent.

i.e., $\qquad \alpha_1 X_1 + \alpha_2 X_2 + ... + \alpha_n X_n = 0 \qquad$ and not all $\alpha_i = 0 \qquad (2)$

$\therefore \qquad A.(\alpha_1 X_1 + \alpha_2 X_2 + ... + \alpha_n X_n) = A.(0)$

i.e., $\qquad \alpha_1 AX_1 + \alpha_2 AX_2 + ... + \alpha_n AX_n = 0$

i.e., $\qquad \alpha_1 \lambda_1 X_1 + \alpha_2 \lambda_2 X_2 + ... + \alpha_n \lambda_n X_n = 0 \qquad\qquad\qquad (3)$

Now multiplying (2) by $-\lambda_1$ and adding with (3) we get

$$\alpha_2(\lambda_2 - \lambda_1)X_2 + \alpha_3(\lambda_3 - \lambda_1)X_3 + ... + \alpha_n(\lambda_n - \lambda_1)X_n = 0 \qquad\qquad (4)$$

Since by assumption any set of n-1 or lesser number of eigen vectors are linearly independent and $\lambda_i \neq \lambda_j$ for $i \neq j$, from (4) we get, $\alpha_2 = \alpha_3 = ... = \alpha_n = 0$.

Substituting in (2) we get, $\alpha_1 = 0$.

Hence $X_1, X_2, ..., X_n$ are linearly independent.

(5) The Sum of the eigen values of a matrix A is equal to the trace of A and the product of the eigen values of A is equal to the determinant of A.

Proof: Let A be a matrix of order n. The eigen values are given by the characteristic equation $|A-\lambda I| = 0$. This equation may be written as

$$\lambda^n - D_1\lambda^{n-1} + D_2\lambda^{n-2} - \ldots +(-1)^n D_n = 0 \qquad (1)$$

where D_i is the sum of all the minors of A of order i, whose prinicipal diagonals lie along the principal diagonals of A. Let $\lambda_1, \lambda_2, \ldots, \lambda_n$ be the eigen values of the matrix A. They are the roots of the equation (1)

$\therefore$ The sum of the roots $= \lambda_1+\lambda_2+\ldots+\lambda_n = -(-D_1)/1$

$$= D_1$$
$$= a_{11} + a_{22} + \ldots + a_{nn}$$
$$= \text{Trace of A.}$$

The product of the roots $= \lambda_1\lambda_2\ldots\lambda_n = (-1)^n.(-1)^n.D_n/1$

$$= D_n$$
$$= |A|$$

Note:

If A is singular, then $|A| = 0$. i.e, $\lambda_1\lambda_2\ldots\lambda_n = 0$. $\therefore$ atleast one of the eigen values of A is zero. Conversely if an eigen value of A is zero, then A is a singular matrix.

(6) The eigen values of a square matrix A and its transpose A^T are the same.

Proof: Let $A = (a_{ij})$

The characteristic polynomial of A

$$= |A - \lambda I| = \begin{vmatrix} a_{11} - \lambda & a_{12} & \cdots & a_{1n} \\ a_{21} & a_{22} - \lambda & \cdots & a_{2n} \\ \cdots & \cdots & \cdots & \cdots \\ a_{n1} & a_{n2} & \cdots & a_{nn} - \lambda \end{vmatrix}$$

$$= \begin{vmatrix} a_{11} - \lambda & a_{21} & \cdots & a_{n1} \\ a_{12} & a_{22} - \lambda & \cdots & a_{n2} \\ \cdots & \cdots & \cdots & \cdots \\ a_{1n} & a_{2n} & \cdots & a_{nn} - \lambda \end{vmatrix} \quad \text{by interchanging rows and columns}$$

$$= |A^T - \lambda I|$$
$$= \text{The characteristic polynomial of } A^T.$$

$\therefore$ The characteristic equations of A and A^T are the same.

(7) If $\lambda_1, \lambda_2, \ldots, \lambda_n$ are the eigen values of a matrix A, then

   (i) $k\lambda_1, k\lambda_2, \ldots, k\lambda_n$ are the eigen values of the matrix kA, where k is a non-zero scalar.

   (ii) $\lambda_1^p, \lambda_2^p, \lambda_3^p, \cdots, \lambda_n^p$ are the eigen values of A^p, where p is a positive integer.

(iii) $1/\lambda_1, 1/\lambda_2, \ldots, 1/\lambda_n$ are the eigen values of the inverse matrix A^{-1}, provided $\lambda_i \neq 0$ for all i.

Proof:

(i)   Let X_i be an eigen vector corresponding to an eigen value λ_i of the matrix A.

    Then $A X_i = \lambda_I X_i$

    $\therefore$ $k(A X_i) = k(\lambda_i X_i)$

    i.e., $(kA) X_i = (k\lambda_i) X_i$

    i.e., $k\lambda_i$ is an eigen value of kA and the corresponding eigen vector is X_i, which is same as that of λ_i.

(ii)   We have $A X_i = \lambda_i X_i$ (1)

    $\therefore$ $A (A X_i) = A (\lambda_i X_i)$

    i.e., $(AA) Xi = \lambda_i (A X_i)$

    i.e., $A^2 X_i = \lambda_i (\lambda_i X_i)$ using (1)

              $A^2 X_i = \lambda_i^2 X_i$

    Similarly, $A^3 X_i = \lambda_i^3 X_i$

    And in general $A^P X_i = \lambda_i^P X_i$

    $\therefore$ λ_i^p is an eigen value of A^P and the corresponding eigen vector is X_i, same as that of λ_i

(iii)   We have $A X_i = \lambda_i X_i$ and $\lambda_i \neq 0$

    $A^{-1}(A X_i) = A^{-1}(\lambda_i X_i)$

    i.e., $(A^{-1}A)X_i = \lambda_i(A^{-1}X_i)$

    i.e., $I X_i = \lambda_i(A^{-1}X_i)$

    i.e., $X_i = \lambda_i(A^{-1}X_i)$

    i.e., $A^{-1}X_i = \dfrac{1}{\lambda_i} X_i$ as $\lambda_i \neq 0$

    $\therefore \dfrac{1}{\lambda_i}$ is an eigen value of A^{-1} and the corresponding eigen vector is X_i, same as that of λ_I

(8) The eigen values of a real symmetric matrix are real.

Proof: Let λ be an eigen value of the real symmetric matrix A and X be the corresponding eigen vector. Let $\overline{X}^T$ be the transpose of the complex conjugate of X and $\overline{\lambda}$ be the complex conjugate of λ (λ may be a complex number)

  We have $AX = \lambda X$

  $\therefore$ $\overline{X}^T AX = \overline{X}^T \lambda X$

  i.e., $\overline{X}^T AX = \lambda \overline{X}^T X$ (1)

Taking the complex conjugate on both sides

$$X^T \overline{A} \, \overline{X} = \overline{\lambda} X^T \overline{X}$$

i.e., $X^T A \overline{X} = \overline{\lambda} X^T \overline{X}$ (Since $\overline{A} = A$, as A is real)

Taking transpose on both sides

$$\overline{X}^T A^T X = \overline{\lambda} \, \overline{X}^T X$$ (Since $(AB)^T = B^T A^T$ and $(X^T)^T = X$)

i.e., $\overline{X}^T A X = \overline{\lambda} \, \overline{X}^T X$ (Since $A^T = A$, as A is symmetric)

i.e., $\lambda \overline{X}^T X = \overline{\lambda} \, \overline{X}^T X$ using (1)

i.e., $(\lambda - \overline{\lambda}) \overline{X}^T X = 0$

$\overline{X}^T X$ is a sum of squares and is positive.

$\therefore (\lambda - \overline{\lambda}) = 0$

i.e., $\lambda = \overline{\lambda}$

i.e., λ is real

Hence all the eigen values of A are real.

(9) If λ is an eigen value of an orthogonal matrix A, then $\dfrac{1}{\lambda}$ is also an eigen value of A.

Proof: If λ is an eigen value of A, then $\dfrac{1}{\lambda}$ is an eigen value of A^{-1}. Since A is

orthogonal, $A^{-1} = A^T$. $\therefore \dfrac{1}{\lambda}$ is an eigen value of A^T. But the matrices A and A^T have

the same eigen values. Hence $\dfrac{1}{\lambda}$ is also an eigen value of A.

(10) The eigen vectors corresponding to distinct eigen values of a real symmetric matrix are othogonal.

Proof: Let λ_1 and λ_2 be any two distinct eigen values of the real symmetric matrix A and X_1, X_2 be the corresponding eigen vectors respectively.

Then $AX_1 = \lambda_1 X_1$ (1)

and $AX_2 = \lambda_2 X_2$ (2)

Using (1) $X_2^T (AX_1) = X_2^T (\lambda_1 X_1)$

i.e., $X_2^T AX_1 = \lambda_1 X_2^T X_1$

Taking transpose on both sides

$$X_1^T A^T X_2 = \lambda_1 X_1^T X_2$$

i.e., $X_1^T AX_2 = \lambda_1 X_1^T X_2$ $(\because A^T = A)$ (3)

Using (2) $X_1^T (AX_2) = X_1^T (\lambda_2 X_2)$

i.e., $X_1^T AX_2 = \lambda_2 X_1^T X_2$ (4)

From (3) and (4) $\lambda_1 X_1^T X_2 = \lambda_2 X_1^T X_2$

$\qquad$ i.e., $(\lambda_1 - \lambda_2)X_1^T X_2 = 0$

Since $\lambda_1 \neq \lambda_2$, $\qquad$ $X_1^T X_2 = 0$

i.e., the eigen vectors X_1 and X_2 are orthogonal.

(11) The eigen values of an triangular matrix are the diagonal elements.

(12) The eigen values of a diagonal matrix are the diagonal elements.

(13) The eigen values of the matrix A and the matrix $P^{-1}AP$ are the same.
$\qquad$ (i.e., similar matrices have the same eigen values)

Proof: Let $B = P^{-1}AP$

$$B - \lambda I = P^{-1}AP - \lambda I$$
$$= P^{-1}AP - \lambda P^{-1}P$$
$$= P^{-1}AP - P^{-1}\lambda P$$
$$= P^{-1}(A - \lambda I) P$$

$$\therefore |B - \lambda I| = |P^{-1}| |A - \lambda I| |P|$$
$$= |A - \lambda I| |P^{-1}| |P|$$
$$= |A - \lambda I| |P^{-1}P|$$
$$= |A - \lambda I| |I|$$
$$= |A - \lambda I|$$

$\therefore$ A and B have the same characteristic equation. Hence they have the same eigen values.

Example 1.2.5 Find the eigen values of $4A^2$ and A^{-1}, if $A = \begin{bmatrix} 4 & 1 \\ 3 & 2 \end{bmatrix}$

Solution: Eigen values of A are given by the characteristic equation $|A - \lambda I| = 0$.

$\qquad$ i.e., $\begin{vmatrix} 4-\lambda & 1 \\ 3 & 2-\lambda \end{vmatrix} = 0$

Expanding, we get,

$\qquad$ $(4 - \lambda)(2 - \lambda) - 3 = 0$

i.e., $\qquad$ $\lambda^2 - 6\lambda + 5 = 0$

i.e., $\qquad$ $(\lambda-5)(\lambda-1) = 0$

i.e., $\qquad$ $\lambda = 1, 5$

$\therefore$ Eigen values of A are 1 and 5.

Hence eigen values of A^2 are the squares of 1 and 5. i.e., 1^2 and 5^2

i.e., eigen values of A^2 are 1 and 25.

$\therefore$ eigen values of $4A^2$ are 4 and 100.

Eigen values of A^{-1} are the reciprocals of the eigen values of A. i.e., the reciprocals of 1 and 5. i.e., 1 and 1/5.

Example 1.2.6 Find the sum and product of the eigen values of the matrix

$$A = \begin{bmatrix} 6 & -2 & 2 \\ -2 & 3 & -1 \\ 2 & -1 & 3 \end{bmatrix}$$

Solution:

Sum of the eigen values of A = trace of A

$$= \text{Sum of the diagonal elements.}$$
$$= 6 + 3 + 3$$
$$= 12$$

Product of the eigen values of A = |A|

$$= \begin{vmatrix} 6 & -2 & 2 \\ -2 & 3 & -1 \\ 2 & -1 & 3 \end{vmatrix}$$

$$= 6(9-1) + 2(-6 + 2) + 2(2 - 6)$$
$$= 6(8) + 2(-4) + 2(-4)$$
$$= 48 - 8 - 8$$
$$= 32.$$

Example 1.2.7 Find the eigen values and eigen vectors of adj A, given that the matrix

Solution: $A = \begin{bmatrix} 2 & 2 & 1 \\ 1 & 3 & 1 \\ 1 & 2 & 2 \end{bmatrix}$

The characteristic equation of the matrix A is

$$|A - \lambda I| = \begin{vmatrix} 2-\lambda & 2 & 1 \\ 1 & 3-\lambda & 1 \\ 1 & 2 & 2-\lambda \end{vmatrix} = 0$$

i.e., $(2 - \lambda) [(3 - \lambda)(2 - \lambda) -2] -2[2 - \lambda -1] + 1[2 - (3 - \lambda)] = 0$

i.e., $(2 - \lambda) [\lambda^2 - 5\lambda + 4] -2[1 - \lambda] + (-1 + \lambda) = 0$

i.e., $(2 - \lambda) [\lambda^2 - 5\lambda + 4] + 3(\lambda -1) = 0$

i.e., $(2 - \lambda) [(\lambda -1)(\lambda - 4)] + 3(\lambda -1) = 0$

$(\lambda -1) [(2 - \lambda) (\lambda - 4) +3] = 0$

$(\lambda -1) [-\lambda^2 + 6\lambda - 5] = 0$

$- (\lambda -1) (\lambda -1)(\lambda-5) = 0$

The eigen values of A are $\lambda = 1, 1, 5$.

Eigen vectors of A are given by the system of equations $(A - \lambda I) X = 0$.

i.e.,
$$\begin{bmatrix} 2-\lambda & 2 & 1 \\ 1 & 3-\lambda & 1 \\ 1 & 2 & 2-\lambda \end{bmatrix} \begin{bmatrix} x_1 \\ x_2 \\ x_3 \end{bmatrix} = \begin{bmatrix} 0 \\ 0 \\ 0 \end{bmatrix}$$

Case 1. $\lambda = 1$

$$\begin{bmatrix} 1 & 2 & 1 \\ 1 & 2 & 1 \\ 1 & 2 & 1 \end{bmatrix} \sim \begin{bmatrix} 1 & 2 & 1 \\ 0 & 0 & 0 \\ 0 & 0 & 0 \end{bmatrix}$$

$\therefore \qquad x_1 + 2x_2 + x_3 = 0$ (only one equation)

Values of any two variables can be chosen arbitrarily. Choose $x_1 = s$, $x_2 = t$.

Then $x_3 = -s - 2t$. The eigen vectors corresponding to $\lambda = 1$ are given by $\begin{bmatrix} s \\ t \\ -s-2t \end{bmatrix}$

Put $s = 0$, $t = 1$. Then one eigen vector corresponding to $\lambda = 1$ is $\begin{bmatrix} 0 \\ 1 \\ -2 \end{bmatrix}$

Put $s = 1$, $t = 0$. Then we get another linearly independent eigen vector $\begin{bmatrix} 1 \\ 0 \\ -1 \end{bmatrix}$

Case 2. $\lambda = 5$

$$\begin{bmatrix} -3 & 2 & 1 \\ 1 & -2 & 1 \\ 1 & 2 & -3 \end{bmatrix} \sim \begin{bmatrix} 1 & 2 & -3 \\ 0 & -4 & 4 \\ 0 & 0 & 0 \end{bmatrix}$$

$\therefore \qquad x_1 + 2x_2 - 3x_3 = 0$ \hfill (1)

$\qquad\qquad -4x_2 + 4x_3 = 0$ \hfill (2)

$x_3 = 1$, $x_2 = 1$, $x_1 = 1$.

$\therefore$ The eigen vector is $\begin{bmatrix} 1 \\ 1 \\ 1 \end{bmatrix}$

The eigen values of A are 1, 1, 5

$\therefore$ The eigen values of A^{-1} are 1, 1, 1/5

i.e., the eigen values of $\dfrac{\text{adj } A}{|A|}$ are 1, 1, 1/5

$\therefore$ The eigen values of adj A are $|A|$, $|A|$, $\dfrac{|A|}{5}$

$$|A| = \begin{vmatrix} 2 & 2 & 1 \\ 1 & 3 & 1 \\ 1 & 2 & 2 \end{vmatrix} = 2(6-2) - 2(2-1) + 1(2-3)$$

$$= 8 - 2 - 1$$
$$= 5.$$

Hence the eigen values of adj A are 5, 5, 1.

Eigen vectors of adj A are same as those of $|A|.A^{-1}$.

But eigen vectors of $|A|.A^{-1}$ are same as those of A^{-1}, which are same as those of A.

$\therefore$ Eigen vectors of adj A are $\begin{bmatrix} 0 \\ 1 \\ -2 \end{bmatrix}$, $\begin{bmatrix} 1 \\ 0 \\ -1 \end{bmatrix}$, $\begin{bmatrix} 1 \\ 1 \\ 1 \end{bmatrix}$

EXERCISE 1.2

PART – A

1. Define characteristic equation of a matrix.
2. Define eigen values and eigen vectors of a matrix.
3. Show that the eigen values of a triangular matrix are equal to the elements of its main diagonal.
4. Prove that A and A^T have the same eigen values.
5. If λ is an eigen value of a non-singular matrix A, prove that $\dfrac{|A|}{\lambda}$ is an eigen value of adj A.
6. If $A = \begin{bmatrix} 4 & 1 \\ 3 & 2 \end{bmatrix}$, find the eigen values of A^5.
7. Find the eigen values of $A = \begin{bmatrix} 2 & 3 \\ 0 & 4 \end{bmatrix}$.
8. Find the sum and product of the eigen values of the matrix $A = \begin{bmatrix} 1 & 1 & 1 \\ 1 & 2 & 2 \\ 1 & 2 & 3 \end{bmatrix}$
9. If $A = \begin{bmatrix} 3 & 1 & 4 \\ 0 & 2 & 6 \\ 0 & 0 & 5 \end{bmatrix}$, find the eigen values of A^3 and A^{-1}.
10. If the eigen values of a matrix A are 2, 3, 4 find the eigen values of adj A.

11. If two eigen values of $A = \begin{bmatrix} 6 & -2 & 2 \\ -2 & 3 & -1 \\ 2 & -1 & 3 \end{bmatrix}$ are 2 and 8, find the third eigen value.

12. Find the product of the eigen values of the matrix $A = \begin{bmatrix} 1 & 0 & 0 \\ 0 & 3 & -1 \\ 0 & -1 & 3 \end{bmatrix}$

PART – B

Find the eigen values and the eigen vectors of the following matrices.

13. $\begin{bmatrix} 2 & 0 & -1 \\ 0 & 2 & 0 \\ -1 & 0 & 2 \end{bmatrix}$ 14. $\begin{bmatrix} 7 & -2 & -2 \\ -2 & 1 & 4 \\ -2 & 4 & 1 \end{bmatrix}$ 15. $\begin{bmatrix} 11 & -4 & -7 \\ 7 & -2 & -5 \\ 10 & -4 & -6 \end{bmatrix}$

16. $\begin{bmatrix} 6 & -6 & 5 \\ 14 & -13 & 10 \\ 7 & -6 & 4 \end{bmatrix}$ 17. $\begin{bmatrix} -15 & 4 & 3 \\ 10 & -12 & 6 \\ 20 & -4 & 2 \end{bmatrix}$ 18. $\begin{bmatrix} 2 & 2 & 1 \\ 1 & 3 & 1 \\ 1 & 2 & 2 \end{bmatrix}$ 19. $\begin{bmatrix} 3 & 10 & 5 \\ -2 & -3 & -4 \\ 3 & 5 & 7 \end{bmatrix}$

20. Show that the eigen vectors of the matrix $A = \begin{bmatrix} a & b \\ -b & a \end{bmatrix}$ are $\begin{bmatrix} 1 \\ i \end{bmatrix}$ and $\begin{bmatrix} 1 \\ -i \end{bmatrix}$

21. Verify that the sum of the eigen values of A equals the trace of A and that their

    product equals | A |, for the matrix $A = \begin{bmatrix} -2 & 2 & -3 \\ 2 & 1 & -6 \\ -1 & -2 & 0 \end{bmatrix}$

22. Find the eigen values and eigen vectors of adj A, given $A = \begin{bmatrix} 2 & 0 & -1 \\ 0 & 2 & 0 \\ -1 & 0 & 2 \end{bmatrix}$

23. Verify that the eigen vectors of the real symmetric matrix $A = \begin{bmatrix} 3 & -1 & 1 \\ -1 & 5 & -1 \\ 1 & -1 & 3 \end{bmatrix}$ are

    orthogonal in pairs.

24. Show that the matrix $A = \dfrac{1}{3} \begin{bmatrix} -1 & 2 & 2 \\ 2 & -1 & 2 \\ 2 & 2 & -1 \end{bmatrix}$ is orthogonal. Also verify that $1/\lambda$ is an

    eigen value of A, if λ is an eigen value and that the eigen values of A are of unit
    modulus.

25. Verify that the eigen values of A^2 and A^{-1} are respectively the squares and reciprocals

of the eigen values of A, given that $A = \begin{bmatrix} 3 & 1 & 4 \\ 0 & 2 & 6 \\ 0 & 0 & 5 \end{bmatrix}$

1.3 CAYLEY – HAMILTON THEOREM

Theorem 1.3.1 Every square matrix satisfies its own characteristic equation.

Proof: Let A be any square matrix of order n and $A = (a_{ij})$.
Let $|A - \lambda I| = a_0 \lambda^n + a_1 \lambda^{n-1} + \dots + a_{n-1} \lambda + a_n = 0$ (1)
be the characteristic equation of A.
We have to prove that $a_0 A^n + a_1 A^{n-1} + \dots + a_{n-1} A + a_n I = 0$.
Let $B = adj(A - \lambda I)$ (2).
The elements of B are cofactors of the elements of $|A - \lambda I|$. Therefore the elements of B
will be polynomials in λ of degree at most n-1. Therefore B can be written as
$B = B_0 \lambda^{n-1} + B_1 \lambda^{n-2} + \dots + B_{n-2} \lambda + B_{n-1} = 0$ (3)
where $B_0, B_1, \dots, B_{n-1}$ are matrices of order n, whose elements are dependent on the
elements of A. We have $(A - \lambda I) . adj(A - \lambda I) = |A - \lambda I|. I$ (4)
Using (1), (2) and (3) in (4), we get,
$(A - \lambda I)(B_0 \lambda^{n-1} + B_1 \lambda^{n-2} + \dots + B_{n-2} \lambda + B_{n-1}) = (a_0 \lambda^n + a_1 \lambda^{n-1} + \dots + a_{n-1} \lambda + a_n).I$
Equating the coefficients of various powers of λ on both sides, we get
$$-B_0 = a_0 I$$
$$AB_0 - B_1 = a_1 I$$
$$AB_1 - B_2 = a_2 I$$

$$AB_{n-2} - B_{n-1} = a_{n-1} I$$
$$AB_{n-1} = a_n I$$
Now pre-multiplying these equations by $A^n, A^{n-1}, \dots, A, I$ respectively and adding, we
get
$a_0 A^n + a_1 A^{n-1} + \dots + a_{n-1} A + a_n I = -A^n B_0 + (A^n B_0 - A^{n-1} B_1) + (A^{n-1} B_1 - A^{n-2} B_2) + \text{----}$
$$+ (A^2 B_{n-2} - AB_{n-1}) + AB_{n-1}$$
$$= 0$$
Hence A satisfies its characteristic equation.

Note:

(i)   Cayley-Hamilton theorem can be used for the computation of the inverse of
large matrices. We have $a_0 A^n + a_1 A^{n-1} + \dots + a_{n-1} A + a_n I = 0$
Multiplying by A^{-1}, we get
$a_0 A^{n-1} + a_1 A^{n-2} + \dots + a_{n-1} I + a_n A^{-1} = 0$
Hence $A^{-1} = (-1/a_n) [a_0 A^{n-1} + a_1 A^{n-2} + \dots + a_{n-1} I]$

1.3.1   REDUCTION OF A REAL MATRIX TO A DIAGONAL FORM

If D is a diagonal matrix, then the eigen values of D are its diagonal elements. Given any real square matrix A, if we can find a matrix M such that $M^{-1}AM = D$, a diagonal matrix, then A and D are similar. Further the diagonal elements of D are the eigen values of A.

Definition 1.3.1 The process of finding a matrix M such that $M^{-1}AM = D$, where D is a diagonal matrix, is called diagonalisation of the real matrix A.

Note:

(i)   Let A be a real square matrix of order n. The matrix M which diagonlises A is called the **modal matrix of A** and the resulting diagonal matrix D is called the **spectral matrix** of A. The following theorem provides a method of computing the modal matrix M of a given square matrix A, whose eigen values are distinct.

Theorem 1.3.2 If A is a real square matrix with distinct eigen values and M is the matrix whose columns are the eigen vectors of A, then A can be diagonalised by the similarity transformation $M^{-1}AM = D$, where D is a diagonal matrix whose diagonal elements are the eigen values of A.

Proof: Let λ_1, λ_2 ... λ_n be the distinct eigen values of A and X_1, X_2 ... X_n the corresponding eigen vectors. Then $A X_i = \lambda_i X_i$ (i = 1, 2 ... n) and X_1, X_2 ... X_n are linearly independent. Let M be the matrix whose columns are X_1, X_2 ... X_n.
i.e., $M = [X_1, X_2 ... X_n]$. As X_1, X_2 ... X_n are linearly independent, M is a non-singular matrix. Hence its inverse M^{-1} exists.
Now $AM = [AX_1, AX_2 ... A X_n]$
$$= [\lambda_1 X_1, \lambda_2 X_2 ... \lambda_n X_n]$$

$$= [X_1 \quad X_2 \quad \cdots \quad X_n] \begin{bmatrix} \lambda_1 & 0 & 0 & 0 & \cdots & 0 \\ 0 & \lambda_2 & 0 & 0 & \cdots & 0 \\ \cdots & \cdots & \cdots & \cdots & \cdots & \cdots \\ 0 & 0 & 0 & 0 & \cdots & \lambda_n \end{bmatrix}$$

i.e., $AM = M.D$
Multiplying both sides by M^{-1}, we get
$$M^{-1}AM = M^{-1}(MD)$$
$$= (M^{-1}M)D$$
$$= ID$$
$$M^{-1}AM = D$$

D is diagonal matrix whose diagonal elements are λ_1, λ_2 ...λ_n which are the eigen values of A.

Note:

(i)   The above diagonalisation process holds good even if two or more eigen values of A are equal, provided all the eigen vectors of A are linearly independent.

1.3.2 COMPUTATION OF THE POWERS OF A MATRIX

The diagonal form D of a matrix A can be used to compute the powers of A. A non singular matrix M can be found such that $D = M^{-1}AM$

Then $A = MDM^{-1}$

$$A^2 = (MDM^{-1})(MDM^{-1})$$
$$= MD^2M^{-1}$$

Similarly, $A^3 = MD^3M^{-1}$

In general, $A^k = MD^kM^{-1}$

If $D = \begin{bmatrix} \lambda_1 & 0 & 0 & 0 & \cdots & 0 \\ 0 & \lambda_{21} & 0 & 0 & \cdots & 0 \\ \cdots & \cdots & \cdots & \cdots & \cdots & \cdots \\ 0 & 0 & 0 & 0 & \cdots & \lambda_n \end{bmatrix}$ then $D^k = \begin{bmatrix} \lambda_1^k & 0 & 0 & 0 & \cdots & 0 \\ 0 & \lambda_2^k & 0 & 0 & \cdots & 0 \\ \cdots & \cdots & \cdots & \cdots & \cdots & \cdots \\ 0 & 0 & 0 & 0 & \cdots & \lambda_n^k \end{bmatrix}$

$\therefore A^k = M \begin{bmatrix} \lambda_1^k & 0 & 0 & 0 & \cdots & 0 \\ 0 & \lambda_2^k & 0 & 0 & \cdots & 0 \\ \cdots & \cdots & \cdots & \cdots & \cdots & \cdots \\ 0 & 0 & 0 & 0 & \cdots & \lambda_n^k \end{bmatrix} M^{-1}$ where M is the model matrix of A

Example 1.3.1 Verify Cayley-Hamilton theorem and find the inverse of

$$A = \begin{bmatrix} 3 & -4 & 2 \\ -2 & 1 & 0 \\ -1 & -1 & 1 \end{bmatrix}$$

Solution:

The characteristic equation of A is

$$\begin{vmatrix} 3-\lambda & -4 & 2 \\ -2 & 1-\lambda & 0 \\ -1 & -1 & 1-\lambda \end{vmatrix} = 0$$

i.e., $(3 - \lambda)[(1 - \lambda)(1 - \lambda) - 0] + 4[-2(1 - \lambda) - 0] + 2[2 + 1-\lambda] = 0$

i.e., $(3 - \lambda)(1 - 2\lambda + \lambda^2) - 8(1 - \lambda) + 2(3 - \lambda) = 0$

i.e., $3 - 6\lambda + 3\lambda^2 - \lambda + 2\lambda^2 - \lambda^3 - 8 + 8\lambda + 6 - 2\lambda = 0$

i.e., $-\lambda^3 + 5\lambda^2 - \lambda + 1 = 0$

$$\text{i.e., } \lambda^3 - 5\lambda^2 + \lambda - 1 = 0$$

Cayley-Hamilton theorem states that $A^3 - 5A^2 + A - I = 0$ (1)

To verify this compute A^2 and A^3

$$A^2 = \begin{bmatrix} 3 & -4 & 2 \\ -2 & 1 & 0 \\ -1 & -1 & 1 \end{bmatrix} \begin{bmatrix} 3 & -4 & 2 \\ -2 & 1 & 0 \\ -1 & -1 & 1 \end{bmatrix} = \begin{bmatrix} 15 & -18 & 8 \\ -8 & 9 & -4 \\ -2 & 2 & -1 \end{bmatrix}$$

$$A^3 = A.A^2 = \begin{bmatrix} 3 & -4 & 2 \\ -2 & 1 & 0 \\ -1 & -1 & 1 \end{bmatrix} \begin{bmatrix} 15 & -18 & 8 \\ -8 & 9 & -4 \\ -2 & 2 & -1 \end{bmatrix} = \begin{bmatrix} 73 & -86 & 38 \\ -38 & 45 & -20 \\ -9 & 11 & -5 \end{bmatrix}$$

$$A^3 - 5A^2 + A - I = \begin{bmatrix} 73 & -86 & 38 \\ -38 & 45 & -20 \\ -9 & 11 & -5 \end{bmatrix} - \begin{bmatrix} 75 & -90 & 40 \\ -40 & 45 & -20 \\ -10 & 10 & -5 \end{bmatrix}$$

$$+ \begin{bmatrix} 3 & -4 & 2 \\ -2 & 1 & 0 \\ -1 & -1 & 1 \end{bmatrix} - \begin{bmatrix} 1 & 0 & 0 \\ 0 & 1 & 0 \\ 0 & 0 & 1 \end{bmatrix}$$

$$= \begin{bmatrix} 0 & 0 & 0 \\ 0 & 0 & 0 \\ 0 & 0 & 0 \end{bmatrix}$$

Thus Cayley-Hamilton theorem is verified

Multiplying (1) by A^{-1}, we get

$$A^2 - 5A + I - A^{-1} = 0$$

$\therefore$ $A^{-1} = A^2 - 5A + I$

$$A^{-1} = \begin{bmatrix} 15 & -18 & 8 \\ -8 & 9 & -4 \\ -2 & 2 & -1 \end{bmatrix} - 5 \begin{bmatrix} 3 & -4 & 2 \\ -2 & 1 & 0 \\ -1 & -1 & 1 \end{bmatrix} + \begin{bmatrix} 1 & 0 & 0 \\ 0 & 1 & 0 \\ 0 & 0 & 1 \end{bmatrix}$$

$$= \begin{bmatrix} 1 & 2 & -2 \\ 2 & 5 & -4 \\ 3 & 7 & -5 \end{bmatrix}$$

Example 1.3.2 Given $A = \begin{bmatrix} 1 & 0 & 3 \\ 2 & 1 & -1 \\ 1 & -1 & 1 \end{bmatrix}$, use Cayley-Hamilton theorem to find the inverse of

A and also find A^4

Solution:

The characteristic equation of A is

$$\begin{vmatrix} 1-\lambda & 0 & 3 \\ 2 & 1-\lambda & -1 \\ 1 & -1 & 1-\lambda \end{vmatrix} = 0$$

i.e., $(1-\lambda)[(1-\lambda)(1-\lambda)-1] + 3[-2-(1-\lambda)] = 0$

i.e., $(1-\lambda)^3 - (1-\lambda) - 6 - 3 + 3\lambda = 0$

i.e., $1 - 3\lambda + 3\lambda^2 - \lambda^3 - 1 + \lambda - 9 + 3\lambda = 0$

i.e., $-\lambda^3 + 3\lambda^2 + \lambda - 9 = 0$

i.e., $\lambda^3 - 3\lambda^2 - \lambda + 9 = 0$

By Cayley-Hamilton theorem, $A^3 - 3A^2 - A + 9I = 0$

Multiplying by A^{-1}, $A^2 - 3A - I + 9A^{-1} = 0$

$\therefore \qquad A^{-1} = \dfrac{1}{9}[-A^2 + 3A + I]$

$$A^2 = \begin{bmatrix} 1 & 0 & 3 \\ 2 & 1 & -1 \\ 1 & -1 & 1 \end{bmatrix}\begin{bmatrix} 1 & 0 & 3 \\ 2 & 1 & -1 \\ 1 & -1 & 1 \end{bmatrix} = \begin{bmatrix} 4 & -3 & 6 \\ 3 & 2 & 4 \\ 0 & -2 & 5 \end{bmatrix}$$

$$A^{-1} = \frac{1}{9}\begin{bmatrix} -4 & 3 & -6 \\ -3 & -2 & -4 \\ 0 & 2 & -5 \end{bmatrix} + \begin{bmatrix} 3 & 0 & 9 \\ 6 & 3 & -3 \\ 3 & -3 & 3 \end{bmatrix} + \begin{bmatrix} 1 & 0 & 0 \\ 0 & 1 & 0 \\ 0 & 0 & 1 \end{bmatrix}$$

$$= \frac{1}{9}\begin{bmatrix} 0 & 3 & 3 \\ 3 & 2 & -7 \\ 3 & -1 & -1 \end{bmatrix}$$

To find A^4:

We have $A^3 - 3A^2 - A + 9I = 0$

i.e., $A^3 = 3A^2 + A - 9I$ \hfill (1)

Multiplying (1) by A, we get,

$A^4 = 3A^3 + A^2 - 9A$

$= 3(3A^2 + A - 9I) + A^2 - 9A$ \qquad using (1)

$= 10A^2 - 6A - 27I$

$$= 10\begin{bmatrix} 4 & -3 & 6 \\ 3 & 2 & 4 \\ 0 & -2 & 5 \end{bmatrix} - 6\begin{bmatrix} 1 & 0 & 3 \\ 2 & 1 & -1 \\ 1 & -1 & 1 \end{bmatrix} - 27\begin{bmatrix} 1 & 0 & 0 \\ 0 & 1 & 0 \\ 0 & 0 & 1 \end{bmatrix}$$

$$= \begin{bmatrix} 7 & -30 & 42 \\ 18 & -13 & 46 \\ -6 & -14 & 17 \end{bmatrix}$$

Example 1.3.4 Given $A = \begin{bmatrix} 5 & 3 \\ 1 & 3 \end{bmatrix}$, find A^n using Cayley-Hamilton theorem. Hence find A^4.

Solution:

The Characteristic equation of A is $\begin{bmatrix} 5 - \lambda & 3 \\ 1 & 3 - \lambda \end{bmatrix} = 0$

$$\text{i.e.,} \quad (5 - \lambda)(3 - \lambda) - 3 = 0$$
$$\text{i.e.,} \quad \lambda^2 - 8\lambda + 12 = 0$$
$$\text{i.e.,} \quad (\lambda - 2)(\lambda - 6) = 0$$

$\therefore$ The eigen values of A are $\lambda = 2, 6$.

By Cayley-Hamilton theorem $A^2 - 8A + 12I = 0$ (1)

When λ^n is divided by $\lambda^2 - 8\lambda + 12$, let the quotient be $Q(\lambda)$ and the remainder be $a\lambda + b$.

Then $\lambda^n = (\lambda^2 - 8\lambda + 12) Q(\lambda) + a\lambda + b$ (2)

Put $\lambda = 2$ in (2). $2^n = 2a + b$ (3)

Put $\lambda = 6$ in (2). $6^n = 6a + b$ (4)

Solving (3) and (4)

$$a = \frac{6^n - 2^n}{4}, \quad b = \frac{3 \cdot 2^n - 6^n}{2}$$

Put $\lambda = A$ in (2)

$$A^n = (A^2 - 8A + 12I) Q(A) + aA + bI$$
$$= aA + bI \quad \text{using (1)}$$

$$= \left(\frac{6^n - 2^n}{4} \right) \begin{bmatrix} 5 & 3 \\ 1 & 3 \end{bmatrix} + \left(\frac{3 \cdot 2^n - 6^n}{2} \right) \begin{bmatrix} 1 & 0 \\ 0 & 1 \end{bmatrix}$$

When $n = 4$ $A^4 = \left(\frac{6^4 - 2^4}{4} \right) \begin{bmatrix} 5 & 3 \\ 1 & 3 \end{bmatrix} + \left(\frac{3 \cdot 2^4 - 6^4}{2} \right) \begin{bmatrix} 1 & 0 \\ 0 & 1 \end{bmatrix}$

$$= 320 \begin{bmatrix} 5 & 3 \\ 1 & 3 \end{bmatrix} + (-624) \begin{bmatrix} 1 & 0 \\ 0 & 1 \end{bmatrix}$$

$$= \begin{bmatrix} 976 & 960 \\ 320 & 336 \end{bmatrix}$$

Example 1.3.5 Given the matrix $A = \begin{bmatrix} 1 & 1 & 1 \\ 1 & 2 & -3 \\ 2 & -1 & 3 \end{bmatrix}$,

compute $A^8 - 6A^7 + 5A^6 + 11A^5 + A^4 - 6A^3 + 7A^2 + 12A + I$

Solution:

The characteristic equation of A is $\begin{bmatrix} 1-\lambda & 1 & 1 \\ 1 & 2-\lambda & -3 \\ 2 & -1 & 3-\lambda \end{bmatrix} = 0$

i.e., $(1 - \lambda) [(2 - \lambda) (3 - \lambda) -3] -1[3 - \lambda + 6] + 1[-1 - 2(2 - \lambda)] = 0$

i.e., $(1 - \lambda) (6 - 5\lambda + \lambda^2 - 3) -1(9 - \lambda) + 1(2\lambda - 5) = 0$

i.e., $(1 - \lambda) (3 - 5\lambda + \lambda^2) - 9 + \lambda + 2\lambda - 5 = 0$

i.e., $3 - 5\lambda + \lambda^2 - 3\lambda + 5\lambda^2 - \lambda^3 - 9 + \lambda + 2\lambda - 5 = 0$

i.e., $-\lambda^3 + 6\lambda^2 - 5\lambda - 1 = 0$

i.e., $\lambda^3 - 6\lambda^2 + 5\lambda + 11 = 0$

By Cayley-Hamilton theorem we have, $A^3 - 6A^2 + 5A + 11.I = 0$ \hfill (1)

Now, $A^8 - 6A^7 + 5A^6 + 11A^5 + A^4 - 6A^3 + 7A^2 + 12A + I$

$= A^5 [A^3 - 6A^2 + 5A + 11.I] + A[A^3 - 6A^2 + 5A + 11.I] + 2A^2 + A + I$

$\qquad = A^5.0 + A.0 + 2A^2 + A + I \quad$ using (1)

$\qquad = 2A^2 + A + I$ \hfill (2)

Now $A^2 = \begin{bmatrix} 1 & 1 & 1 \\ 1 & 2 & -3 \\ 2 & -1 & 3 \end{bmatrix} \begin{bmatrix} 1 & 1 & 1 \\ 1 & 2 & -3 \\ 2 & -1 & 3 \end{bmatrix} = \begin{bmatrix} 4 & 2 & 1 \\ -3 & 8 & -14 \\ 7 & -3 & 14 \end{bmatrix}$

∴ Substituting in (2), the given polynomial in A

$= 2\begin{bmatrix} 4 & 2 & 1 \\ -3 & 8 & -14 \\ 7 & -3 & 14 \end{bmatrix} + \begin{bmatrix} 1 & 1 & 1 \\ 1 & 2 & -3 \\ 2 & -1 & 3 \end{bmatrix} + \begin{bmatrix} 1 & 0 & 0 \\ 0 & 1 & 0 \\ 0 & 0 & 1 \end{bmatrix} = \begin{bmatrix} 10 & 5 & 3 \\ -5 & 19 & -31 \\ 16 & -7 & 32 \end{bmatrix}$

Example 1.3.6 Find the similarity transformation, which diagonalises $A = \begin{bmatrix} 1 & 1 & 1 \\ 0 & 2 & 1 \\ -4 & 4 & 3 \end{bmatrix}$

Solution:

The characteristic equation of A is $\begin{bmatrix} 1-\lambda & 1 & 1 \\ 0 & 2-\lambda & 1 \\ -4 & 4 & 3-\lambda \end{bmatrix} = 0$

i.e., $(1 - \lambda) [(2 - \lambda) (3 - \lambda) - 4] -1[0 + 4] +1[0 + 4(2 - \lambda)] = 0$

i.e., $(1 - \lambda)(\lambda^2 - 5\lambda + 6 - 4) - 4 + 8 - 4\lambda = 0$

i.e., $(1 - \lambda)(\lambda^2 - 5\lambda + 2) + 4 - 4\lambda = 0$

i.e., $(1 - \lambda)(\lambda^2 - 5\lambda + 2 + 4) = 0$

$$\text{i.e.,} \quad (\lambda - 1)(\lambda^2 - 5\lambda + 6) = 0$$
$$\text{i.e.,} \quad (\lambda - 1)(\lambda - 2)(\lambda - 3) = 0$$

$\therefore$ The eigen values of A are $\lambda = 1, 2, 3$.

The eigen vectors are given by $\begin{bmatrix} 1-\lambda & 1 & 1 \\ 0 & 2-\lambda & 1 \\ -4 & 4 & 3-\lambda \end{bmatrix} \begin{bmatrix} x_1 \\ x_2 \\ x_3 \end{bmatrix} = \begin{bmatrix} 0 \\ 0 \\ 0 \end{bmatrix}$

Case 1. $\lambda = 1$

$$\begin{bmatrix} 0 & 1 & 1 \\ 0 & 1 & 1 \\ -4 & 4 & 2 \end{bmatrix} \sim \begin{bmatrix} -4 & 4 & 2 \\ 0 & 1 & 1 \\ 0 & 0 & 0 \end{bmatrix}$$

$$-4x_1 + 4x_2 + 2x_3 = 0$$
$$x_2 + x_3 = 0$$

A solution is, $x_3 = 2$, $x_2 = -2$, $x_1 = -1$

$\therefore$ An eigen vector $X_1 = \begin{bmatrix} -1 \\ -2 \\ 2 \end{bmatrix}$

Case 2. $\lambda = 2$

$$\begin{bmatrix} -1 & 1 & 1 \\ 0 & 0 & 1 \\ -4 & 4 & 1 \end{bmatrix} \sim \begin{bmatrix} -1 & 1 & 1 \\ 0 & 0 & 1 \\ 0 & 0 & 0 \end{bmatrix}$$

$$-x_1 + x_2 + x_3 = 0$$
$$x_3 = 0$$

A solution is, $x_3 = 0$, $x_2 = 1$, $x_1 = 1$

$\therefore$ An eigen vector $X_2 = \begin{bmatrix} 1 \\ 1 \\ 0 \end{bmatrix}$

Case 3. $\lambda = 3$

$$\begin{bmatrix} -2 & 1 & 1 \\ 0 & -1 & 1 \\ -4 & 4 & 0 \end{bmatrix} \sim \begin{bmatrix} -2 & 1 & 1 \\ 0 & -1 & 1 \\ 0 & 0 & 0 \end{bmatrix}$$

$$-2x_1 + x_2 + x_3 = 0$$
$$-x_2 + x_3 = 0$$

A solution of these equations is $x_3 = 1$, $x_2 = 1$, $x_1 = 1$

$\therefore$ An eigen vector $X_3 = \begin{bmatrix} 1 \\ 1 \\ 1 \end{bmatrix}$

Hence the modular matrix $M = \begin{bmatrix} -1 & 1 & 1 \\ -2 & 1 & 1 \\ 2 & 0 & 1 \end{bmatrix}$

$|M| = -1(1-0) -1(-2-2) + 1(-2) = -1 + 4 - 2 = 1$

Cofactor matrix $C = \begin{bmatrix} 1 & 4 & -2 \\ -1 & -3 & 2 \\ 0 & -1 & 1 \end{bmatrix}$

$\text{Adj}(M) = C^T = \begin{bmatrix} 1 & -1 & 0 \\ 4 & -3 & -1 \\ -2 & 2 & 1 \end{bmatrix}$

$M^{-1} = \dfrac{\text{Adj}(M)}{|M|} = \begin{bmatrix} 1 & -1 & 0 \\ 4 & -3 & -1 \\ -2 & 2 & 1 \end{bmatrix}$

$\therefore$ The similarity transformation is $M^{-1}AM = D$, where $D = \begin{bmatrix} 1 & 0 & 0 \\ 0 & 2 & 0 \\ 0 & 0 & 3 \end{bmatrix}$

We verify this:

$AM = \begin{bmatrix} 1 & 1 & 1 \\ 0 & 2 & 1 \\ -4 & 4 & 3 \end{bmatrix}\begin{bmatrix} -1 & 1 & 1 \\ -2 & 1 & 1 \\ 2 & 0 & 1 \end{bmatrix} = \begin{bmatrix} -1 & 2 & 3 \\ -2 & 2 & 3 \\ 2 & 0 & 3 \end{bmatrix}$

$M^{-1}AM = \begin{bmatrix} 1 & -1 & 0 \\ 4 & -3 & -1 \\ -2 & 2 & 1 \end{bmatrix}\begin{bmatrix} -1 & 2 & 3 \\ -2 & 2 & 3 \\ 2 & 0 & 3 \end{bmatrix} = \begin{bmatrix} 1 & 0 & 0 \\ 0 & 2 & 0 \\ 0 & 0 & 3 \end{bmatrix} = D$

Now $A^4 = MD^4M^{-1} = \begin{bmatrix} -1 & 1 & 1 \\ -2 & 1 & 1 \\ 2 & 0 & 1 \end{bmatrix}\begin{bmatrix} 1 & 0 & 0 \\ 0 & 16 & 0 \\ 0 & 0 & 81 \end{bmatrix} = \begin{bmatrix} 1 & -1 & 0 \\ 4 & -3 & -1 \\ -2 & 2 & 1 \end{bmatrix}$

$= \begin{bmatrix} -1 & 16 & 81 \\ -2 & 16 & 81 \\ 2 & 0 & 81 \end{bmatrix}\begin{bmatrix} 1 & -1 & 0 \\ 4 & -3 & -1 \\ -2 & 2 & 1 \end{bmatrix} = \begin{bmatrix} -99 & 115 & 65 \\ -100 & 116 & 65 \\ -160 & 160 & 81 \end{bmatrix}$

1.3.3   PROPERTIES OF ORTHOGONAL MATRICES

(1)   If A is orthogonal , then $|A| = \pm 1$

Proof: Let A be orthogonal. Then, we have, $A.A^T = A^T.A = I$.

$$\therefore \qquad |AA^T| = |I|$$

i.e., $|A|.|A^T| = 1$

i.e., $|A|.|A| = 1$, since $|A| = |A^T|$

i.e., $|A|^2 = 1$

i.e., $|A| = \pm 1$.

(2)   If A is orthogonal, $A^{-1} = A^T$.

(3)   If A and B are orthogonal, then AB is orthogonal.

Proof: If A is orthogonal, $AA^T = A^TA = I$ (1)

If B is orthogonal, $BB^T = B^TB = I$ (2)

Now $(AB)(AB)^T = (AB)(B^TA^T)$

$\qquad\qquad\qquad = A(BB^T)A^T$

$\qquad\qquad\qquad = A(I)A^T \quad$ by (2)

$\qquad\qquad\qquad = AA^T$

$\qquad\qquad\qquad = I \qquad\qquad$ by (1)

$\therefore$ AB is orthogonal.

(4)   If A is orthogonal, A^T and A^{-1} are orthogonal.

Proof: If A is orthogonal $A.A^T = A^TA = I$

But $A = (A^T)^T$

$\therefore (A^T)^TA^T = A^T(A^T)^T = I$

i.e., A^T is orthogonal.

When A is orthogonal, $A^{-1} = A^T \quad \therefore$ We get A^{-1} is orthogonal.

(5)   The eigen values of an orthogonal matrix have absolute value 1.

(6)   If λ is an eigen value of an orthogonal matrix A, then $\dfrac{1}{\lambda}$ is also an eigen value of A.

1.3.4   DIAGONALISATION BY ORTHOGONAL TRANSFORMATION

Let A be a real symmetric matrix. Then eigen vectors of A corresponding to distinct eigen values are linearly independent and pairwise orthogonal. If we divide each element of an eigen vector X_i by the square root of the sum of the squares of all the elements of X_i, we get a normalized eigen vector of A. The modal matrix N formed with the normalized eigen vectors of A is called the **normalized modal matrix.** The normalized modal matrix N is an orthogonal matrix and hence $N^{-1} = N^T$.

Thus the similarity transformation $M^{-1}AM = D$ becomes $N^TAN = D$. This transformation of the real symmetric matrix A into a diagonal matrix D by means of the normalized modal matrix N is called diagonalisation by orthogonal transformation or **orthogonal reduction**.

Example 1.3.7 Diagonalise the matrix $A = \begin{bmatrix} 8 & -6 & 2 \\ -6 & 7 & -4 \\ 2 & -4 & 3 \end{bmatrix}$ by orthogonal transformation.

Solution:

The characteristic equation of A is $\begin{vmatrix} 8-\lambda & -6 & 2 \\ -6 & 7-\lambda & -4 \\ 2 & -4 & 3-\lambda \end{vmatrix} = 0$

i.e, $(8 - \lambda)[(7 - \lambda)(3 - \lambda) - 16] + 6[-6(3 - \lambda) + 8] + 2[24 - 2(7 - \lambda)] = 0$

i.e., $(8 - \lambda)[21 - 10\lambda + \lambda^2 - 16] + 6[-18 + 6\lambda + 8] + 2[10 + 2\lambda] = 0$

i.e., $(8 - \lambda)[\lambda^2 - 10\lambda + 5] + 6[6\lambda - 10] + 2[10 + 2\lambda] = 0$

i.e., $8\lambda^2 - 80\lambda + 40 - \lambda^3 + 10\lambda^2 - 5\lambda + 36\lambda - 60 + 4\lambda + 20 = 0$

i.e., $-\lambda^3 + 18\lambda^2 - 45\lambda = 0$

i.e., $\lambda^3 - 18\lambda^2 + 45\lambda = 0$

i.e., $\lambda(\lambda^2 - 18\lambda + 45) = 0$

i.e., $\lambda(\lambda - 3)(\lambda - 15) = 0$

$\therefore$ The eigen values are $\lambda = 0, 3, 15$.

Case 1. $\lambda = 0$

The corresponding eigen vector is given by $\begin{bmatrix} 8 & -6 & 2 \\ -6 & 7 & -4 \\ 2 & -4 & 3 \end{bmatrix}\begin{bmatrix} x_1 \\ x_2 \\ x_3 \end{bmatrix} = \begin{bmatrix} 0 \\ 0 \\ 0 \end{bmatrix}$

$\begin{bmatrix} 8 & -6 & 2 \\ -6 & 7 & -4 \\ 2 & -4 & 3 \end{bmatrix} \sim \begin{bmatrix} 2 & -4 & 3 \\ 0 & -5 & 5 \\ 0 & 0 & 0 \end{bmatrix}$

$\therefore 2x_1 - 4x_2 + 3x_3 = 0$

$-5x_2 + 5x_3 = 0$

A solution is, $x_3 = 2, x_2 = 2, x_1 = 1$.

$\therefore$ Eigen vector corresponding to $\lambda = 0$ is $\begin{bmatrix} 1 \\ 2 \\ 2 \end{bmatrix}$

Case 2. $\lambda = 3$

The corresponding eigen vector is given by $\begin{bmatrix} 5 & -6 & 2 \\ -6 & 4 & -4 \\ 2 & -4 & 0 \end{bmatrix} \begin{bmatrix} x_1 \\ x_2 \\ x_3 \end{bmatrix} = \begin{bmatrix} 0 \\ 0 \\ 0 \end{bmatrix}$

$$\begin{bmatrix} 5 & -6 & 2 \\ -6 & 4 & -4 \\ 2 & -4 & 0 \end{bmatrix} \sim \begin{bmatrix} 2 & -4 & 0 \\ 0 & -8 & -4 \\ 0 & 0 & 0 \end{bmatrix}$$

$$\therefore \quad 2x_1 - 4x_2 = 0$$
$$-8x_2 - 4x_3 = 0$$

A solution is, $x_3 = -2$, $x_2 = 1$, $x_1 = 2$.

$\therefore$ Eigen vector corresponding to $\lambda = 3$ is $\begin{bmatrix} 2 \\ 1 \\ -2 \end{bmatrix}$

Case 3. $\lambda = 15$

The corresponding eigen vector is given by $\begin{bmatrix} -7 & -6 & 2 \\ -6 & -8 & -4 \\ 2 & -4 & -12 \end{bmatrix} \begin{bmatrix} x_1 \\ x_2 \\ x_3 \end{bmatrix} = \begin{bmatrix} 0 \\ 0 \\ 0 \end{bmatrix}$

$$\begin{bmatrix} -7 & -6 & 2 \\ -6 & -8 & -4 \\ 2 & -4 & -12 \end{bmatrix} \sim \begin{bmatrix} 2 & -4 & -12 \\ 0 & -20 & -40 \\ 0 & 0 & 0 \end{bmatrix}$$

$$\therefore \quad 2x_1 - 4x_2 - 12x_3 = 0$$
$$-20x_2 - 40x_3 = 0$$

A solution is, $x_3 = 1$, $x_2 = -2$, $x_1 = 2$.

$\therefore$ Eigen vector corresponding to $\lambda = 15$ is $\begin{bmatrix} 2 \\ -2 \\ 1 \end{bmatrix}$

The modal matrix $M = \begin{bmatrix} 1 & 2 & 2 \\ 2 & 1 & -2 \\ 2 & -2 & 1 \end{bmatrix}$

Normalizing each column vector, we get the normalized modal

matrix $N = \begin{bmatrix} 1/3 & 2/3 & 2/3 \\ 2/3 & 1/3 & -2/3 \\ 2/3 & -2/3 & 1/3 \end{bmatrix}$

∴ The orthogonal transformation that diagonalises A is $N^T AN = D$ where

$$D = \begin{bmatrix} 0 & 0 & 0 \\ 0 & 3 & 0 \\ 0 & 0 & 15 \end{bmatrix}$$

To verify $N^T AN = D$

$$N^T A = \begin{bmatrix} 1/3 & 2/3 & 2/3 \\ 2/3 & 1/3 & -2/3 \\ 2/3 & -2/3 & 1/3 \end{bmatrix} \begin{bmatrix} 8 & -6 & 2 \\ -6 & 7 & -4 \\ 2 & -4 & 3 \end{bmatrix} = \begin{bmatrix} 0 & 0 & 0 \\ 2 & 1 & -2 \\ 10 & -10 & 5 \end{bmatrix}$$

$$N^T AN = \begin{bmatrix} 0 & 0 & 0 \\ 2 & 1 & -2 \\ 10 & -10 & 5 \end{bmatrix} \begin{bmatrix} 1/3 & 2/3 & 2/3 \\ 2/3 & 1/3 & -2/3 \\ 2/3 & -2/3 & 1/3 \end{bmatrix} = \begin{bmatrix} 0 & 0 & 0 \\ 0 & 3 & 0 \\ 0 & 0 & 15 \end{bmatrix}$$

EXERCISE 1.3

PART – A

1. State Cayley Hamilton Theorem.
2. What are the applications of Cayley Hamilton Theorem?
3. What can you say about the eigen values of a matrix and its diagonal matrix?
4. When do you say that two matrices are similar?
5. State any property of similar matrices
6. What is meant by diagonalisation of a matrix?
7. Define similarity transformation.
8. How will you find A^k using a similarity transformation?
9. What type of matrices can be diagonalised using orthogonal transformation?

10. Verify Cayley Hamilton Theorem for the matrix $A = \begin{bmatrix} 2 & 3 \\ 1 & 4 \end{bmatrix}$

11. Find the inverse of $A = \begin{bmatrix} 2 & 3 \\ 3 & 5 \end{bmatrix}$ using Cayley Hamilton theorem.

12. Find A^3 using Cayley Hamilton theorem, given $A = \begin{bmatrix} -1 & 3 \\ 2 & 4 \end{bmatrix}$.

13. Define modal matrix of a given square matrix A.
14. Define normalised modal matrix of A.

15. Find the modal matrix that diagonalises the matrix $A = \begin{bmatrix} 5 & 3 \\ 1 & 3 \end{bmatrix}$.

PART – B

16. Verify Cayley Hamilton theorem for $A = \begin{bmatrix} 1 & 0 & 3 \\ 2 & 1 & -1 \\ 1 & -1 & 1 \end{bmatrix}$. Hence find A^{-1} and A^4.

17. Verify Cayley Hamilton theorem and find A^{-1}, given $A = \begin{bmatrix} 1 & 2 & -2 \\ 2 & 5 & -4 \\ 3 & 7 & -5 \end{bmatrix}$

18. Using Cayley Hamilton theorem find A^{-1} when $A = \begin{bmatrix} 1 & 2 & 3 \\ 0 & 1 & 2 \\ 0 & 0 & 1 \end{bmatrix}$

19. Using Cayley Hamilton theorem, find A^4 when $A = \begin{bmatrix} 2 & -2 & 1 \\ 0 & 1 & 2 \\ 1 & 0 & 1 \end{bmatrix}$

20. Find the characteristic equation of $A = \begin{bmatrix} 1 & 2 \\ 4 & 3 \end{bmatrix}$ and hence evaluate A^3 and A^{-1}.

21. Show that the matrix $A = \begin{bmatrix} a & b \\ c & d \end{bmatrix}$ satisfies its characteristic equation. Hence find A^{-1}

22. If $A = \begin{bmatrix} 5 & 3 \\ 1 & 3 \end{bmatrix}$, obtain A^n in terms of A. Hence find A^4.

23. Find A^n, using Cayley Hamilton theorem, when $A = \begin{bmatrix} 7 & 3 \\ 2 & 6 \end{bmatrix}$, hence find A^3.

24. Obtain the matrix $A^5 - 25A^2 + 122A$ when $A = \begin{bmatrix} 0 & 0 & 2 \\ 2 & 1 & 0 \\ -1 & -1 & 3 \end{bmatrix}$.

25. Using Cayley Hamilton theorem, compute $A^8 - 5A^7 + 7A^6 - 3A^5 + A^4 - 5A^3 + 8A^2 - 2A$
    $+ I$, if the matrix $A = \begin{bmatrix} 2 & 1 & 1 \\ 0 & 1 & 0 \\ 1 & 1 & 2 \end{bmatrix}$.

26. Using Cayley Hamilton theorem, compute $A^6 - 5A^5 + 8A^4 - 2A^3 - 9A^2 + 31A - 36\,I$, if
    the matrix $A = \begin{bmatrix} 1 & 0 & 3 \\ 2 & 1 & -1 \\ 1 & -1 & 1 \end{bmatrix}$.

27. Diagonalise the following matrices by similarity transformation.

(i) $\begin{bmatrix} 1 & 0 & 0 \\ 0 & 3 & -1 \\ 0 & -1 & 3 \end{bmatrix}$ (ii) $\begin{bmatrix} 2 & 2 & 1 \\ 1 & 3 & 1 \\ 1 & 2 & 2 \end{bmatrix}$ (iii) $\begin{bmatrix} 1 & 1 & 1 \\ 0 & 2 & 1 \\ -4 & 4 & 3 \end{bmatrix}$

(iv) $\begin{bmatrix} 11 & -4 & -7 \\ 7 & -2 & -5 \\ 10 & -4 & -6 \end{bmatrix}$ (v) $\begin{bmatrix} 2 & 2 & -7 \\ 2 & 1 & 2 \\ 0 & 1 & -3 \end{bmatrix}$ (vi) $\begin{bmatrix} 6 & -2 & 2 \\ -2 & 3 & 1 \\ 2 & -1 & 3 \end{bmatrix}$

28. Diagonalise the matrix $A = \begin{bmatrix} 1 & 1 & 3 \\ 1 & 5 & 1 \\ 3 & 1 & 1 \end{bmatrix}$ and hence find A^4.

29. Diagonalise the matrix $A = \begin{bmatrix} 3 & -1 & 1 \\ -1 & 5 & -1 \\ 1 & -1 & 3 \end{bmatrix}$ and hence find A^4.

30. Find the matrix M which transforms $\begin{bmatrix} 3 & 1 & 4 \\ 2 & 2 & 4 \\ 4 & 1 & 3 \end{bmatrix}$ to diagonal form.

31. Diagonalise the following matrices by means of an orthogonal transformation.

(i) $\begin{bmatrix} 10 & -2 & -5 \\ -2 & 2 & 5 \\ -5 & 3 & 5 \end{bmatrix}$ (ii) $\begin{bmatrix} 2 & -1 & 1 \\ -1 & 2 & -1 \\ 1 & -1 & 2 \end{bmatrix}$ (iii) $\begin{bmatrix} 2 & 0 & 4 \\ 0 & 6 & 0 \\ 4 & 0 & 2 \end{bmatrix}$

(iv) $\begin{bmatrix} 1 & -1 & 0 \\ -1 & 2 & 1 \\ 0 & 1 & 1 \end{bmatrix}$ (v) $\begin{bmatrix} 1 & 1 & 1 \\ 0 & 2 & 1 \\ -4 & 4 & 3 \end{bmatrix}$

1.4   REDUCTION OF A QUADRATIC FORM TO CANONOCAL FORM

Definition 1.4.1 A homogeneous polynomial of the second degree in any number of variables is called a **quadratic form**.

For example $10x_1^2 + 2x_2^2 + 5x_3^2 + 6x_2x_3 - 10x_3x_1 - 4x_1x_2$ (1)

is a quadratic form in the three variables x_1, x_2 and x_3.

The general quadratic form Q in the n variables $x_1, x_2, ..., x_n$ is

$$Q = c_{11}x_1^2 + c_{12}x_1x_2 + \cdots + c_{1n}x_1x_n$$
$$+ c_{21}x_2x_1 + c_{22}x_2^2 + \cdots + c_{2n}x_2x_n$$
$$+ \cdots \quad \cdots \cdots \cdots \cdots \quad \cdots$$
$$+ c_{n1}x_nx_1 + c_{n2}x_nx_2 + \cdots + c_{nn}x_n^2$$

i.e. $Q = \sum\limits_{j=1}^{n}\sum\limits_{i=1}^{n} c_{ij}x_ix_j$

In general c_{ij} and c_{ji} need not be equal. The coefficient of x_ix_j in Q is $c_{ij} + c_{ji}$. If we define $a_{ij} = \dfrac{1}{2}(c_{ij} + c_{ji})$ then $a_{ii} = c_{ii}$, $a_{ij} = a_{ji}$ and $a_{ij} + a_{ji} = 2a_{ij} = c_{ij} + c_{ji}$.

$\therefore$ $Q = \sum\limits_{j=1}^{n}\sum\limits_{i=1}^{n} a_{ij}x_ix_j$ where $a_{ij} = a_{ji} = \dfrac{1}{2}(c_{ij} + c_{ji})$

$\therefore$ The matrix $A = [a_{ij}]$ is symmetric and in matrix form, $Q = X^T A X$ where $X = \begin{bmatrix} x_1 \\ x_2 \\ \vdots \\ x_n \end{bmatrix}$

The matrix A is called the matrix of the quadratic form Q. To determine A, the coefficient of x_i^2 is placed in the a_{ii} position and $\dfrac{1}{2}$(coefficient of $x_i x_j$) is placed in each of the a_{ij} and a_{ji} positions.

For example, the matrix of the quadratic form (1) is $\begin{bmatrix} 10 & -2 & -5 \\ -2 & 2 & 3 \\ -5 & 3 & 5 \end{bmatrix}$

If $Q = X^T A X$. then $|A|$ is called the **determinant or modulus of Q.** The rank r of the matrix A is called **the rank of the quadratic form Q.** If A is singular, the quadratic form Q is called **singular.** Otherwise, it is **non-singular.**

1.4.1 LINEAR TRANSFORMATION OF A QUADRATIC FORM

$Q = X^T A X$ be a quadratic form in the n variables x_1, x_2, ..., x_n. P be a non-singular matrix of order n. The transformation $X = PY$ that transforms the variable $X = [x_1, x_2, ..., x_n]^T$ to the variable $Y = [y_1, y_2, ..., y_n]^T$ is called a **non-singular linear transformation.** By this transformation, the quadratic form is transformed as follows.

$$Q = X^T A X$$
$$= (PY)^T A (PY)$$
$$= Y^T (P^T A P) Y$$
$$= Y^T B Y \quad \text{where } B = P^T A P$$
$$B^T = (P^T A P)^T = P^T A^T P = P^T A P = B \quad (\text{Since A is symmetric, } A^T = A)$$

$\therefore$ B is a symmetric matrix.

Hence $Y^T B Y$ is a quadratic form in the variables y_1, y_2, ..., y_n. Thus under the linear transformation $X = PY$ the quadratic form $X^T A X$ is transformed into a quadratic form $Y^T B Y$ where $B = P^T A P$.

1.4.2 CANONICAL FORM OF A QUADRATIC FORM

If P in the linear transformation $X = PY$ is chosen such that $B = P^T A P$ is a diagonal matrix

of the form $\begin{bmatrix} \lambda_1 & 0 & 0 & 0 & \cdots & 0 \\ 0 & \lambda_2 & 0 & 0 & \cdots & 0 \\ \cdots & \cdots & \cdots & \cdots & \cdots & \cdots \\ 0 & 0 & 0 & 0 & \cdots & \lambda_n \end{bmatrix}$, then the quadratic form $Q = X^T A X$ gets reduced to as $Q = Y^T BY$

$$= [y_1, y_2, \cdots, y_n] \begin{bmatrix} \lambda_1 & 0 & 0 & 0 & \cdots & 0 \\ 0 & \lambda_2 & 0 & 0 & \cdots & 0 \\ \cdots & \cdots & \cdots & \cdots & \cdots & \cdots \\ 0 & 0 & 0 & 0 & \cdots & \lambda_n \end{bmatrix} \begin{bmatrix} y_1 \\ y_2 \\ \vdots \\ y_n \end{bmatrix}$$

$$= \lambda_1 y_1^2 + \lambda_2 y_1^2 + \cdots + \lambda_n y_n^2$$

This form of Q is called the **canonical form**. It is also called the sum of the square form of Q.

1.4.3 ORTHOGONAL REDUCTION OF A QUADRATIC FORM TO THE CANONICAL FORM

The transformation X = PY is called an **orthogonal transformation** if P is an orthogonal matrix. If the transformation X =PY where P is an orthogonal matrix, transform the quadratic form Q to the canonical form, then Q is said be reduced to the canonical form by an orthogonal transformation.

Let $Q = X^T AX$. If N is the normalized modal matrix of A, then N is an orthogonal matrix and $N^T AN = D$ is a diagonal matrix with eigen values of A as diagonal elements. Thus the orthogonal transformation X = NY will reduce the quadratic form $Q = X^T AX$ to the canonical form $Y^T DY$ where $D = N^T AN$ is the diagonal matrix with the eigen values of A as the diagonal elements.

1.4.4 NATURE OF QUADRATIC FORMS

Definition: 1.4.2 The quadratic form $Q = X^T AX$ in n variables is said to be

(i)   Positive definite if $X^T AX > 0$ $\forall X \neq 0$

(ii)   Negative definite if $X^T AX < 0$ $\forall X \neq 0$

(iii)   Positive semidefinite if $X^T AX \geq 0$ $\forall X$

(iv)   Negative semidefinite if $X^T AX \leq 0$ $\forall X$

(v)   Indefinite in all other cases. i.e., $X^T AX$ takes positive and negative values.

1.4.5 PROCEDURE TO CHECK FOR THE NATURE OF A QUADRATIC FORM

In the quadratic form $Q = X^T AX$, let the rank of the matrix A be r. When Q is reduced to the canonical form, it will contain only r terms. These r terms in the canonical form may be positive or negative. The number of positive terms in the canonical form is called the **index of the quadratic form**, denoted by p. The difference p-(r-p) between the number of positive terms and the number of negative terms is called the **signature of the quadratic form**, denoted by s.
Thus s = p - (r - p) = 2p - r.

The quadratic form $Q = X^T A X$ is

(i)   Positive definite, if $r = n$ and $p = n$
        i.e., if all the eigen values of A are positive

(ii)   Negative definite, if $r = n$ and $p = 0$
        i.e., if all the eigen values of A are negative.

(iii)   Positive semidefinite if $r < n$ and $p = r$
        i.e., if all the eigen values of A are non-negative and atleast one eigen value is zero.

(iv)   Negative semidefinite if $r < n$ and $p = 0$
        i.e., if all the eigen values of A are non-positive and atleast one eigen value is zero.

(v)   Indefinite in all other cases.
        i.e., if A has positive as well as negative eigen values.

1.4.6   ALTERNATIVE METHOD TO FIND THE NATURE OF A QUADRATIC FORM

Let $Q = X^T A X$ where $A = (a_{ij})$ is a real symmetric matrix of order n. The following method is used to determine the nature of this quadratic form without reducing it to the canonical form. Let $D_1 = |a_{11}|$

$$D_2 = \begin{vmatrix} a_{11} & a_{12} \\ a_{21} & a_{22} \end{vmatrix} \qquad D_3 = \begin{vmatrix} a_{11} & a_{12} & a_{13} \\ a_{21} & a_{22} & a_{23} \\ a_{31} & a_{32} & a_{33} \end{vmatrix} \quad \text{etc}$$

and $D_n = |A|$.

$D_1, D_2 \ldots D_n$ are called the **principal sub-determinants** or **principal minors of A**.

The quadratic form $Q = X^T A X$ is

(i)   Positive definite, if $D_1, D_2 \ldots D_n$ are all positive.
        i.e. $D_n > 0$ for all n.

(ii)   Negative definite, if $D_1, D_3, D_5 \ldots$ are all negative and $D_2, D_4, D_6 \ldots$ are all positive.
        i.e. $(-1)^n D_n > 0$ for all n.

(iii)   Positive semidefinite, if $D_n \geq 0$ and atleast one principal minor is zero.

(iv)   Negative semidefinite, if $(-1)^n D_n \geq 0$ and atleast one principal minor is zero.

(v)   Indefinite in all other cases.

1.4.7   SIMULTANEOUS REDUCTION OF A PAIR OF QUADRATIC FORMS TO CANONICAL FORMS

Theorem 1.4.1 If A and B are two symmetric matrices such that the roots of the equation $|A - \lambda B| = 0$ are all distinct, then there exists a matrix P such that $P^T A P$ and $P^T B P$ are both diagonal matrices.

Let $X^T AX$ and $X^T BX$ be the two quadratic forms. They can be simultaneously reduced to canonical forms by the same linear transformation using the above theorem. The procedure for this is given below.

(1) Form the characteristic equation $|A - \lambda B| = 0$ let the roots of this equation (eigen values) be $\lambda_1, \lambda_2 \ldots \lambda_n$.

(2) Find the eigen vectors $X_1, X_2 \ldots X_n$ corresponding to $\lambda_1, \lambda_2, \ldots, \lambda_n$ solving the equations $(A - \lambda_i B)X_i = 0$ $i = 1, 2, \ldots, n$.

(3) Form the matrix P whose column vectors are $X_1, X_2 \ldots X_n$. Then $X = PY$ is the linear transformation that transforms the quadratic forms to canonical forms.

(4) Compute $P^T AP$ and $P^T BP$, which are diagonal matrices.

(5) The quadratic forms corresponding to these diagonal matrices are the required canonical forms.

Example 1.4.1 Reduce the quadratic form $8x_1^2 + 7x_2^2 + 3x_3^2 - 12x_1x_2 + 4x_1x_3 - 8x_2x_3$ to the canonical form by an orthogonal transformation. Find also the rank, index, signature and the nature of the quadratic form.

Solution:

The matrix of the quadratic form is $A = \begin{bmatrix} 8 & -6 & 2 \\ -6 & 7 & -4 \\ 2 & -4 & 3 \end{bmatrix}$

The eigen values of this matrix are 0, 3 and 15 (refer example 1.19) and the

corresponding eigen vectors are $X_1 = \begin{bmatrix} 1 \\ 2 \\ 2 \end{bmatrix}$, $X_2 = \begin{bmatrix} 2 \\ 1 \\ -2 \end{bmatrix}$, $X_3 = \begin{bmatrix} 2 \\ -2 \\ 1 \end{bmatrix}$

The normalized modal matrix is $N = \begin{bmatrix} 1/3 & 2/3 & 2/3 \\ 2/3 & 1/3 & -2/3 \\ 2/3 & -2/3 & 1/3 \end{bmatrix}$

and $N^T AN = D = \begin{bmatrix} 0 & 0 & 0 \\ 0 & 3 & 0 \\ 0 & 0 & 15 \end{bmatrix}$

Now the orthogonal transformation $X = NY$ will reduce the given quadratic form to the canonical form $0y_1^2 + 3y_2^2 + 15y_3^2$.

Also rank = 2, index = 2, signature = 2. The quadratic form is positive semidefinite.

Example 1.4.2 Find the orthogonal transformation which reduces the quadratic form $2x_1^2 + 2x_2^2 + 2x_3^2 - 2x_1x_2 - 2x_2x_3 + 2x_1x_3$ into the canonical form. Determine the rank, index, signature and the nature of the quadratic form.

Solution:

The matrix of the quadratic form is $A = \begin{bmatrix} 2 & -1 & 1 \\ -1 & 2 & -1 \\ 1 & -1 & 2 \end{bmatrix}$

The characteristic equation of A is $\begin{vmatrix} 2-\lambda & -1 & 1 \\ -1 & 2-\lambda & -1 \\ 1 & -1 & 2-\lambda \end{vmatrix} = 0$

Expanding $\lambda^3 - 6\lambda^2 + 9\lambda - 4 = 0$

$\lambda = 1$ is a root

Dividing $\lambda^3 - 6\lambda^2 + 9\lambda - 4$ by $\lambda - 1$,

$$\begin{array}{r} 1 \quad -6 \quad 9 \quad -4 \\ 0 \quad 1 \quad -5 \quad 4 \\ \hline 1 \quad -5 \quad 4 \quad \underline{|0} \end{array}$$

The remaining roots are given by $\lambda^2 - 5\lambda + 4 = 0$

$$\lambda^2 - 5\lambda + 4 = (\lambda - 1)(\lambda - 4) = 0$$

i.e., $\lambda = 1, 4$

$\therefore$ The eigen values of A are $\lambda = 4, 1, 1$

Case 1. $\lambda = 4$

The eigen vectors are given by $\begin{bmatrix} 2-4 & -1 & 1 \\ -1 & 2-4 & -1 \\ 1 & -1 & 2-4 \end{bmatrix} \begin{bmatrix} x_1 \\ x_2 \\ x_3 \end{bmatrix} = \begin{bmatrix} 0 \\ 0 \\ 0 \end{bmatrix}$

$$\begin{bmatrix} -2 & -1 & 1 \\ -1 & -2 & -1 \\ 1 & -1 & -2 \end{bmatrix} \sim \begin{bmatrix} 1 & -1 & -2 \\ 0 & -3 & -3 \\ 0 & 0 & 0 \end{bmatrix}$$

$$\therefore \quad x_1 - x_2 - 2x_3 = 0$$
$$-3x_2 - 3x_3 = 0$$

A solution is $x_3 = 1, x_2 = -1, x_1 = 1$.

$\therefore$ The corresponding eigen vector is $X_1 = \begin{bmatrix} 1 \\ -1 \\ 1 \end{bmatrix}$

Case 2. $\lambda = 1$

The eigen vectors are given by $\begin{bmatrix} 2-1 & -1 & 1 \\ -1 & 2-1 & -1 \\ 1 & -1 & 2-1 \end{bmatrix} \begin{bmatrix} x_1 \\ x_2 \\ x_3 \end{bmatrix} = \begin{bmatrix} 0 \\ 0 \\ 0 \end{bmatrix}$

$$\begin{bmatrix} 1 & -1 & 1 \\ -1 & 1 & -1 \\ 1 & -1 & 1 \end{bmatrix} \sim \begin{bmatrix} 1 & -1 & 1 \\ 0 & 0 & 0 \\ 0 & 0 & 0 \end{bmatrix}$$

$$\therefore \quad x_1 - x_2 + x_3 = 0$$

Put $x_3 = 0$. We get $x_1 = x_2 = 1$. Let $x_1 = x_2 = 1$

$\therefore$ The eigen vector corresponding to $\lambda = 1$ is $X_2 = \begin{bmatrix} 1 \\ 1 \\ 0 \end{bmatrix}$

X_1 and X2 are orthogonal as $X_1.X_2 = 1 \cdot 0 + (-1) \cdot 1 + 1 \cdot 1 = 0$.

To find another vector $X_3 = \begin{bmatrix} a \\ b \\ c \end{bmatrix}$ corresponding to $\lambda = 1$ such that it is orthogonal to both

X_1 and X_2 and satisfies $x_1 - x_2 + x_3 = 0$

i.e., $\quad X_1.X_3 = 0, \quad X_2.X_3 = 0$ and $a - b + c = 0$

i.e., $\quad 1.a - 1.b + 1.c = 0, \ 1.a + 1.b + 0.c = 0$ and $a - b + c = 0$.

i.e., $\quad a - b + c = 0$ and $\quad a + b = 0$

i.e., $\quad a = -b$ and $\quad c = 2b$

Put $b = 1$, so that $a = -1, c = 2$

$\therefore \quad X_3 = \begin{bmatrix} -1 \\ 1 \\ 2 \end{bmatrix}$

The model matrix is $\begin{bmatrix} 1 & 1 & -1 \\ -1 & 1 & 1 \\ 1 & 0 & 2 \end{bmatrix}$

Hence the normalized model matrix is $N = \begin{bmatrix} 1/\sqrt{3} & 1/\sqrt{2} & -1/\sqrt{6} \\ -1/\sqrt{3} & 1/\sqrt{2} & 1/\sqrt{6} \\ 1/\sqrt{3} & 0 & 2/\sqrt{6} \end{bmatrix}$

$\therefore$ The required orthogonal transformation is $X = NY$

i.e., $\quad \begin{bmatrix} x_1 \\ x_2 \\ x_3 \end{bmatrix} = \begin{bmatrix} 1/\sqrt{3} & 1/\sqrt{2} & -1/\sqrt{6} \\ -1/\sqrt{3} & 1/\sqrt{2} & 1/\sqrt{6} \\ 1/\sqrt{3} & 0 & 2/\sqrt{6} \end{bmatrix} \begin{bmatrix} y_1 \\ y_2 \\ y_3 \end{bmatrix}$

i.e., $\quad x_1 = \dfrac{1}{\sqrt{3}} y_1 + \dfrac{1}{\sqrt{2}} y_2 - \dfrac{1}{\sqrt{6}} y_3$

$\quad x_2 = -\dfrac{1}{\sqrt{3}} y_1 + \dfrac{1}{\sqrt{2}} y_2 + \dfrac{1}{\sqrt{6}} y_3$

$$x_3 = \frac{1}{\sqrt{3}} y_1 + \frac{2}{\sqrt{6}} y_3$$

The canonical form of the given quadratic form is $\lambda_1 y_1^2 + \lambda_2 y_2^2 + \lambda_3 y_3^2$

i.e., $\quad Q = 4y_1^2 + y_2^2 + y_3^2$

Rank of the quadratic form = 3, index = 3, signature = 3. The quadratic form is positive definite.

Example 1.4.3 Reduce the quadratic form $10x_1^2 + 2x_2^2 + 5x_3^2 + 6x_2x_3 - 10x_3x_1 - 4x_1x_2$ to a canonical form by orthogonal reduction. Find also a set of non-zero values of x_1, x_2, x_3 which will make the quadratic form zero.

Solution:

The matrix of the quadratic form is $A = \begin{bmatrix} 10 & -2 & -5 \\ -2 & 2 & 3 \\ -5 & 3 & 5 \end{bmatrix}$

The characteristic equation is $|A - \lambda I| = 0$ i.e. $\begin{vmatrix} 10-\lambda & -2 & -5 \\ -2 & 2-\lambda & 3 \\ -5 & 3 & 5-\lambda \end{vmatrix} = 0$

$$(10 - \lambda)(\lambda^2 - 7\lambda + 10 - 9) + 2(-10 + 2\lambda + 15) - 5(-6 + 10 - 5\lambda) = 0$$

i.e., $10\lambda^2 - 70\lambda + 10 - \lambda^3 + 7\lambda^2 - \lambda + 10 + 4\lambda - 20 + 25\lambda = 0$

i.e., $-\lambda^3 + 17\lambda^2 - 42\lambda = 0$

i.e., $\lambda^3 - 17\lambda^2 + 42\lambda = 0$

i.e., $\lambda(\lambda - 3)(\lambda - 14) = 0$

$$\lambda = 0, 3, 14.$$

$\therefore$ The eigen values are 0, 3 and 14.

The eigen vectors are given by

$$\begin{bmatrix} 10-\lambda & -2 & -5 \\ -2 & 2-\lambda & 3 \\ -5 & 3 & 5-\lambda \end{bmatrix} \begin{bmatrix} x_1 \\ x_2 \\ x_3 \end{bmatrix} = \begin{bmatrix} 0 \\ 0 \\ 0 \end{bmatrix}$$

Case 1. $\lambda = 0$

$$\begin{bmatrix} 10 & -2 & -5 \\ -2 & 2 & 3 \\ -5 & 3 & 5 \end{bmatrix} \sim \begin{bmatrix} -2 & 2 & 3 \\ 0 & 8 & 10 \\ 0 & 0 & 0 \end{bmatrix}$$

$$\therefore -2x_1 + 2x_2 + 3x_3 = 0$$
$$8x_2 + 10x_3 = 0$$

A solution is $x_3 = 4$, $x_2 = -5$, $x_1 = 1$.

∴ The eigen vector corresponding to $\lambda = 0$ is $\begin{bmatrix} 1 \\ -5 \\ 4 \end{bmatrix}$

Case 2. $\lambda = 3$

$$\begin{bmatrix} 7 & -2 & -5 \\ -2 & -1 & 3 \\ -5 & 3 & 2 \end{bmatrix} \sim \begin{bmatrix} -2 & -1 & 3 \\ 0 & -11 & 11 \\ 0 & 0 & 0 \end{bmatrix}$$

$$\therefore \quad -2x_1 - x_2 + 3x_3 = 0$$
$$-11x_2 + 11x_3 = 0$$

A solution is $x_3 = 1$, $x_2 = 1$, $x_1 = 1$.

∴ The eigen vector corresponding to $\lambda = 3$ is $\begin{bmatrix} 1 \\ 1 \\ 1 \end{bmatrix}$

Case 3. $\lambda = 14$

The eigen vectors are given by

$$\begin{bmatrix} -4 & -2 & -5 \\ -2 & -12 & 3 \\ -5 & 3 & -9 \end{bmatrix} \sim \begin{bmatrix} -2 & -12 & 3 \\ 0 & 22 & -11 \\ 0 & 0 & 0 \end{bmatrix}$$

$$\therefore \quad -2x_1 - 12x_2 + 3x_3 = 0$$
$$22x_2 - 11x_3 = 0$$

A solution is $x_3 = 2$, $x_2 = 1$, $x_1 = -3$.

∴ The eigen vector corresponding to $\lambda = 14$ is $\begin{bmatrix} -3 \\ 1 \\ 2 \end{bmatrix}$

The modal matrix is $M = \begin{bmatrix} 1 & 1 & -3 \\ -5 & 1 & 1 \\ 4 & 1 & 2 \end{bmatrix}$

∴ The normalized modal matrix is $N = \begin{bmatrix} 1/\sqrt{42} & 1/\sqrt{3} & -3/\sqrt{14} \\ -5/\sqrt{42} & 1/\sqrt{3} & 1/\sqrt{14} \\ 4/\sqrt{42} & 1/\sqrt{3} & 2/\sqrt{14} \end{bmatrix}$

Also $\quad N^T A N = D = \begin{bmatrix} 0 & 0 & 0 \\ 0 & 3 & 0 \\ 0 & 0 & 14 \end{bmatrix}$

The orthogonal transformation $X = NY$ transforms the given quadratic form to the canonical form $0.y_1^2 + 3.y_2^2 + 14.y_3^2$

i.e., $\qquad Q = 3.y_2^2 + 14.y_3^2$

When $y_2 = 0$, $y_3 = 0$ and y_1 = arbitrary, the canonical form of the quadratic form is zero.

Putting $y_1 = \sqrt{42}$, $y_2 = 0$, $y_3 = 0$ in the orthogonal transformation $X = NY$

i.e., $\qquad x_1 = \dfrac{1}{\sqrt{42}} y_1 + \dfrac{1}{\sqrt{3}} y_2 - \dfrac{3}{\sqrt{14}} y_3$

$$x_2 = -\dfrac{5}{\sqrt{42}} y_1 + \dfrac{1}{\sqrt{3}} y_2 + \dfrac{1}{\sqrt{14}} y_3$$

$$x_3 = \dfrac{4}{\sqrt{42}} y_1 + \dfrac{1}{\sqrt{3}} y_2 + \dfrac{2}{\sqrt{14}} y_3$$

we get $x_1 = 1$, $x_2 = -5$, $x_3 = 4$.

Example 1.4.4 Determine the nature of the quadratic form

$x_1^2 + 2x_2^2 + 3x_3^2 + 2x_1x_2 + 2x_2x_3 - 2x_3x_1$ without reducing it to canonical form.

Solution:

Matrix of the quadratic form is $A = \begin{bmatrix} 1 & 1 & -1 \\ 1 & 2 & 1 \\ -1 & 1 & 3 \end{bmatrix}$

$$D_1 = |1| = 1 \quad D_2 = \begin{vmatrix} 1 & 1 \\ 1 & 2 \end{vmatrix} = 2 - 1 = 1$$

$$D_3 = \begin{vmatrix} 1 & 1 & -1 \\ 1 & 2 & 1 \\ -1 & 1 & 3 \end{vmatrix} = 1(6-1) - 1(3+1) - 1(1+2)$$

$$= 6 - 4 - 3$$

$$= -1$$

$D_1 > 0$, but $D_3 < 0$

$\therefore$ The quadratic form is indefinite.

Example 1.4.5 Reduce the quadratic forms $3x_1^2 + 6x_2^2 + 2x_3^2 - 4x_2x_3 + 8x_1x_2$ and $5x_1^2 + 5x_2^2 + x_3^2 - 2x_2x_3 + 8x_1x_2$ simultaneously to canonical forms by a real non-singular transformation.

Solution:

The matrices of the quadratic forms are $A = \begin{bmatrix} 3 & 4 & 0 \\ 4 & 6 & -2 \\ 0 & -2 & 2 \end{bmatrix}$ and $B = \begin{bmatrix} 5 & 4 & 0 \\ 4 & 5 & -1 \\ 0 & -1 & 1 \end{bmatrix}$

The characteristic equation is $|A - \lambda B| = 0$.

i.e., $\begin{vmatrix} 3-5\lambda & 4-4\lambda & 0 \\ 4-4\lambda & 6-5\lambda & -2+\lambda \\ 0 & -2+\lambda & 2-\lambda \end{vmatrix} = 0$

i.e., $(3 - 5\lambda) [(6 - 5\lambda)(2 - \lambda) - (-2 + \lambda)^2] - (4 - 4\lambda)[(4 - 4\lambda)(2 - \lambda) - 0] = 0$

i.e., $(3 - 5\lambda)(2 - \lambda)[6 - 5\lambda - (2 - \lambda)] - 4(1 - \lambda).4(1 - \lambda)(2 - \lambda) = 0$

i.e., $(2 - \lambda)(3 - 5\lambda)(4 - 4\lambda) - 16(1 - \lambda)^2(2 - \lambda) = 0$

i.e., $4(2 - \lambda)(1 - \lambda)[3 - 5\lambda - 4(1 - \lambda)] = 0$

i.e., $4(2 - \lambda)(1 - \lambda)(-1 - \lambda) = 0$

$\lambda = 1, 2, -1$

The eigen values are $\lambda = 2, 1, -1$

The eigen vectors are given by $\begin{bmatrix} 3-5\lambda & 4-4\lambda & 0 \\ 4-4\lambda & 6-5\lambda & -2+\lambda \\ 0 & -2+\lambda & 2-\lambda \end{bmatrix} \begin{bmatrix} x_1 \\ x_2 \\ x_3 \end{bmatrix} = \begin{bmatrix} 0 \\ 0 \\ 0 \end{bmatrix}$

When $\lambda = 2$, the eigen vector is given by

$-7x_1 - 4x_2 \quad = 0$

$-4x_1 - 4x_2 \quad = 0$

$x_1 = 0, x_2 = 0, x_3 = 1$ is a solution.

$\therefore$ The eigen vector corresponding to $\lambda = 2$ is $X_1 = \begin{bmatrix} 0 \\ 0 \\ 1 \end{bmatrix}$

When $\lambda = 1$, the eigen vector is given by

$-2x_1 \quad = 0$

$x_2 - x_3 \quad = 0$

$-x_2 + x_3 \quad = 0$

$x_1 = 0, x_2 = 1, x_3 = 1$ is a solution.

$\therefore$ The eigen vector corresponding to $\lambda = 1$ is $X_2 = \begin{bmatrix} 0 \\ 1 \\ 1 \end{bmatrix}$.

When $\lambda = -1$, the eigen vector is given by

$8x_1 + 8x_2 \quad = 0$

$8x_1 + 11x_2 - 3x_3 \quad = 0$

$-3x_2 + 3x_3 \quad = 0$

Solving $x_3 = x_2$ and $x_1 = -x_2$

Let $x_2 = -1$, so that $x_3 = -1$ and $x_1 = 1$.

$\therefore$ The eigen vector corresponding to $\lambda = -1$ is $X_3 = \begin{bmatrix} 1 \\ -1 \\ -1 \end{bmatrix}$

The matrix P whose column vectors are X_1, X_2 and X_3 is $\begin{bmatrix} 0 & 0 & 1 \\ 0 & 1 & -1 \\ 1 & 1 & -1 \end{bmatrix}$. P is a real non-singular matrix.

$$P^T AP = \begin{bmatrix} 0 & 0 & 1 \\ 0 & 1 & -1 \\ 1 & 1 & -1 \end{bmatrix} \begin{bmatrix} 3 & 4 & 0 \\ 4 & 6 & -2 \\ 0 & -2 & 2 \end{bmatrix} \begin{bmatrix} 0 & 0 & 1 \\ 0 & 1 & -1 \\ 1 & 1 & -1 \end{bmatrix} = \begin{bmatrix} 2 & 0 & 0 \\ 0 & 4 & 0 \\ 0 & 0 & -1 \end{bmatrix}$$

$$P^T BP = \begin{bmatrix} 0 & 0 & 1 \\ 0 & 1 & -1 \\ 1 & 1 & -1 \end{bmatrix} \begin{bmatrix} 5 & 4 & 0 \\ 4 & 5 & -1 \\ 0 & -1 & 1 \end{bmatrix} \begin{bmatrix} 0 & 0 & 1 \\ 0 & 1 & -1 \\ 1 & 1 & -1 \end{bmatrix} = \begin{bmatrix} 1 & 0 & 0 \\ 0 & 4 & 0 \\ 0 & 0 & 1 \end{bmatrix}$$

$\therefore$ The corresponding canonical forms are $2y_1^2 + 4y_2^2 - y_3^2$ and $y_1^2 + 4y_2^2 + y_3^2$

The non-singular linear transformation that reduces the given quadratic forms to the above canonical forms is X = PY

i.e., $\begin{bmatrix} x_1 \\ x_2 \\ x_3 \end{bmatrix} = \begin{bmatrix} 0 & 0 & 1 \\ 0 & 1 & -1 \\ 1 & 1 & -1 \end{bmatrix} \begin{bmatrix} y_1 \\ y_2 \\ y_3 \end{bmatrix}$

i.e., $x_1 = y_3$

$x_2 = y_2 - y_3$

$x_3 = y_1 + y_2 - y_3$

EXERCISE 1.4

PART – A

1. Define quadratic form.

2. Write the matrix of the quadratic form $4x_1^2 + 2x_2^2 - 3x_3^2 + 2x_1x_2 + 4x_3x_1$

3. Write down the matrix of the quadratic form $x_1^2 + 2x_2^2 - 7x_3^2 - 4x_1x_2 + 8x_1x_3 + 5x_2x_3$

4. Write down the quadratic form corresponding to the symmetric matrix $\begin{bmatrix} 1 & 2 \\ 2 & -4 \end{bmatrix}$.

5. Define rank of a quadratic form.

6. When do you say that a quadratic form is singular ?

7. What is meant by canonical form of a quadratic form ?

8. If the transformation X = PY reduces the quadratic form $X^T AX$ to the canonical form $Y^T BY$, what is B?

9. Determine the orthogonal transformation that reduces the quadratic form $X^T AX$ to the canonical form.

10. Define index and signature of a quadratic form.

11. Find the rank, index and signature of the quadratic form $x_1^2 + x_2^2 - 4x_3^2$.

12. State the conditions for a quadratic form $X^T AX$ in n variables to be positive semidefinite.

13. What is the nature of the quadratic form $x_1^2 - 2x_2^2 + 5x_3^2$.

14. Find the nature of the quadratic form $2x^2 + 2xy + 3y^2$

15. Find the nature of the quadratic form $10x^2 + 2y^2 + 5z^2 + 6yz - 10zx - 4xy$.

PART – B

16. Reduce the following quadratic form to a canonical form by orthogonal reduction and find the rank, index, signature and the nature of the quadratic form : $-x^2 + y^2 + 4yz + 4zx$.

17. Reduce the following quadratic form to a canonical form: $8x^2 + 7y^2 + 3z^2 - 12xy - 8yz + 4zx$.

18. Reduce the quadratic form $3x_1^2 + 5x_2^2 + 3x_3^2 - 2x_2x_3 + 2x_3x_1 - 2x_1x_2$ to a canonical form by orthogonal reduction.

19. Find the orthogonal transformation which will reduce the quadratic form: $2x_1x_2 + 2x_1x_3 - 2x_2x_3$ to the canonical form and hence find its nature.

20. Show that the quadratic form: $2x_1^2 + x_2^2 + x_3^2 + 2x_1x_2 - 2x_1x_3 - 4x_2x_3$ is indefinite and find two sets of values of x_1, x_2, x_3 for which the form assumes positive and negative values.

21. Reducing the quadratic form: $8x_1^2 + 7x_2^2 + 3x_3^2 - 12x_1x_2 - 8x_2x_3 + 4x_3x_1$ to the canonical form; find a set of non-zero values of x_1, x_2, x_3 that will make the quadratic form zero.

22. Determine the nature of the following quadratic forms without reducing it to canonical form

    (i) $6x_1^2 + 3x_2^2 + 14x_3^2 + 4x_2x_3 + 18x_3x_1 + 4x_1x_2$

    (ii) $x_1^2 + 2x_2^2 + x_3^2 - 2x_1x_2 + 2x_2x_3$

    (iii) $x_1^2 + 2x_2^2 + 3x_3^2 + 2x_2x_3 - 2x_3x_1 + 2x_1x_2$

23. Reduce the given quadratic forms $6x_1^2 + 3x_2^2 + 14x_3^2 + 4x_2x_3 + 18x_3x_1 + 4x_1x_2$ and $2x_1^2 + 5x_2^2 + 4x_1x_2 + 2x_3x_1$ simultaneously to canonical forms by a real non-singular transformation.

24. Find the real non-singular transformation that reduces the quadratic forms $2x_1^2 + 2x_2^2 + 3x_3^2 + 2x_1x_2 - 4x_2x_3 - 4x_3x_1$ and $2x_2x_3 - 2x_1x_2 - x_2^2$ simultaneously to the canonical forms.

ANSWERS

EXERCISE 1.1

(2) n　　　　　　(3) 4　　　　　(4) $\begin{vmatrix} a_1 & b_1 & c_1 \\ a_2 & b_2 & c_2 \\ a_3 & b_3 & c_3 \end{vmatrix} = 0$

(10) $a = 8$　　　(14) $a = -4, b = 6$

(15) No unique solution for any value of λ

(16). $\lambda \neq 1$, $\mu =$ any value

(17) $\lambda = 2$, $\mu = -3$.　　　　(18) $\lambda = 3$

(19) 2　　　　(20) 2　　　　(21) 3

(22) $2x_1 + x_2 - x_3 = 0$

(23) $3x_1 - x_2 + x_3 = 0$

(24) $x_1 - x_2 + x_3 - x_4 = 0$

(28) Inconsistent　　　(29) consistent; $x = -1, y = 1, z = 2$

(30) Consistent; $x = 2s - 1, y = 3 - 2s, z = s$

(31) $a = 8, b = 15$ infinite number of solutions, $a = 8, b \neq 15$ no solution, $a \neq 8$ unique solution

(32) $a + 2b - c = 0$

EXERCISE 1.2

(6) 1, 3125　　　　　　(7) 2, 4　　　(8) 6, 1

(9) the eigen values of A^3 are 27, 8, 125, the eigen values of A^{-1} are 1/3, ½, 1/5

(10) 12, 8, 6　(11) 2　　(12) 8

(13) $\lambda = 1, 2, 3$ $\begin{bmatrix} 1 \\ 0 \\ 1 \end{bmatrix}$, $\begin{bmatrix} 0 \\ 1 \\ 0 \end{bmatrix}$, $\begin{bmatrix} 1 \\ 0 \\ -1 \end{bmatrix}$

(14) $\lambda = 3, -3, 9$ $\begin{bmatrix} 1 \\ 1 \\ 1 \end{bmatrix}$, $\begin{bmatrix} 0 \\ -1 \\ 1 \end{bmatrix}$, $\begin{bmatrix} 2 \\ -1 \\ -1 \end{bmatrix}$

(15) $\lambda = 0, 1, 2$ $\begin{bmatrix} 1 \\ 1 \\ 1 \end{bmatrix}$, $\begin{bmatrix} 1 \\ -1 \\ 2 \end{bmatrix}$, $\begin{bmatrix} 2 \\ 1 \\ 2 \end{bmatrix}$

(16) $\lambda = -1, -1, -1$ $\begin{bmatrix} -5 \\ 0 \\ 7 \end{bmatrix}$, $\begin{bmatrix} 6 \\ 7 \\ 0 \end{bmatrix}$, $\begin{bmatrix} 0 \\ 5 \\ 6 \end{bmatrix}$ and general form of eigen vector is $\begin{bmatrix} \dfrac{6}{7}k_1 - \dfrac{5}{7}k_2 \\ k_1 \\ k_2 \end{bmatrix}$

(17) $\lambda = -10, -20, 5$ $\begin{bmatrix} 1 \\ 2 \\ -1 \end{bmatrix}$, $\begin{bmatrix} 2 \\ -1 \\ 2 \end{bmatrix}$, $\begin{bmatrix} 1 \\ 2 \\ 4 \end{bmatrix}$

(18) $\lambda = 1, 1, 5$ $\begin{bmatrix} -2s-t \\ s \\ t \end{bmatrix}\begin{bmatrix} 1 \\ 1 \\ 1 \end{bmatrix}$ or $\begin{bmatrix} 1 \\ 0 \\ -1 \end{bmatrix}$, $\begin{bmatrix} -2 \\ 1 \\ 0 \end{bmatrix}$, $\begin{bmatrix} 1 \\ 1 \\ 1 \end{bmatrix}$

(19) $\lambda = 2, 2, 3$ $\begin{bmatrix} 5 \\ 2 \\ -5 \end{bmatrix}$, $\begin{bmatrix} 5 \\ 2 \\ -5 \end{bmatrix}$, $\begin{bmatrix} 1 \\ 1 \\ -2 \end{bmatrix}$

(21) $\lambda = -3, -3, 5$ (22) $\lambda = 6, 3, 2$ $\begin{bmatrix} 1 \\ 0 \\ 1 \end{bmatrix}$, $\begin{bmatrix} 0 \\ 1 \\ 0 \end{bmatrix}$, $\begin{bmatrix} 1 \\ 0 \\ -1 \end{bmatrix}$

(23) $\lambda = 2, 3, 6$ $\begin{bmatrix} 1 \\ 0 \\ -1 \end{bmatrix}$, $\begin{bmatrix} 1 \\ 1 \\ 1 \end{bmatrix}$, $\begin{bmatrix} 1 \\ -2 \\ 1 \end{bmatrix}$

EXERCISE 1.3

(11) $\begin{bmatrix} 5 & -3 \\ -3 & 2 \end{bmatrix}$ (12) $\begin{bmatrix} -19 & 57 \\ 38 & 76 \end{bmatrix}$ (15) $M = \begin{bmatrix} 1 & 3 \\ -1 & 1 \end{bmatrix}$

(16) $A^{-1} = \dfrac{1}{9}\begin{bmatrix} 0 & 3 & 3 \\ 3 & 2 & -7 \\ 3 & -1 & -1 \end{bmatrix}$ $A^4 = \begin{bmatrix} 7 & -30 & 42 \\ 18 & -13 & 46 \\ -6 & -14 & 17 \end{bmatrix}$

(17) $A^{-1} = \begin{bmatrix} 3 & -4 & 2 \\ -2 & 1 & 0 \\ -1 & -1 & 1 \end{bmatrix}$

(18) $\begin{bmatrix} 1 & -2 & 1 \\ 0 & 1 & -2 \\ 0 & 0 & 1 \end{bmatrix}$ (19) $\begin{bmatrix} 10 & -34 & -31 \\ 24 & -19 & 10 \\ 17 & -24 & -7 \end{bmatrix}$

(20) $\lambda^2 - 4\lambda - 5 = 0$, $A^3 = \begin{bmatrix} 41 & 42 \\ 84 & 83 \end{bmatrix}$ $A^{-1} = -\dfrac{1}{5} \begin{bmatrix} 3 & -2 \\ -4 & 1 \end{bmatrix}$

(21) $\dfrac{1}{ad - bc} \begin{bmatrix} d & -b \\ -c & a \end{bmatrix}$

(22) $A^n = \dfrac{6^n - 2^n}{4} \begin{bmatrix} 5 & 3 \\ 1 & 3 \end{bmatrix} + \dfrac{3.2^n - 6^n}{2} \begin{bmatrix} 1 & 0 \\ 0 & 1 \end{bmatrix}$ $A^4 = \begin{bmatrix} 976 & 960 \\ 320 & 336 \end{bmatrix}$

(23) $A^n = \dfrac{9^n - 4^n}{5} \begin{bmatrix} 7 & 3 \\ 2 & 6 \end{bmatrix} + \dfrac{9.4^n - 4.9^n}{5} \begin{bmatrix} 1 & 0 \\ 0 & 1 \end{bmatrix}$; $\begin{bmatrix} 463 & 399 \\ 266 & 330 \end{bmatrix}$

(24) $\begin{bmatrix} -34 & 0 & -20 \\ -20 & -54 & 0 \\ 10 & 10 & -74 \end{bmatrix}$ (25) $\begin{bmatrix} 8 & 5 & 5 \\ 0 & 3 & 0 \\ 5 & 5 & 8 \end{bmatrix}$ (26) $\begin{bmatrix} 0 & 0 & 0 \\ 0 & 0 & 0 \\ 0 & 0 & 0 \end{bmatrix}$

(27) (i) $M = \begin{bmatrix} 1 & 0 & 0 \\ 0 & 1 & 1 \\ 0 & 1 & -1 \end{bmatrix}$; $\begin{bmatrix} 1 & 0 & 0 \\ 0 & 2 & 0 \\ 0 & 0 & 4 \end{bmatrix}$ (ii) $M = \begin{bmatrix} 0 & 1 & 1 \\ 1 & 0 & 1 \\ -2 & -1 & 1 \end{bmatrix}$; $\begin{bmatrix} 1 & 0 & 0 \\ 0 & 1 & 0 \\ 0 & 0 & 5 \end{bmatrix}$

(iii) $M = \begin{bmatrix} -1 & 1 & 1 \\ -2 & 1 & 1 \\ 2 & 0 & 1 \end{bmatrix}$; $\begin{bmatrix} 1 & 0 & 0 \\ 0 & 2 & 0 \\ 0 & 0 & 3 \end{bmatrix}$ (iv) $M = \begin{bmatrix} 1 & 1 & -2 \\ 1 & -1 & -1 \\ 1 & 2 & -2 \end{bmatrix}$; $\begin{bmatrix} 0 & 0 & 0 \\ 0 & 1 & 0 \\ 0 & 0 & 2 \end{bmatrix}$

(v) $M = \begin{bmatrix} 1 & 5 & 3 \\ -4 & 6 & -2 \\ -1 & 1 & 2 \end{bmatrix}$; $\begin{bmatrix} 1 & 0 & 0 \\ 0 & 3 & 0 \\ 0 & 0 & -4 \end{bmatrix}$ (vi) $M = \begin{bmatrix} 2 & 1 & 1 \\ -1 & 0 & 2 \\ 1 & -2 & 0 \end{bmatrix}$; $\begin{bmatrix} 8 & 0 & 0 \\ 0 & 2 & 0 \\ 0 & 0 & 2 \end{bmatrix}$

(28) $\begin{bmatrix} -2 & 0 & 0 \\ 0 & 3 & 0 \\ 0 & 0 & 6 \end{bmatrix}$; $A^4 = \begin{bmatrix} 251 & 475 & 235 \\ 475 & 1051 & 475 \\ 235 & 475 & 251 \end{bmatrix}$

(29) $\begin{bmatrix} 2 & 0 & 0 \\ 0 & 3 & 0 \\ 0 & 0 & 6 \end{bmatrix}$; $A^4 = \begin{bmatrix} 251 & -405 & 235 \\ -405 & 891 & -405 \\ 235 & -405 & 251 \end{bmatrix}$

(30) $M = \begin{bmatrix} -4 & 1 & 1 \\ -4 & -6 & 1 \\ 5 & 1 & 1 \end{bmatrix}$

(31) (i) $N = \begin{bmatrix} 1/\sqrt{42} & 1/\sqrt{3} & -3/\sqrt{14} \\ -5/\sqrt{42} & 1/\sqrt{3} & 1/\sqrt{14} \\ 4/\sqrt{42} & 1/\sqrt{3} & 2/\sqrt{14} \end{bmatrix}$ $D = \begin{bmatrix} 0 & 0 & 0 \\ 0 & 3 & 0 \\ 0 & 0 & 14 \end{bmatrix}$

(ii) $N = \begin{bmatrix} 1/\sqrt{3} & 1/\sqrt{2} & -1/\sqrt{6} \\ -1/\sqrt{3} & 1/\sqrt{2} & 1/\sqrt{6} \\ 1/\sqrt{3} & 0 & 2/\sqrt{6} \end{bmatrix}$ $D = \begin{bmatrix} 4 & 0 & 0 \\ 0 & 1 & 0 \\ 0 & 0 & 1 \end{bmatrix}$

(iii) $N = \begin{bmatrix} 1/\sqrt{2} & 1/\sqrt{2} & 0 \\ 0 & 0 & 1 \\ -1/\sqrt{2} & 1/\sqrt{2} & 0 \end{bmatrix}$ $D = \begin{bmatrix} -2 & 0 & 0 \\ 0 & 6 & 0 \\ 0 & 0 & 6 \end{bmatrix}$

(iv) $N = \begin{bmatrix} 1/\sqrt{3} & 1/\sqrt{2} & -1/\sqrt{6} \\ 1/\sqrt{3} & 0 & 2/\sqrt{6} \\ -1/\sqrt{3} & 1/\sqrt{2} & 1/\sqrt{6} \end{bmatrix}$ $D = \begin{bmatrix} 0 & 0 & 0 \\ 0 & 1 & 0 \\ 0 & 0 & 3 \end{bmatrix}$

(v) $N = \begin{bmatrix} 1/3 & 1/\sqrt{2} & 1/\sqrt{3} \\ 2/3 & 1/\sqrt{2} & 1/\sqrt{3} \\ -2/3 & 0 & 1/\sqrt{3} \end{bmatrix}$ $D = \begin{bmatrix} 1 & 0 & 0 \\ 0 & 2 & 0 \\ 0 & 0 & 3 \end{bmatrix}$

EXERCISE 1.4

(2) $\begin{bmatrix} 4 & 1 & 2 \\ 1 & 2 & 0 \\ 2 & 0 & -3 \end{bmatrix}$ (3) $\begin{bmatrix} 1 & -2 & 4 \\ -2 & 2 & 2.5 \\ 4 & 2.5 & -7 \end{bmatrix}$

(4) $x_1^2 - 4x_2^2 + 4x_1 x_2$ (8) $B = P^T A P$

(11) Rank = 3, Index = 2, Signature = 1   (13) Indefinite

(14) Positive definite

(15) eigen values are 0 ,3, 14 $\therefore$ Positive semi definite

(16) $N = \begin{bmatrix} 2/3 & 1/3 & 2/3 \\ -2/3 & 2/3 & 1/3 \\ 1/3 & 2/3 & -2/3 \end{bmatrix}$ Q.F $= 0.y_1^2 + 3y_2^2 - 3y_3^2$

Rank = 2, Index = 1, Signature = 0, Q.F is indefinite

(17) $N = \begin{bmatrix} 1/3 & 2/3 & 2/3 \\ 2/3 & 1/3 & -2/3 \\ 2/3 & -2/3 & 1/3 \end{bmatrix}$ Q.F $= 0.y_1^2 + 3y_2^2 + 15y_3^2$

(18) $N = \begin{bmatrix} 1/\sqrt{2} & 1/\sqrt{3} & 1/\sqrt{6} \\ 0 & 1/\sqrt{3} & -2/\sqrt{6} \\ -1/\sqrt{2} & 1/\sqrt{3} & 1/\sqrt{6} \end{bmatrix}$ Q.F $= 2y_1^2 + 3y_2^2 + 6y_3^2$

(19) $X = NY$ where $N = \begin{bmatrix} 1/\sqrt{2} & 1/\sqrt{6} & -1/\sqrt{3} \\ 0 & 2/\sqrt{6} & 1/\sqrt{3} \\ 1/\sqrt{2} & -1/\sqrt{6} & 1/\sqrt{3} \end{bmatrix}$ Q.F $= y_1^2 + y_2^2 - 2y_3^2$ Q.F indefinite.

(20) Q.F $= -y_1^2 + y_2^2 + 4y_3^2$

$x_1 = \dfrac{2}{\sqrt{6}} y_2 + \dfrac{1}{\sqrt{3}} y_3, \quad x_2 = \dfrac{1}{\sqrt{2}} y_1 - \dfrac{1}{\sqrt{6}} y_2 + \dfrac{1}{\sqrt{3}} y_3, \quad x_3 = \dfrac{1}{\sqrt{2}} y_1 + \dfrac{1}{\sqrt{6}} y_2 - \dfrac{1}{\sqrt{3}} y_3$

When $x_1 = 0$, $x_2 = 1$, $x_3 = 1$, Q.F is negative

When $x_1 = 3$, $x_2 = 0$, $x_3 = 0$, Q.F is positive

(21) $N = \begin{bmatrix} 1/3 & 2/3 & 2/3 \\ 2/3 & 1/3 & -2/3 \\ 2/3 & -2/3 & 1/3 \end{bmatrix}$; $3y_2^2 + 15y_3^2$; $x_1 = 1$, $x_2 = 2$, $x_3 = 2$.

(22) (i) Positive definite   (ii) Positive semidefinite   (iii) Indefinite

(23) $y_1^2 + y_2^2 + y_3^2$, $-y_1^2 + 5y_2^2 + y_3^2$

(24) $P = \begin{bmatrix} 1 & 0 & 1 \\ 0 & 1 & 1 \\ 1 & 1 & 1 \end{bmatrix}$, $y_1^2 + y_2^2 + y_3^2$, $y_2^2 - y_3^2$

CHAPTER 2

THREE DIMENSIONAL ANALYTICAL GEOMETRY

2.0   INTRODUCTION

The three dimensional analytical geometry or solid geometry is a natural extension of the analytical geometry of the plane. To represent a point in a plane we need only two coordinates, whereas to represent a point in space we require three coordinates.

Let XOX′, YOY′ and ZOZ′ be three mutually perpendicular straight lines in space intersecting at a point O (Fig. 2.1). Let P be any point in space. Through P draw planes parallel to the planes YOZ, ZOX and XOY (Fig. 2.2). Let these planes meet the lines XOX′, YOY′ and ZOZ′ at A, B and C respectively. The lengths OA, OB and OC are called the **Cartesian coordinates** of the point P. If OA= x, OB=y, OC=z then the point P is denoted by (x, y, z). The point O is called the **origin** and the lines X′OX, Y′OY and Z′OZ are called the **rectangular Cartesian coordinate axes** and are referred as x, y and z-axes. These three coordinate axes determine three mutually

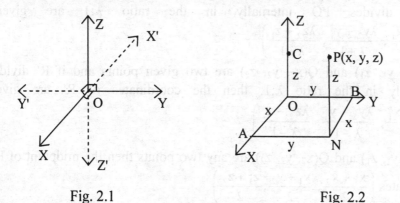

Fig. 2.1   Fig. 2.2

perpendicular planes XOY, YOZ and ZOX, called **coordinate planes** and are referred as xy, yz and zx planes. These planes divide the whole space into eight compartments, which are known as eight **octants**.

The coordinates of a point P in space can be determined using the following construction. Draw PN perpendicular to the plane XOY. From N, the foot of the perpendicular PN, draw the line NA in the plane XOY, parallel to y- axis and meeting the x-axis at A. Then OA = x, AN = y and NP = z. If NB is drawn in the plane XOY, parallel to x-axis and meeting y-axis at B, the OB = y and BN = x. Draw PC perpendicular to z-axis. Then OC = NP = z. PA, PB are also perpendiculars to x and y axes respectively.

The coordinates of the origin O are (0, 0, 0) and those of the points A, B, C are (x, 0, 0), (0, y, 0), (0, 0, z). For any point on the XOY plane, the z-coordinate is equal to zero. Similarly for any point on the YOZ plane, the x-coordinates is equal to zero and for any point on the ZOX plane the y-coordinate is equal to zero.

If x, y, z are measured along OX, OY, OZ then they are all positive. If x, y, z are measured along OX', OY', OZ' then they are all negative. The signs of the coordinates of a point determine the particular octant in which it lies. For any point in the octant OXYZ, all the coordinates are positive.

Given the axes OX and OY, the axis OZ is taken perpendicular to both OX and OY such that the axes OX, OY and OZ form a right-handed system. A right-handed screw when rotated from OX to OY would be driven from O to Z.

Note:
The formulae given below follow directly, as extensions of the corresponding formulae in the two dimensional analytical geometry.

(i)   The distance between two points $P(x_1, y_1, z_1)$ and $Q(x_2, y_2, z_2)$ is

$$PQ = \sqrt{(x_1 - x_2)^2 + (y_1 - y_2)^2 + (z_1 - z_2)^2}$$

(ii)   Distance of the point P(x, y, z) from the origin is $OP = \sqrt{x^2 + y^2 + z^2}$

(iii)   If $P(x_1, y_1, z_1)$ and $Q(x_2, y_2, z_2)$ are two given points, the coordinates of the point R which divides PQ internally in the ratio $\lambda:1$ are given by

$$\left(\frac{\lambda x_2 + x_1}{\lambda + 1}, \frac{\lambda y_2 + y_1}{\lambda + 1}, \frac{\lambda z_2 + z_1}{\lambda + 1} \right)$$

(iv)   If $P(x_1, y_1, z_1)$ and $Q(x_2, y_2, z_2)$ are two given points and if R' divides PQ externally in the ratio $\lambda:1$, then the coordinates of R' are given by

$$\left(\frac{\lambda x_2 - x_1}{\lambda - 1}, \frac{\lambda y_2 - y_1}{\lambda - 1}, \frac{\lambda z_2 - z_1}{\lambda - 1} \right)$$

(v)   If $P(x_1, y_1, z_1)$ and $Q(x_2, y_2, z_2)$ are any two points then the midpoint of PQ has coordinates $\left(\dfrac{x_1 + x_2}{2}, \dfrac{y_1 + y_2}{2}, \dfrac{z_1 + z_2}{2} \right)$

(vi)   The coordinates of the centroid of a triangle with vertices $P(x_1, y_1, z_1)$,

$Q(x_2, y_2, z_2)$ and $R(x_3, y_3, z_3)$ are $\left(\dfrac{x_1 + x_2 + x_3}{3}, \dfrac{y_1 + y_2 + y_3}{3}, \dfrac{z_1 + z_2 + z_3}{3} \right)$

2.0.1 ANGLE BETWEEN TWO STRAIGHT LINES IN SPACE

Two straight lines on a plane (coplanar) always intersect, either at a finite point or at infinity. Two straight lines intersecting at infinity are parallel lines and the angle between them is zero or 180° according as they have the same direction or not. In general two straight lines in space need not intersect. Such non-intersecting lines are called **skew lines**. To find the angle between two skew lines l and m, take any point P in space draw two lines PL and PM through P parallel to l and m. Then the angle between the skew lines l and m is the angle θ, the angle between the intersecting lines PL and PM.

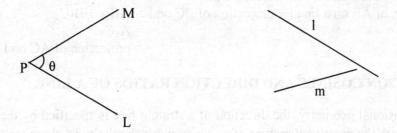

Fig. 2.3

2.0.2 PROJECTIONS

Definition 2.0.1 The **projection** of a point A on a straight-line l in space is the foot of the perpendicular drawn from the point P on the line l. The projection of a finite line segment AB on a line l is the line segment A_1B_1 where A_1 and B_1 are projections of the points A and B on the line l.

Note:

(i)   If θ is the angle between the line segment AB and a line l, then the projection of AB on l is AB cos θ.

To find the projection of AB on a line l, through A draw a line AL parallel to l.

Let $B\hat{A}L = \theta$ (Fig. 2.4). Let M be the foot of the perpendicular drawn from B on AL.

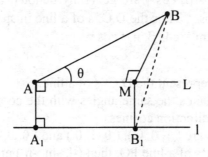

Fig. 2.4

If A_1 and B_1 are projections of the points A and B on the line l, then we have,

Projection of AB on l = A_1B_1 = AM

$$= AB \cos \theta \quad \text{(from } \Delta BAM)$$

(ii)   The algebraic sum of the projections of the line segments of a broken line in space, taken in order, on a straight line is equal to the projection of the closing line segment on that line i.e., if ABC is a broken line consisting of two segments AB and BC, then projection of AB on a line l + projection of BC on l = projection of AC on l.

If projection of AB on a line l is A_1B_1, the projection of BC on l is B_1C_1 and the projection of AC on l is A_1C_1 then, we get

projection of AB on a line l + projection of BC on l = $A_1B_1 + B_1C_1$

$$= A_1C_1$$

$$= \text{projection of AC on l}$$

2.1   DIRECTION COSINES AND DIRECTION RATIOS OF A LINE

In the two dimensional geometry, the direction of a straight line is specified by the angle θ that it makes with the positive direction of x-axis or equivalently by its slope $\tan \theta$. But the direction of a line in space is specified by the three angles α, β and γ that the line makes with the positive direction of x, y and z axes or equivalently by the three quantities $\cos \alpha$, $\cos \beta$ and $\cos \gamma$. This is because the direction of a line in space is fixed when we are given the angles that the line makes with the positive directions of the coordinate axes. Otherwise, there is a unique line through a given point in space making angles α, β and γ with the positive directions of the coordinate axes. In this respect the following definition of direction cosines is important in the study of the analytical geometry of three dimensions.

Definition 2.1.1 The three angles α, β and γ that a line makes with the positive direction of x, y and z-axes are called the **direction angles** of the line. $\cos \alpha$, $\cos \beta$ and $\cos \gamma$ are called the **direction cosines (D.C.'s)** of the line.

The direction cosines $\cos \alpha$, $\cos \beta$, $\cos \gamma$ are generally denoted by l, m, n respectively. i.e., $l = \cos \alpha$, $m = \cos \beta$, $n = \cos \gamma$. Thus the D.C.'s of a line in space are specified by the triplet (l, m, n) where $l = \cos \alpha$, $m = \cos \beta$, $n = \cos \gamma$.

Note:

(i)   Not all triplets (l, m, n) represent the D.C.'s of a line.

(ii)   A set of parallel lines makes the same angles with the coordinate axes and hence they will have the same direction cosines.

(iii)   D.C.'s of x, y and z axes are (1, 0, 0), (0, 1, 0) and (0, 0, 1)

(iv)   If (l, m, n) are the D.C.'s of a line PQ, then (-l, -m, -n) are the D.C.'s of the line QP.

2.1.1   DIRECTION COSINES OF LINES THROUGH THE ORIGIN

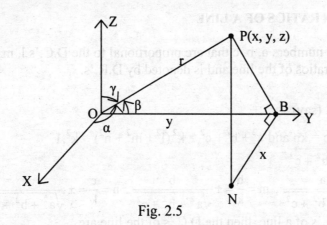

Fig. 2.5

Let P(x, y, z) be any point on a line through the origin. Let OP = r. Draw PN perpendicular to XOY plane and NB parallel to x-axis meeting OY at B. Then PB is perpendicular to OY. If OP makes angles α, β and γ with x, y and z axes, then D.C.'s of OP are (cos α, cos β, cos γ).

In $\triangle POB$, $P\hat{O}B = \beta$ and $P\hat{B}O = 90°$

From $\triangle POB$,

$\therefore$ $\cos\beta = \dfrac{OB}{OP} = \dfrac{OB}{r} = \dfrac{y}{r}$. Similarly, $\cos\alpha = \dfrac{x}{r}, \cos\gamma = \dfrac{z}{r}$

$\therefore$ D. C.'s of line OP are (l, m, n) = $\left(\dfrac{x}{r}, \dfrac{y}{r}, \dfrac{z}{r}\right)$

Note:
(i)   If (l, m, n) are the D.C.'s of line OP and OP = r, then P is (rl, rm, rn)

Theorem: If l, m, n are the D.C.'s of a line, then $l^2 + m^2 + n^2 = 1$.

Proof: If P(x, y, z) is any point on the line through the origin, parallel to the given line and OP = r, then, $l = \cos\alpha = \dfrac{x}{r}, m = \cos\beta = \dfrac{y}{r}$ and $n = \cos\gamma = \dfrac{z}{r}$

Now,   OP $= \sqrt{x^2 + y^2 + z^2}$

i.e., $OP^2 = x^2 + y^2 + z^2$

i.e., $r^2 = x^2 + y^2 + z^2$

i.e., $\left(\dfrac{x}{r}\right)^2 + \left(\dfrac{y}{r}\right)^2 + \left(\dfrac{z}{r}\right)^2 = 1$

i.e., $\cos^2\alpha + \cos^2\beta + \cos^2\gamma = 1$.

i.e., $l^2 + m^2 + n^2 = 1$.

2.1.2 DIRECTION RATIOS OF A LINE

Definition 2.1.2 Three numbers a, b, c that are proportional to the D.C.'s l, m, n of a line are called the direction ratios of the line and is denoted by D.R.'s.

Thus $\dfrac{a}{l} = \dfrac{b}{m} = \dfrac{c}{n} = k$ (say)

i.e., $a = kl$, $b = km$, $c = kn$ and $a^2 + b^2 + c^2 = k^2(l^2 + m^2 + n^2) = k^2 \cdot 1$.

$\therefore \quad k = \pm\sqrt{a^2 + b^2 + c^2}$

$\therefore \quad l = \dfrac{a}{k} = \pm\dfrac{a}{\sqrt{a^2 + b^2 + c^2}}, \quad m = \dfrac{b}{k} = \pm\dfrac{b}{\sqrt{a^2 + b^2 + c^2}}, \quad n = \dfrac{c}{k} = \pm\dfrac{c}{\sqrt{a^2 + b^2 + c^2}}$

If a, b, c are the D.R.'s of a line, then the D.C.'s of the line are

$$\pm\dfrac{a}{\sqrt{a^2 + b^2 + c^2}}, \quad \pm\dfrac{b}{\sqrt{a^2 + b^2 + c^2}}, \quad \pm\dfrac{c}{\sqrt{a^2 + b^2 + c^2}}$$

where the same sign positive or negative is to be chosen throughout.

Note:

The D.R.'s of two parallel lines are proportional.

2.1.3 DIRECTION COSINES OF THE LINE JOINING TWO POINTS

Let $P(x_1, y_1, z_1)$ and $Q(x_2, y_2, z_2)$ be two given points. Let PQ make angles α, β, γ with the coordinate axes. To find (l, m, n) the D.C.'s of PQ.

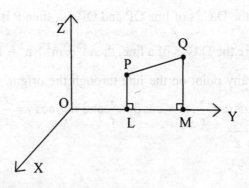

Fig. 2.6

Then $l = \cos\alpha$, $m = \cos\beta$ and $n = \cos\gamma$. Draw PL and QM perpendicular to OY.

Then OL = y_1, OM = y_2 $\therefore$ LM = $y_2 - y_1$.

Also LM = projection of PQ on OY.

$\qquad$ = PQ cos β (as β is the angle between PQ and OY)

i.e., $y_2 - y_1$ = PQ. m

$\therefore$ m = $\dfrac{y_2 - y_1}{PQ}$ where PQ = $\sqrt{(x_2 - x_1)^2 + (y_2 - y_1)^2 + (z_2 - z_1)^2}$

Similarly $l = \dfrac{x_2 - x_1}{PQ}$, $n = \dfrac{z_2 - z_1}{PQ}$

$\therefore$ D.C.'s of PQ are $\left(\dfrac{x_2 - x_1}{PQ}, \dfrac{y_2 - y_1}{PQ}, \dfrac{z_2 - z_1}{PQ} \right)$

Note:

The D.R.'s of PQ are $(x_2 - x_1, y_2 - y_1, z_2 - z_1)$

2.1.4 ANGLE BETWEEN TWO LINES

Let the D.C.'s of the two given lines be (l_1, m_1, n_1) and (l_2, m_2, n_2). Through O draw lines OA and OB parallel to the two given lines. Then the angle between OA and OB is θ, the angle between the given lines. Further choose the points A and B such that OA = OB = 1. Then the coordinates of A are (rl_1, rm_1, rn_1) where r = OA = 1. i.e., the coordinates of A are (l_1, m_1, n_1). Similarly coordinates of B are (l_2, m_2, n_2).

From $\triangle$AOB, $AB^2 = OA^2 + OB^2 - 2 \cdot OA \cdot OB \cdot \cos \theta$

i.e., $(l_1 - l_2)^2 + (m_1 - m_2)^2 + (n_1 - n_2)^2 = 1^2 + 1^2 - 2 \cdot 1 \cdot 1 \cdot \cos \theta$.

i.e., $l_1^2 + m_1^2 + n_1^2 + l_2^2 + m_2^2 + n_2^2 - 2(l_1 l_2 + m_1 m_2 + n_1 n_2) = 2 - 2 \cos \theta$.

i.e., $1 + 1 - 2(l_1 l_2 + m_1 m_2 + n_1 n_2) = 2 - 2 \cos \theta$.

$\therefore$ $\cos \theta = l_1 l_2 + m_1 m_2 + n_1 n_2$.

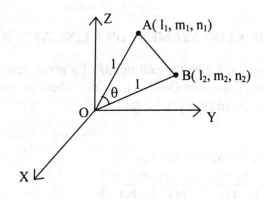

Fig. 2.7

Note:

(i) $\sin\theta = \sqrt{(l_1 m_2 - m_1 l_2)^2 + (m_1 n_2 - n_1 m_2)^2 + (n_1 l_2 - l_1 n_2)^2}$

Now $\sin^2\theta = 1 - \cos^2\theta$

$\qquad\qquad = 1.1 - \cos^2\theta$

$\qquad\qquad = \left(l_1^2 + m_1^2 + n_1^2\right) + \left(l_2^2 + m_2^2 + n_2^2\right) - \left(l_1 l_2 + m_1 m_2 + n_1 n_2\right)^2$

$\qquad\qquad = (l_1 m_2 - m_1 l_2)^2 + (m_1 n_2 - n_1 m_2)^2 + (n_1 l_2 - l_1 n_2)^2$

$\therefore \sin\theta = \pm\sqrt{(l_1 m_2 - m_1 l_2)^2 + (m_1 n_2 - n_1 m_2)^2 + (n_1 l_2 - l_1 n_2)^2}$

(ii)   If the direction ratios of the two lines are (a_1, b_1, c_1) and (a_2, b_2, c_2) then

$$\cos\theta = \pm\frac{(a_1 a_2 + b_1 b_2 + c_1 c_2)}{\sqrt{a_1^2 + b_1^2 + c_1^2}\ \sqrt{a_2^2 + b_2^2 + c_2^2}}$$

$$\sin\theta = \frac{\pm\sqrt{(a_1 b_2 - b_1 a_2)^2 + (b_1 c_2 - c_1 b_2)^2 + (c_1 a_2 - a_1 c_2)^2}}{\sqrt{a_1^2 + b_1^2 + c_1^2}\ \sqrt{a_2^2 + b_2^2 + c_2^2}}$$

2.1.5   CONDITIONS FOR PERPENDICULARITY AND PARALLELISM OF LINES

1. Let the D.C.'s of the two given lines be (l_1, m_1, n_1) and (l_2, m_2, n_2). If the two lines are perpendicular, $\theta = 90°$ or $\cos\theta = 0°$. i.e., $l_1 l_2 + m_1 m_2 + n_1 n_2 = 0$.

If the two lines are parallel, then $l_1 = l_2$, $m_1 = m_2$ and $n_1 = n_2$. i.e., $\dfrac{l_1}{l_2} = \dfrac{m_1}{m_2} = \dfrac{n_1}{n_2} = 1$

2. Let the D.R.'s of the two given lines be (a_1, b_1, c_1) and (a_2, b_2, c_2). If the two lines are perpendicular then $\theta = 90°$ or $\cos\theta = 0°$. i.e., $a_1 a_2 + b_1 b_2 + c_1 c_2 = 0$. If the two lines are parallel then the D.R.'s are proportional. i.e., $\dfrac{a_1}{a_2} = \dfrac{b_1}{b_2} = \dfrac{c_1}{c_2}$

2.1.6   PROJECTION OF A LINE SEGMENT ON A LINE WITH D.C.'S (l, m, n)

Let $P(x_1, y_1, z_1)$ and $Q(x_2, y_2, z_2)$ be the end points of a given line segment and θ be the angle between PQ and the lines with D.C.'s (l, m, n). Then the projection of PQ on the line is $PQ \cos\theta = l(x_2 - x_1) + m(y_2 - y_1) + n(z_2 - z_1)$.

Proof:

Now D.R.'s of PQ are $(x_2 - x_1, y_2 - y_1, z_2 - z_1)$.

$\therefore$ D.C.'s of PQ are $\left(\dfrac{x_2 - x_1}{PQ}, \dfrac{y_2 - y_1}{PQ}, \dfrac{z_2 - z_1}{PQ}\right)$

$\left(\because\ \sqrt{(x_2 - x_1)^2 + (y_2 - y_1)^2 + (z_2 - z_1)^2} = PQ\ \right)$

Since θ is the angle between the lines with D.C.'s $\left(\dfrac{x_2 - x_1}{PQ}, \dfrac{y_2 - y_1}{PQ}, \dfrac{z_2 - z_1}{PQ}\right)$ and

(l, m, n), we have,

$$\cos \theta = \frac{x_2 - x_1}{PQ}.l + \frac{y_2 - y_1}{PQ}.m + \frac{z_2 - z_1}{PQ}.n$$

i.e., $PQ \cos \theta = l(x_2 - x_1) + m(y_2 - y_1) + n(z_2 - z_1)$.

i.e.,   the projection of PQ on the line with D.C.'s (l, m, n) is

$PQ \cos \theta = l(x_2 - x_1) + m(y_2 - y_1) + n(z_2 - z_1)$.

Example 2.1.1 If (2, -1, 3) are the D.R.'s of a line find the D.C.'s.

Solution:

D.C.'s are $\left(\dfrac{2}{\sqrt{2^2 + (-1)^2 + 3^2}}, \dfrac{-1}{\sqrt{2^2 + (-1)^2 + 3^2}}, \dfrac{3}{\sqrt{2^2 + (-1)^2 + 3^2}}\right)$

i.e., $\left(\dfrac{2}{\sqrt{14}}, \dfrac{-1}{\sqrt{14}}, \dfrac{3}{\sqrt{14}}\right)$

Example 2.1.2 Find the D.C.'s of the line joining the points P (1, 3, 4) and Q (-2, 3, -5).

Solution:
D.R.'s of PQ are (-2 -1, 3 - 3, -5 - 4)
i.e.,   (-3, 0, -9)

$\therefore$ D.C.'s are $\left(\dfrac{-3}{\sqrt{90}}, 0, \dfrac{-9}{\sqrt{90}}\right)$

Example 2.1.3 Find the projection of the line segment joining the points P(2, 3, 1) and Q(4, -1, 3) on the line with D.R.'s (2, 3, 6).

Solution:

The D.C.'s of the given line are $\left(\dfrac{2}{7}, \dfrac{3}{7}, \dfrac{6}{7}\right)$

Projection of the line segment PQ on the line

$$= l(x_2 - x_1) + m(y_2 - y_1) + n(z_2 - z_1)$$
$$= (2/7)(4 - 2) + (3/7)(-1-3) + (6/7)(3 - 1).$$
$$= (2/7)(2) + (3/7)(-4) + (6/7)(2).$$
$$= \frac{4}{7} - \frac{12}{7} + \frac{12}{7} = \frac{4}{7}$$

Example 2.1.4 Find the angle subtended at the origin by the line segment joining the points P (3, 4, -2) and Q (7, 1, 2).

Solution:
D.R.'s of the line OP are (3, 4, -2)
D.R.'s of the line OQ are (7, 1, 2).
Angle subtended by PQ at the origin = Angle between OP and OQ
$$= \theta$$

where $\cos\theta = \dfrac{3(7) + 4(1) + (-2)(2)}{\sqrt{3^2 + 4^2 + (-2)^2} . \sqrt{7^2 + 1^2 + 2^2}}$

$$= \dfrac{21 + 4 - 4}{\sqrt{29} . \sqrt{54}} = \dfrac{21}{\sqrt{29} . \sqrt{54}}$$

$$\therefore \theta = \cos^{-1}\left(\dfrac{21}{\sqrt{29} . \sqrt{54}}\right)$$

Example 2.1.5 The projections of a line segment on the axes are 6, 3, 2. Find the length of the segment.

Solution:
Let λ be the length of the line segment and let α, β and γ be the angles made by the line with the x, y and z axes. Then the projection of the line segment on the axes are $\lambda \cos\alpha$, $\lambda \cos\beta$, $\lambda \cos\gamma$.
$\therefore \lambda \cos\alpha = 6$, $\lambda \cos\beta = 3$, $\lambda \cos\gamma = 2$.
Squaring and adding $\lambda^2 (\cos^2\alpha + \cos^2\beta + \cos^2\gamma) = 6^2 + 3^2 + 2^2$.

$$\text{i.e.,} \qquad \lambda^2 . 1 = 49$$
$$\lambda = 7$$

Example 2.1.6 Prove that the points A(3, 1, 3), B(1, -2, -1) and C(-1, -5, -5) are collinear.

Solution:
D.R.'s of line AB are (1 - 3,-2 - 1,-1 - 3)
i.e.,   (-2, -3, -4)
i.e.,   (2, 3, 4)

$\therefore$ D.C.'s of line AB are $\left(\dfrac{+2}{\sqrt{29}}, \dfrac{+3}{\sqrt{29}}, \dfrac{+4}{\sqrt{29}}\right)$

D.R.'s of line AC are (-1- 3, -5 -1, -5 - 3)
i.e.,   (-4, -6, -8)
i.e.,   (2, 3, 4)

$\therefore$ D.C.'s of line AC are $\left(\dfrac{+2}{\sqrt{29}}, \dfrac{+3}{\sqrt{29}}, \dfrac{+4}{\sqrt{29}}\right)$

Since D.C.'s of AB and AC are same, AB is parallel to AC and hence the points A, B, C are collinear.

Example 2.1.7 A (13, 2, -2), B (2, 5, 3), C (3, -6, 2) and D (7, 9, 1) are four points in space. Prove that they are coplanar.

Solution:
If the four points are to be coplanar, the lines AB, AC and AD must lie in a plane. If AB, AC and AD lie in a plane, they will be perpendicular to the normal to this plane. Let (a, b, c) be the D.R.'s of this normal.
Now D.R.'s of AB are (2 -13, 5 - 2, 3 + 2)   i.e., (-11, 3, 5)
D.R.'s of AC are (3 -13, -6 - 2, 2 + 2)   i.e., (-10, -8, 4)
D.R.'s of AD are (7 -13, 9 - 2, 1 + 2)   i.e., (-6, 7, 3)
Since AB, AC and AD are all perpendicular to the normal with D.R.'s (a, b, c), we have,

$$-11a + 3b + 5c = 0 \qquad\qquad (1)$$
$$-10a - 8b + 4c = 0 \qquad\qquad (2)$$
$$-6a + 7b + 3c = 0 \qquad\qquad (3)$$

AB, AC and AD lie in a plane, if the system of equations (1), (2), (3) has a solution for a, b, c

$$\begin{vmatrix} -11 & 3 & 5 \\ -10 & -8 & 4 \\ -6 & 7 & 3 \end{vmatrix} = -11(-24-28) - 3(-30+24) + 5(-70-48)$$

$$= 11(52) + 18 - 5(118)$$
$$= 572 + 18 - 590$$
$$= 590 - 590$$
$$= 0$$

∴ The system of equations is consistent.
∴ AB, AC and AD lie in a plane.
i.e., A, B, C and D are coplanar.

Note: Four points $A(x_1, y_1, z_1)$, $B(x_2, y_2, z_2)$, $C(x_3, y_3, z_3)$ and $D(x_4, y_4, z_4)$ are coplanar if

$$\begin{vmatrix} x_2 - x_1 & y_2 - y_1 & z_2 - z_1 \\ x_3 - x_1 & y_3 - y_1 & z_3 - z_1 \\ x_4 - x_1 & y_4 - y_1 & z_4 - z_1 \end{vmatrix} = 0$$

Example 2.1.8 If two pairs of opposite edges of a tetrahedron are at right angles, prove that so is the third pair.

Solution:

Let OABC be the tetrahedron such that OA is perpendicular to BC and OB is perpendicular to CA. We have to prove that OC is perpendicular to AB. With out loss of generality take the point O as origin. Let (x_1, y_1, z_1), (x_2, y_2, z_2), (x_3, y_3, z_3) be the coordinates of the points A, B, C respectively. The D.R.'s of OA are (x_1, y_1, z_1) and those of BC are $(x_2 - x_3, y_2 - y_3, z_2 - z_3)$. Since OA perpendicular to BC, we have,

$$x_1(x_2 - x_3) + y_1(y_2 - y_3) + z_1(z_2 - z_3) = 0 \qquad (1)$$

Similarly, as OB is perpendicular to CA, we have

$$x_2(x_3 - x_1) + y_2(y_3 - y_1) + z_2(z_3 - z_1) = 0 \qquad (2)$$

Adding (1) and (2) we have

$$x_3(x_2 - x_1) + y_3(y_2 - y_1) + z_3(z_2 - z_1) = 0 \qquad (3)$$

(x_3, y_3, z_3) are the D.R.'s of OC.

$(x_2 - x_1, y_2 - y_1, z_2 - z_1)$ are the D.R.'s of AB.

∴   (3) implies that OC is perpendicular to AB.

Example 2.1.9 Show that the angles between the four diagonals of a rectangular parallelopiped are $\cos^{-1}((\pm a^2 \pm b^2 \pm c^2)/(a^2+b^2+c^2))$, a, b, c being the edges of the parallelopiped

Solution:

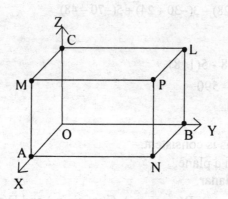

Fig. 2.8

Take one of the vertices O of the parallelopiped as origin and the three rectangular faces OAB, OBC and OCA as the three rectangular coordinate planes. Let OA = a, OB = b, OC = c. The four diagonals of the parallelopiped are OP, AL, BM and CN.

The coordinates of A, B, C are (a, 0, 0), (0, b, 0), (0, 0, c).

The coordinates of L, M, N are (0, b, c), (a, 0, c), (a, b, 0).

The coordinates of O, P are (0, 0, 0), (a, b, c).

The diagonals are OP, CN, AL and BM.

The D.C.'s of OP are $\left(\dfrac{a}{\sqrt{a^2+b^2+c^2}}, \dfrac{b}{\sqrt{a^2+b^2+c^2}}, \dfrac{c}{\sqrt{a^2+b^2+c^2}}\right)$

The D.C.'s of CN are $\left(\dfrac{a}{\sqrt{a^2+b^2+c^2}}, \dfrac{b}{\sqrt{a^2+b^2+c^2}}, \dfrac{-c}{\sqrt{a^2+b^2+c^2}}\right)$

The D.C.'s of AL are $\left(\dfrac{-a}{\sqrt{a^2+b^2+c^2}}, \dfrac{b}{\sqrt{a^2+b^2+c^2}}, \dfrac{c}{\sqrt{a^2+b^2+c^2}}\right)$

The D.C.'s of BM are $\left(\dfrac{a}{\sqrt{a^2+b^2+c^2}}, \dfrac{-b}{\sqrt{a^2+b^2+c^2}}, \dfrac{c}{\sqrt{a^2+b^2+c^2}}\right)$

The angle between OP and CN is given by

$$\cos\theta = \frac{a^2+b^2-c^2}{\sqrt{a^2+b^2+c^2}\,\sqrt{a^2+b^2+c^2}}$$

$$\text{i.e.,} \quad \theta = \cos^{-1}\left(\frac{a^2+b^2-c^2}{a^2+b^2+c^2}\right)$$

Similarly the angles between the remaining five pairs of diagonals are,

$$\cos^{-1}\left(\frac{a^2-b^2+c^2}{a^2+b^2+c^2}\right), \quad \cos^{-1}\left(\frac{-a^2+b^2+c^2}{a^2+b^2+c^2}\right), \quad \cos^{-1}\left(\frac{a^2-b^2-c^2}{a^2+b^2+c^2}\right),$$

$$\cos^{-1}\left(\frac{-a^2+b^2-c^2}{a^2+b^2+c^2}\right), \quad \cos^{-1}\left(\frac{-a^2-b^2+c^2}{a^2+b^2+c^2}\right)$$

Note:

    If OABC is a cube with side a then $\cos\theta = 1/3$

Example 2.1.10 A line makes angel α, β, γ and δ with the four diagonals of a cube. Prove that $\cos^2\alpha + \cos^2\beta + \cos^2\gamma + \cos^2\delta = 4/3$. (Refer Example 2.1.9 with a = b = c)

Solution:

    D.C.'s of OP are $\left(\dfrac{1}{\sqrt{3}}, \dfrac{1}{\sqrt{3}}, \dfrac{1}{\sqrt{3}}\right)$

    D.C.'s of AL are $\left(\dfrac{-1}{\sqrt{3}}, \dfrac{1}{\sqrt{3}}, \dfrac{1}{\sqrt{3}}\right)$

    D.C.'s of BM are $\left(\dfrac{1}{\sqrt{3}}, \dfrac{-1}{\sqrt{3}}, \dfrac{1}{\sqrt{3}}\right)$

D.C.'s of CN are $\left(\dfrac{1}{\sqrt{3}}, \dfrac{1}{\sqrt{3}}, \dfrac{-1}{\sqrt{3}}\right)$

Let the D.C.'s of the given line be (l, m, n). Then $l^2 + m^2 + n^2 = 1$. As the line makes angles α, β, γ and δ with OP, AL, BM, CN we have

$$\cos\alpha = \frac{l}{\sqrt{3}} + \frac{m}{\sqrt{3}} + \frac{n}{\sqrt{3}}$$

$$\cos\beta = -\frac{l}{\sqrt{3}} + \frac{m}{\sqrt{3}} + \frac{n}{\sqrt{3}}$$

$$\cos\gamma = \frac{l}{\sqrt{3}} - \frac{m}{\sqrt{3}} + \frac{n}{\sqrt{3}}$$

$$\cos\delta = \frac{l}{\sqrt{3}} + \frac{m}{\sqrt{3}} - \frac{n}{\sqrt{3}}$$

$\therefore \cos^2\alpha + \cos^2\beta + \cos^2\gamma + \cos^2\delta$

$= (1/3)\,(l + m + n)^2 + 1/3(-l + m + n)^2 + 1/3(l - m + n)^2 + 1/3(l + m - n)^2$

$= (1/3)\,(4l^2 + 4m^2 + 4n^2)$

$= (4/3)\,(l^2 + m^2 + n^2)$

$= 4/3$

Note:

(i) $\cos 2\alpha + \cos 2\beta + \cos 2\delta + \cos 2\gamma + 4/3 = 0$

(ii) $\sin^2\alpha + \sin^2\beta + \sin^2\gamma + \sin^2\delta = \dfrac{8}{3}$

Example 2.1.11 Show that the lines whose direction cosines are given by the equations $2l - m + 2n = 0$ and $mn + nl + lm = 0$ are perpendicular.

Solution:

$2l - m + 2n = 0$ \hfill (1)

$\therefore \qquad m = 2l + 2n$ \hfill (2)

Substituting for m in $mn + nl + lm = 0$

we get, $(2l + 2n)\,n + nl + l\,(2l + 2n) = 0$

i.e., $2n^2 + 5ln + 2l^2 = 0$

i.e., $(2n+l)(n+2l) = 0$

i.e., $2n+l = 0$ or $n+2l = 0$

i.e., $\dfrac{l}{-2} = \dfrac{n}{1}$ or $\dfrac{l}{-1} = \dfrac{n}{2}$ \hfill (Similarly we can eliminate l or n)

Case 1.

$\dfrac{l}{-2} = \dfrac{n}{1} = k$ (say)

i.e., $l = -2k, n = k$

From (1) m = -2k

i.e., $\dfrac{m}{-2} = k$

$\therefore \quad \dfrac{1}{-2} = \dfrac{m}{-2} = \dfrac{n}{1}$

D.R.'s of one line are (-2, -2, 1)

Case 2.

$$\dfrac{1}{-2} = \dfrac{n}{1} = \lambda \text{ (say)}$$

i.e., $\quad 1 = -\lambda, n = 2\lambda$

From (1) m = 2λ

i.e., $\quad \dfrac{m}{2} = \lambda$

$\therefore \quad \dfrac{1}{-1} = \dfrac{m}{2} = \dfrac{n}{2}$

D.R's of the second line are (-1, 2, 2).

If θ is the angel between the two lines, then,

$$\cos\theta = \dfrac{(-2)(-1) + (-2).2 + 1.2}{\sqrt{(-2)^2 + 2^2 + 1^2} \cdot \sqrt{(-1)^2 + 2^2 + 2^2}} = 0$$

$\therefore \quad$ θ = 90°. Hence the lines are perpendicular to each other.

Example 2.1.12 Show that the straight lines whose direction cosines are given by the equations al + bm + cn = 0, $ul^2 + vm^2 + wn^2 = 0$ are perpendicular or parallel according as $a^2(v + w) + b^2(w + u) + c^2(u + v) = 0$ or $a^2/u + b^2/v + c^2/w = 0$.

Solution:
Given al + bm + cn = 0   (1)

$ul^2 + vm^2 + wn^2 = 0$ (2)

Eliminating l between (1) and (2), we get,

$$u\left(\dfrac{bm + cn}{a}\right)^2 + vm^2 + wn^2 = 0$$

$$(b^2u + a^2v)m^2 + 2ubcmn + (c^2u + a^2w)n^2 = 0$$

$$(b^2n + a^2v)\left(\dfrac{m}{n}\right)^2 + 2ubc\left(\dfrac{m}{n}\right) + (c^2u + a^2w) = 0 \qquad (3)$$

If (l_1, m_1, n_1) and (l_2, m_2, n_2) are the D.C.'s of the two lines given by (1) and (2), then $\dfrac{m_1}{n_1}$

and $\dfrac{m_2}{n_2}$ are the roots of (3). If the two lines are parallel, $\dfrac{l_1}{l_2} = \dfrac{m_1}{m_2} = \dfrac{n_1}{n_2}$

Hence $\quad\dfrac{m_1}{m_2}=\dfrac{n_1}{n_2}$

i.e., $\quad\dfrac{m_1}{n_1}=\dfrac{m_2}{n_2}$

i.e., roots of equation (3) are equal

i.e., $\qquad (2ubc)^2 - 4(b^2u + a^2v)(c^2u + a^2w) = 0.$

i.e., $\qquad u^2b^2c^2 - u^2b^2c^2 - uwa^2b^2 - vwa^4 - vua^2c^2 = 0$

i.e., $\qquad uwb^2 + vwa^2 + vuc^2 = 0$

i.e., $\qquad a^2/u + b^2/v + c^2/w = 0$

As $\dfrac{m_1}{n_1}$ and $\dfrac{m_2}{n_2}$ are roots of equation (3)

$$\dfrac{m_1}{n_1}\cdot\dfrac{m_2}{n_2}=\dfrac{c^2u + a^2w}{b^2u + a^2v}$$

i.e., $\quad\dfrac{m_1m_2}{c^2u + a^2w}=\dfrac{n_1n_2}{b^2u + a^2v}$

Similarly eliminating n between (1) and (2), by symmetry, we get,

$$\dfrac{l_1l_2}{b^2w + c^2v}=\dfrac{m_1m_2}{c^2u + a^2w}$$

Thus $\quad\dfrac{l_1l_2}{b^2w + c^2v}=\dfrac{m_1m_2}{c^2u + a^2w}=\dfrac{n_1n_2}{a^2v + b^2u}=k$

If the two lines are perpendicular $l_1l_2 + m_1m_2 + n_1n_2 = 0$.

i.e., $\quad k(b^2w + c^2v) + k(c^2u + a^2w) + k(a^2v + b^2u) = 0$

i.e., $\qquad a^2(v + w) + b^2(w + u) + c^2(u + v) = 0.$

Example 2.1.13 Show that the straight lines whose direction cosines are given by $\quad$ al + bm + cn = 0, fmn + gnl + hlm = 0 are perpendicular if f/a + g/b + h/c = 0 and parallel if $\sqrt{af} \pm \sqrt{bg} \pm \sqrt{ch} = 0$

Solution:

Given $\quad al + bm + cn = 0$ $\hfill(1)$

$\qquad fmn + gnl + hlm = 0$ $\hfill(2)$

Eliminating n between (1) and (2), we get

$$- fm\left(\dfrac{al + bm}{c}\right) - gl\left(\dfrac{al + bm}{c}\right) + hlm = 0$$

i.e., $\quad -aflm - bfm^2 - agl^2 - bglm + chlm = 0.$

i.e., $\quad agl^2 + (af + bg - ch)\,lm + bfm^2 = 0.$

i.e., $\quad ag\left(\dfrac{1}{m}\right)^2 + (af + bg - ch)\left(\dfrac{1}{m}\right) + bf = 0$ $\hfill(3)$

If (l_1, m_1, n_1) and (l_2, m_2, n_2) are the D.C.'s of the two lines given by (1) and (2), then $\dfrac{l_1}{m_1}$ and $\dfrac{l_2}{m_2}$ are the roots of the equation (3).

If the two lines are parallel, $\dfrac{l_1}{l_2} = \dfrac{m_1}{m_2} = \dfrac{n_1}{n_2}$

Hence $\qquad \dfrac{l_1}{l_2} = \dfrac{m_1}{m_2}$

i.e., $\qquad \dfrac{l_1}{m_1} = \dfrac{l_2}{m_2}$

i.e., the roots of equation (3) are equal

i.e., $\qquad (af + bg - ch)^2 - 4ag.bf = 0$

i.e., $\qquad af + bg - ch = \pm 2\sqrt{af.bg}$

i.e., $\qquad af + bg \pm 2\sqrt{af.bg} - ch = 0$

i.e., $\qquad \left(\sqrt{af} \pm \sqrt{bg}\right)^2 - \left(\sqrt{ch}\right)^2 = 0$

i.e., $\qquad \left(\sqrt{af} \pm \sqrt{bg} - \sqrt{ch}\right)\left(\sqrt{af} \pm \sqrt{bg} - \sqrt{ch}\right) = 0$

i.e., $\qquad \sqrt{af} \pm \sqrt{bg} \pm \sqrt{ch} = 0$

Now $\dfrac{l_1}{m_1}$ and $\dfrac{l_2}{m_2}$ are roots of equation (3) implies $\dfrac{l_1}{m_1} \cdot \dfrac{l_2}{m_2} = \dfrac{bf}{ag} = \dfrac{f/a}{g/b}$

i.e., $\qquad \dfrac{l_1 l_2}{f/a} = \dfrac{m_1 m_2}{g/b}$

Similarly eliminating l between (1) and (2), by symmetry we get $\dfrac{m_1 m_2}{g/b} = \dfrac{n_1 n_2}{h/c}$

Thus $\qquad \dfrac{l_1 l_2}{f/a} = \dfrac{m_1 m_2}{g/b} = \dfrac{n_1 n_2}{h/c} = k$

If the two lines are perpendicular, $l_1 l_2 + m_1 m_2 + n_1 n_2 = 0$.

i.e., $\qquad k.\dfrac{f}{a} + k.\dfrac{g}{b} + k.\dfrac{h}{c} = 0$

i.e., $\qquad \dfrac{f}{a} + \dfrac{g}{b} + \dfrac{h}{c} = 0$

EXERCISE 2.1

PART–A

1. Define direction cosines of a line in space.
2. Given direction ratios of a line, how will you find the direction cosines?
3. Define projection of a line segment on a line.

4. What is the projection of the line segment joining $P(x_1, y_1, z_1)$ and $Q(x_2, y_2, z_2)$ on a line with D.C.'s (l, m, n)?

5. Define angle between skew lines.

6. $(2, 3, 6)$ are the D.R.'s of a line. What are the D.C.'s?

7. What are the D.C.'s of lines equally inclined to the axes? How many such lines are there?

8. Find the D.C.'s of the line joining the points $(7, -5, 9)$ and $(5, -3, 8)$

9. Show that the points $(1, -2, 3), (2, 3, -4), (0, -7, 10)$ are collinear.

10. The projections of a line on the coordinate axes are 12, 4, 3. Find the length and D.C.'s of the line.

11. Find the length of the line whose projections on the coordinate axes are 2, 3, 6.

12. Find the D.C.'s of the line which is perpendicular to the lines with D. R.'s $(1, -2, -2), (0, 2, 1)$.

13. Find the D.C.'s of a line perpendicular to two lines whose D. R.'s are $(1, 2, 3)$ and $(-2, 1, 4)$

14. A line makes angles 30° and 60° with the X and Y axes. Find the angle made by the line with the Z axis.

15. A straight line is inclined to the Y and Z axes at angles of 45° and 60°. Find the inclination of the line to the X-axis.

16. If a line makes complementary angles with any two coordinate axes, prove that it is perpendicular to the third axis.

17. If α, β, γ are the angles which a line makes with the axis, prove that $\sin^2\alpha + \sin^2\beta + \sin^2\gamma = 2$.

18. Find the angle between the lines whose D.R.'s are $(1, 2, 1)$ and $(2, -3, 6)$.

19. A, B, C, D are the points $(1, 2, 3), (4, 5, 7), (-4, 3, -6), (2, 9, 2)$ respectively. Prove that AB parallel to CD.

20. A, B, C, D are the points $(k, 3, -1), (3, 5, -3), (1, 2, 3), (3, 5, 7)$ respectively. If AB is perpendicular to CD, find the value of k.

21. Show that the triangle with the vertices at the points $(1, -5, 3), (3, -7, 4)$ and $(2, -3, 5)$ is right angled.

PART- B

22. Show that the figure formed by the points $(4, 0, 1), (3, 2, -1), (5, 4, 0)$ and $(6, 2, 2)$ is a square.

23. Show that the points $(2, 1, 5), (-2, -1, 1), (3, 2, 4)$ and $(7, 4, 8)$ are the vertices of a Parallelogram.

24. Show that the points $(4, -1, -3), (10, 1, -6), (7, 7, -4)$ and $(1, 5, -1)$ are the vertices of a rhombus.

25. Find the angle between two diagonals of a cube.

26. Find the projection of AB on CD if given $A(1, 2, 3), B(-1, 0, 2), C(1, 1, 2)$ and $D(0, 5, 6)$.

27. Find the foot of the perpendicular from $(0, 9, 6)$ on the line joining the points

(1, 2, 3) and (7, -2, 5)

28. Find the projection of A(-3, -16, 6) on the line joining the points B(4, -1, 3) and C (0, 5, -2)

29. Prove that the points (2, 5, 3), (7, 9, 1), (3, -6, 2) and (13, 2, -2) are coplanar and find the D.R.'s of the normal to the plane in which they lie.

30. If (l_1, m_1, n_1) and (l_2, m_2, n_2) are the D.C.'s are two mutually perpendicular lines, show that the D.C.'s of the line perpendicular to both of them are $m_1n_2-m_2n_1$, $n_1l_2-n_2l_1$, $l_1m_2-l_2m_1$.

31. If the distance between two points P and Q is d and the lengths of the projections of PQ on the coordinate planes are d_1, d_2, d_3. Show that $2d^2 = d_1^2 + d_2^2 + d_3^2$.

32. Prove that three concurrent lines with D.C.'s (l_1, m_1, n_1), (l_2, m_2, n_2), (l_3, m_3, n_3) are

coplanar if $\begin{vmatrix} l_1 & m_1 & n_1 \\ l_2 & m_2 & n_2 \\ l_3 & m_3 & n_3 \end{vmatrix} = 0$.

33. Find the angle between the lines whose D. C.'s are given by the equations $l + m + n = 0$, $l^2 + m^2 - n^2 = 0$.

34. Find the angle between the lines whose D.C.'s are given by the equations $3l + m + 5n = 0$, $6mn - 2nl + 5lm = 0$.

35. Prove that the lines whose D.C.'s are given by the equations $m + 2n + l = 0$ and $4lm - 3mn - 4nl = 0$ are perpendicular.

36. Find the angle between the lines whose D.C.'s are given by the equations $l + m + n = 0$, $2lm - mn + 2ln = 0$.

37. Show that the lines whose D.C.'s are given by the equations $l + m + n = 0$ and $al^2 + bm^2 + cn^2 = 0$ are perpendicular if $a + b + c = 0$ and parallel if $ab + bc + ca = 0$.

38. Show that the lines whose D.C.'s are given by the equations $a^2l + b^2m + c^2n = 0$ and $lm + mn + nl = 0$ are perpendicular if $\dfrac{1}{a^2} + \dfrac{1}{b^2} + \dfrac{1}{c^2} = 0$ and parallel if $a \pm b \pm c = 0$.

39. If (l_1, m_1, n_1), (l_2, m_2, n_2), (l_3, m_3, n_3) are the D.C.'s of the three mutually perpendicular lines, prove that the line whose D. R's are

$(l_1 + l_2 + l_3, m_1 + m_2 + m_3, n_1 + n_2 + n_3)$ makes equal angles with them.

40. A plane makes intercepts OA, OB, OC on the axes of coordinates. If OA = a, OB = b, OC = c, find the area of triangle ABC.

41. If in a tetrahedron the sums of the squares of opposite sides are equal, show that its pairs of opposite sides are at right angles.

42. Two edges PQ, RS of a tetrahedron PQRS are perpendicular. Show that the distance between the mid-points of PS and QR is equal to the distance between the midpoints of PR and QS.

2.2   THE PLANE

A plane is a special surface in space. It has the characteristic property that given any two points on it, every point on the straight line joining the two points also lies on it. Given any three points in space, there is a unique plane passing through these three points. Similarly given a straight line and a point not lying on this line, there is a unique plane passing through the point and containing the line. In the following sections we determine different forms of the equation of a plane.

2.2.1   GENERAL FORM OF THE EQUATION OF A PLANE

Theorem 2.2.1 The general equation of the first degree in x, y, z represents a plane.

Proof: The general equation of the first degree in x, y, z is

$$ax + by + cz + d = 0 \qquad (1)$$

where a, b, c, d are constants. Equation (1) represents a surface in space.
Let $P(x_1, y_1, z_1)$ and $Q(x_2, y_2, z_2)$ be any two points on this surface.

Then

$$ax_1 + by_1 + cz_1 + d = 0 \qquad (2)$$
$$ax_2 + by_2 + cz_2 + d = 0 \qquad (3)$$

Let R be any point on the line PQ and let R divide PQ in the ratio $\lambda:1$

Then R is $\left(\dfrac{\lambda x_2 + x_1}{\lambda + 1}, \dfrac{\lambda y_2 + y_1}{\lambda + 1}, \dfrac{\lambda z_2 + z_1}{\lambda + 1} \right)$

Substituting the coordinates of R in (1), we get,

$$a\left(\frac{\lambda x_2 + x_1}{\lambda + 1} \right) + b\left(\frac{\lambda y_2 + y_1}{\lambda + 1} \right) + c\left(\frac{\lambda z_2 + z_1}{\lambda + 1} \right) + d = \frac{1}{\lambda + 1}\left[\lambda(ax_2 + by_2 + cz_2 + d) + (ax_1 + by_1 + cz_1 + d) \right]$$

$$= \frac{1}{\lambda + 1}[\lambda.0 + 0] \text{ using (2) and (3)}$$

$$= 0.$$

∴ R lies on the surface (1)

∴ Equation (1) represents a plane.

Note:
(i)   If $d = 0$ equation (1) becomes $ax + by + cz = 0$, which represents a plane passing through the origin.
(ii)   If the plane $ax + by + cz + d = 0$ passes through the point (x_1, y_1, z_1), then
$ax_1 + by_1 + cz_1 + d = 0$.
i.e., $d = - (ax_1 + by_1 + cz_1)$.
∴ The equation of the plane becomes $ax + by + cz - (ax_1 + by_1 + cz_1) = 0$.
i.e., $a(x - x_1) + b(y - y_1) + c(z - z_1) = 0$.
∴ Equation of a plane through (x_1, y_1, z_1) is
$a(x - x_1) + b(y - y_1) + c(z - z_1) = 0$.

2.2.2   INTERCEPT FORM OF THE EQUATION OF A PLANE

Let a given plane meet the coordinate axes OX, OY, OZ at the points A, B, C respectively. Let OA = a, OB = b, OC = c. Then a, b and c are the x, y and z intercepts of the plane. Coordinates of A, B, C are (a, 0, 0), (0, b, 0), (0, 0, c). The equation of the plane can be determined in terms of the intercepts a, b and c and this equation in terms of a, b and c is called the intercept form of the equation of the plane.

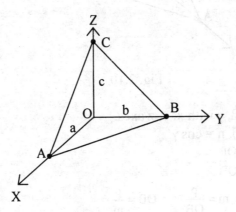

Fig. 2.9

Let the equation of the plane be
$$px + qy + rz + s = 0 \qquad (1)$$
As this equation is satisfied by the coordinates of A, B, C we get,
$$pa + s = 0, \ qb + s = 0, \ ra + s = 0$$
Solving, p = -s/a, q = -s/b, r = -s/c.
∴ Equation (1) becomes
$$-\frac{s}{a}x - \frac{s}{b}y - \frac{s}{c}z + s = 0$$
i.e.,
$$\frac{x}{a} + \frac{y}{b} + \frac{z}{c} = 1$$
This is the equation of the plane in the intercept form.

2.2.3   NORMAL FORM OF THE EQUATION OF A PLANE

Let a given plane be at a distance p from the origin and l, m, n be the D.C.'s of the normal to the plane. Draw the perpendicular OP from O to the plane. Then OP = p and the D.C.'s of line OP are l, m, n. The equation of the plane can be determined in terms of l, m, n and p. This equation of the plane in terms of l. m. n and p is called the normal form of the equation of the plane. Let the plane cut the x, y, z axes at A, B, C respectively.

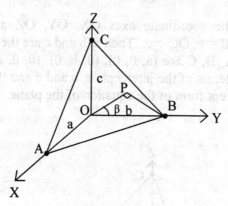

Fig. 2.10

Let OP makes angles α, β, γ with x, y, z axes.

Then $l = \cos \alpha$, $m = \cos \beta$, $n = \cos \gamma$.

From $\triangle$ OPB, $\cos P\hat{O}B = \dfrac{OP}{OB}$

i.e., $\cos \beta = \dfrac{p}{OB}$ i.e., $m = \dfrac{p}{OB}$ $\therefore OB = \dfrac{p}{m}$

Similarly $OA = \dfrac{p}{l}$, $OC = \dfrac{p}{n}$

OA, OB, OC are the x, y, z intercepts of the plane, and hence the equation of the plane in terms of its intercepts is

$$\frac{x}{OA} + \frac{y}{OB} + \frac{z}{OC} = 1$$

i.e., $$\frac{x}{p/l} + \frac{y}{p/m} + \frac{z}{p/n} = 1$$

i.e.,   **lx + my + nz = p**

This is the equation of the plane in the normal form.

Note:

(i)   If the coordinates of the foot of the perpendicular P on the plane

     $lx + my + nz = p$ is (x_1, y_1, z_1) then

     $x_1 = p \cos \alpha = pl$.

     Similarly $y_1 = pm$

     and $z_1 = pn$

     $\therefore$ P is (pl, pm, pn)

(ii)   **Distance of the origin from the plane ax + by + cz + d = 0**

The normal form of the equation of the plane is $lx + my + nz = p$ comparing $ax + by + c = -d$ with $lx + my + nz = p$, we get $\dfrac{l}{a} + \dfrac{m}{b} + \dfrac{n}{c} = \dfrac{p}{-d}$

Each ratio is equal to $\dfrac{\pm\sqrt{l^2 + m^2 + n^2}}{\sqrt{a^2 + b^2 + c^2}}$ i.e., $\dfrac{\pm 1}{\sqrt{a^2 + b^2 + c^2}}$

$\therefore \dfrac{p}{-d} = \dfrac{\pm 1}{\sqrt{a^2 + b^2 + c^2}}$

i.e., $p = \dfrac{\pm d}{\sqrt{a^2 + b^2 + c^2}}$ is the distance of the origin from the plane

$ax + by + cz + d = 0$ or equivalently it is the length of the perpendicular from the origin on the plane.

(iii) $\dfrac{l}{a} = \dfrac{m}{b} = \dfrac{n}{c}$ implies that a, b, c are proportional to l, m, n, the D.C.'s of the

normal to the plane $lx + my + nz = p$ $\therefore$ a, b, c are the D.R.'s of the normal to plane $ax + by + cz + d = 0$

The D.C.'s are $l = \dfrac{\pm a}{\sqrt{a^2 + b^2 + c^2}}$, $m = \dfrac{\pm b}{\sqrt{a^2 + b^2 + c^2}}$, $n = \dfrac{\pm c}{\sqrt{a^2 + b^2 + c^2}}$

2.2.4   ANGLE BETWEEN TWO PLANES

Let the equation of two given planes be $a_1x + b_1y + c_1z + d_1 = 0$ and $a_2x + b_2y + c_2z + d_2 = 0$. The angle between the two planes is equal to the angle between their normals. The D.R.'s of the normals to the planes are (a_1, b_1, c_1) and (a_2, b_2, c_2). If θ is the angle between the planes, then,

$$\cos\theta = \pm\frac{(a_1a_2 + b_1b_2 + c_1c_2)}{\sqrt{a_1^2 + b_1^2 + c_1^2}\,\sqrt{a_2^2 + b_2^2 + c_2^2}}$$

Note:

(i)   If two planes are perpendicular, their normals are also perpendicular and if the planes are parallel, their normals are parallel.

$\therefore$ The planes $a_1x + b_1y + c_1z + d_1 = 0$ and $a_2x + b_2y + c_2z + d_2 = 0$

are perpendicular if $a_1a_2 + b_1b_2 + c_1c_2 = 0$ and are parallel if $\dfrac{a_1}{a_2} = \dfrac{b_1}{b_2} = \dfrac{c_1}{c_2}$.

(ii)   If the equation of a given plane is $ax + by + cz + d = 0$, then the equation of any plane parallel to it is $ax + by + cz + k = 0$.

2.2.5   ANGLE BETWEEN A PLANE AND A LINE

Let $ax + by + cz + d = 0$ be the equation of a given plane and l, m, n be the D.C.'s of a given line. Let θ be the angle between the plane and the line. Then the angle between the normal to the plane and the line is $90 - \theta$. The D.R.'s of the normal to the plane are a, b, c.

$$\therefore \quad \cos(90° - \theta) = \pm \frac{\pm(al + bm + cn)}{\sqrt{a^2 + b^2 + c^2}\sqrt{l^2 + m^2 + n^2}}$$

$$\text{i.e., } \quad \sin\theta = \pm \frac{\pm(al + bm + cn)}{\sqrt{a^2 + b^2 + c^2}\sqrt{l^2 + m^2 + n^2}}$$

Note:

(i)   The given line will be parallel to the plane if it is perpendicular to the normal to the plane. i.e., if $al + mb + cn = 0$.

(ii)   The given line will be perpendicular to the plane, if it is parallel to the normal

to the plane. i.e., if $\dfrac{a}{l} = \dfrac{b}{m} = \dfrac{c}{n}$

(iii)   The general equation of a plane is $ax + by + cz + d = 0$.

i.e., $\quad \dfrac{a}{d}x + \dfrac{b}{d}y + \dfrac{c}{d}z + 1 = 0$

i.e., $\quad px + qy + rz + 1 = 0$

Hence, to determine the equation of a plane, we have to find the three quantities p, q, r. i.e., a plane has three degrees of freedom. Thus a plane can be drawn to satisfy three conditions. There is a unique plane passing through three given points.

2.2.6   EQUATION OF THE PLANE PASSING THROUGH THREE POINTS

Let the three given points be (x_1, y_1, z_1), (x_2, y_2, z_2) and (x_3, y_3, z_3).

The equation of the plane passing through (x_1, y_1, z_1) is

$$a(x - x_1) + b(y - y_1) + c(z - z_1) = 0 \tag{1}$$

If (1) passes through (x_2, y_2, z_2) and (x_3, y_3, z_3) we have

$$a(x_2 - x_1) + b(y_2 - y_1) + c(z_2 - z_1) = 0 \tag{2}$$

$$a(x_3 - x_1) + b(y_3 - y_1) + c(z_3 - z_1) = 0 \tag{3}$$

Eliminating a, b, c from (1), (2) and (3) we get,

$$\begin{vmatrix} x - x_1 & y - y_1 & z - z_1 \\ x_2 - x_1 & y_2 - y_1 & z_2 - z_1 \\ x_3 - x_1 & y_3 - y_1 & z_3 - z_1 \end{vmatrix} = 0$$

This is the required equation of the plane passing through the three given points.

2.2.7   EQUATION OF A PLANE PASSING THROUGH THE INTERSECTION OF TWO GIVEN PLANES

Let the two given planes be $\quad U = a_1x + b_1y + c_1z + d_1 = 0 \tag{1}$

$$V = a_2x + b_2y + c_2z + d_2 = 0 \tag{2}$$

Consider $U + \lambda V = (a_1x + b_1y + c_1z + d_1) + \lambda(a_2x + b_2y + c_2z + d_2) = 0 \tag{3}$

where λ is a constant.

Equation (3) is first degree in x, y, z and hence determines a plane.

If (x_1, y_1, z_1) is a point on the line of intersection of the two planes then it satisfies equations (1) and (2).

∴ $\qquad a_1x_1 + b_1y_1 + c_1z_1 + d_1 = 0$

and $\qquad a_2x_1 + b_2y_1 + c_2z_1 + d_2 = 0$

Also $\quad (a_1x_1 + b_1y_1 + c_1z_1 + d_1) + \lambda(a_2x_1 + b_2y_1 + c_2z_1 + d_2) = 0 + \lambda.0 = 0$

∴ $\qquad (x_1, y_1, z_1)$ lies on the plane determined by equation (3).

Hence all points which are common to (1) and (2) lie on the plane (3) also.

∴ $\qquad U + \lambda V = (a_1x + b_1y + c_1z + d_1) + \lambda(a_2x + b_2y + c_2z + d_2) = 0$

is the equation of a plane passing through the line of intersection of the planes,

$$U = a_1x + b_1y + c_1z + d_1 = 0 \text{ and}$$
$$V = a_2x + b_2y + c_2z + d_2 = 0.$$

2.2.8 DISTANCE OF A POINT P (x_1, y_1, z_1) FROM THE PLANE lx + my + nz = p

$P(x_1, y_1, z_1)$ is the given point and the equation of the given plane is

$$lx + my + nz = p \qquad (1)$$

Consider the plane through P parallel to $lx + my + nz = p$

Its equation is of the form

$$lx + my + nz = p_1 \qquad (2)$$

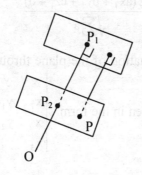

Fig. 2.11

Since $P(x_1, y_1, z_1)$ lies on this plane, we get,

$$lx_1 + my_1 + nz_1 = p_1 \qquad (3)$$

Let the perpendicular OP_1 drawn from the origin on the plane (1) meets the second plane at P_2. Then from (1) and (2) $OP_1 = p$ and $OP_2 = p_1$.

Then the length of the perpendicular from P on the given plane (1) is

$OP_1 - OP_2 = p - p_1$

$\qquad\qquad = p - (lx_1 + my_1 + nz_1)$ (By using (3))

i.e., the distance of the point $P(x_1, y_1, z_1)$ from the plane $lx + my + nz = p$ is

$\qquad p - (lx_1 + my_1 + nz_1)$

2.2.9   DISTANCE OF A POINT $P(x_1, y_1, z_1)$ FROM THE PLANE $ax + by + cz + d = 0$

The equation of the given plane is
$$ax + by + cz + d = 0 \tag{1}$$
The normal form of the equation of the plane is
$$lx + my + nz = p. \tag{2}$$
Comparing the equations (1) and (2) of the plane, we get,

$$\frac{l}{a} = \frac{m}{b} = \frac{n}{c} = \frac{-p}{d} = \frac{\pm\sqrt{l^2 + m^2 + n^2}}{\sqrt{a^2 + b^2 + c^2}} = \frac{\pm 1}{\sqrt{a^2 + b^2 + c^2}}$$

$$\therefore \quad l = \frac{\pm a}{\sqrt{\sum a^2}} \qquad m = \frac{\pm b}{\sqrt{\sum a^2}} \qquad n = \frac{\pm c}{\sqrt{\sum a^2}} \qquad p = \frac{\mp d}{\sqrt{\sum a^2}}$$

$\therefore$ Distance of the point $P(x_1, y_1, z_1)$ from the given plane $= p - (lx_1 + my_1 + nz_1)$

$$= \frac{\mp d}{\sqrt{\sum a^2}} - \left(\frac{\pm ax_1}{\sqrt{\sum a^2}} \pm \frac{by_1}{\sqrt{\sum a^2}} \pm \frac{cz_1}{\sqrt{\sum a^2}} \right)$$

$$= \frac{\mp (ax_1 + by_1 + cz_1 + d)}{\sqrt{\sum a^2}}$$

$$= \frac{\pm (ax_1 + by_1 + cz_1 + d)}{\sqrt{\sum a^2}}$$

Example 2.2.1 Prove that the equation of the plane through three points (x_1, y_1, z_1), (x_2, y_2, z_2), and (x_3, y_3, z_3) can be written in the form

$$\begin{vmatrix} x & y & z & 1 \\ x_1 & y_1 & z_1 & 1 \\ x_2 & y_2 & z_2 & 1 \\ x_3 & y_3 & z_3 & 1 \end{vmatrix} = 0$$

Solution:
Let the equation of the plane through the three given points be
$$ax + by + cz + d = 0 \tag{1}$$
These three given points satisfy this equation.
$$\therefore \ ax_1 + by_1 + cz_1 + d = 0 \tag{2}$$
$$ax_2 + by_2 + cz_2 + d = 0. \tag{3}$$
$$ax_3 + by_3 + cz_3 + d = 0. \tag{4}$$
Elimination of a, b, c, d from equations (1), (2), (3) and (4), we get,

$$\begin{vmatrix} x & y & z & 1 \\ x_1 & y_1 & z_1 & 1 \\ x_2 & y_2 & z_2 & 1 \\ x_3 & y_3 & z_3 & 1 \end{vmatrix} = 0$$

This is the equation of the plane through the three given points.

Note: Four points $A(x_1, y_1, z_1)$, $B(x_2, y_2, z_2)$, $C(x_3, y_3, z_3)$ and $D(x_4, y_4, z_4)$ are coplanar if

$$\begin{vmatrix} x_1 & y_1 & z_1 & 1 \\ x_2 & y_2 & z_2 & 1 \\ x_3 & y_3 & z_3 & 1 \\ x_4 & y_4 & z_4 & 1 \end{vmatrix} = 0$$

Example 2.2.2 Find the ratio in which the plane $ax + by + cz + d = 0$ divides line joining the points $P(x_1, y_1, z_1)$ and $Q(x_2, y_2, z_2)$.

Solution:

Let the line PQ meets the plane at R. Let R divides PQ in the ratio $\lambda:1$.
Then the coordinates of R are

$$\left(\frac{\lambda x_2 + x_1}{\lambda + 1}, \frac{\lambda y_2 + y_1}{\lambda + 1}, \frac{\lambda z_2 + z_1}{\lambda + 1} \right)$$

R lies on the plane $ax + by + cz + d = 0$

$$\therefore a\left(\frac{\lambda x_2 + x_1}{\lambda + 1} \right) + b\left(\frac{\lambda y_2 + y_1}{\lambda + 1} \right) + c\left(\frac{\lambda z_2 + z_1}{\lambda + 1} \right) + d = 0$$

i.e., $(ax_1 + by_1 + cz_1 + d) + \lambda(ax_2 + by_2 + cz_2 + d) = 0$

$$\lambda = -\frac{ax_1 + by_1 + cz_1 + d}{ax_2 + by_2 + cz_2 + d} \tag{1}$$

Thus the given plane divides PQ in the ratio $\lambda:1$ where λ is given by (1)

Note:

If λ is positive, then P and Q lie on opposite sides of the plane and if λ is negative, then P and Q lie on the same side of the plane.
i.e., P and Q lie on the same or opposite sides of the line according as
$ax_1 + by_1 + cz_1 + d$ and $ax_2 + by_2 + cz_2 + d$ have same or opposite signs.

Example 2.2.3 Show that the points $(0, -1, -1)$, $(-4, -4, 4)$, $(4, 5, 1)$ and $(3, 9, 4)$ are coplanar and find the equation of the plane on which they lie.

Solution:

There is a unique plane through the three points $(0, -1, -1)$, $(-4, -4, 4)$, $(4, 5, 1)$.
Equation of any plane through the point $(0, -1, -1)$ is
$$a(x - 0) + b(y + 1) + c(z + 1) = 0 \tag{1}$$
If this plane passes through $(-4, 4, 4)$ and $(4, 5, 1)$, we get,
$$a(-4) + b(4 + 1) + c(4 + 1) = 0$$

i.e., $-4a + 5b + 5c = 0$ (2)

and $a(4 - 0) + b(5 + 1) + c(1 + 1) = 0$

i.e., $4a + 6b + 2c = 0$ (3)

Eliminating a, b, c from (1), (2) and (3), we get,

$$\begin{vmatrix} x & y+1 & z+1 \\ -4 & 5 & 5 \\ 4 & 6 & 2 \end{vmatrix} = 0$$

i.e., $x(-20) - (y + 1)(-28) + (z + 1)(-44) = 0$

i.e., $-20x + 28y - 44z - 16 = 0$

i.e., $5x - 7y + 11z + 4 = 0$ (4)

Substituting the point $(3, 9, 4)$ in (4), we get,

$5(3) - 7(9) + 11(4) + 4 = 15 - 63 + 44 + 4 = 63 - 63 = 0$

∴ The plane passing through the first three points passes through the fourth point also. Hence the four points are coplanar. Equation of the plane containing the four points is $5x - 7y + 11z + 4 = 0$.

Example 2.2.4 Find the equation of the plane which passes through the points $(1, 2, -2)$ and $(3, -2, 6)$ and which is perpendicular to the plane $2x - y - z + 7 = 0$.

Solution:

Equation of any plane through $(1, 2, -2)$ is

$$a(x-1) + b(y-2) + c(z + 2) = 0 \tag{1}$$

If this passes through $(3, -2, 6)$, we get, $a(3-1) + b(-2 - 2) + c(6 + 2) = 0$

i.e., $2a - 4b + 8c = 0$ (2)

If plane (1) is perpendicular to $2x - y - z + 7 = 0$, we get,

$2a - b - c = 0$ (3)

(Using the condition $a_1a_2 + b_1b_2 + c_1c_2 = 0$)

Eliminating a, b, c from (1), (2) and (3), we get,

$$\begin{vmatrix} x-1 & y-2 & z+2 \\ 2 & -4 & 8 \\ 2 & -1 & -1 \end{vmatrix} = 0$$

i.e., $(x-1)12 - (y - 2)(-18) + (z + 2)6 = 0$

i.e., $12x + 18y + 6z - 36 = 0$

i.e., $2x + 3y + z - 6 = 0$

This is the equation of the required plane.

Note: The equation of the plane passing through the points (x_1, y_1, z_1) and (x_2, y_2, z_2) and which is perpendicular to the plane $ax + by + cz + d = 0$ is given by the following determinant.

$$\begin{vmatrix} x & y & z & 1 \\ x_1 & y_1 & z_1 & 1 \\ x_2 & y_2 & z_2 & 1 \\ a & b & c & 1 \end{vmatrix} = 0 \qquad \text{(Prove this)}$$

Example 2.2.5 Find the equation of the plane through (1, 1, 2) and perpendicular to the planes $2x - 2y - 4z = 3$, $3x + y + 6z = 4$.

Solution:
Equation of any plane through the point (1, 1, 2) is
$$a(x-1) + b(y-1) + c(z-2) = 0 \qquad (1)$$
If this plane is perpendicular to the two given planes, we get,
$$2a - 2b - 4c = 0 \qquad (2)$$
and $\quad 3a + b + 6c = 0 \qquad (3)$
Eliminating a, b, c from (1), (2) and (3), we get,

$$\begin{vmatrix} x-1 & y-1 & z-2 \\ 2 & -2 & -4 \\ 3 & 1 & 6 \end{vmatrix} = 0$$

i.e., $(x-1)(-8) - (y-1)(24) + (z-2)(8) = 0$
i.e., $\qquad\qquad -8x - 24y + 8z + 16 = 0$
i.e., $\qquad\qquad x + 3y - z - 2 = 0$
This is the equation of the required plane.

Note: The equation of the plane passing through the point (x_1, y_1, z_1) and which is perpendicular to the planes $a_1 x + b_1 y + c_1 z + d_1 = 0$ and $a_2 x + b_2 y + c_2 z + d_2 = 0$ is

$$\begin{vmatrix} x & y & z & 1 \\ x_1 & y_1 & z_1 & 1 \\ a_1 & b_1 & c_1 & 1 \\ a_2 & b_2 & c_2 & 1 \end{vmatrix} = 0 \qquad \text{(Prove this)}$$

Example 2.2.6 Find the equation of the plane which passes through the point (2, -4, 5) and is parallel to the plane $4x + 2y - 7z + 6 = 0$.

Solution:
Equation of any plane parallel to the plane $4x + 2y - 7z + 6 = 0$ is
$$4x + 2y - 7z + k = 0.$$
If this passes through (2, -4, 5) we have
$$4(2) + 2(-4) - 7(5) + k = 0$$

i.e.,   k = 35.

∴ The equation of the required plane is 4x + 2y - 7z + 35 = 0.

Example 2.2.7 Find the equation of the plane which passes through (4, 5, -6) and is perpendicular to the line joining the points (5, 2, 3) and (1, 6, 4).

Solution:

D.R.'s of the line joining the points P(5, 2, 3) and Q(1, 6, 4) are (1-5, 6-2, 4-3)

i.e.,   (-4, 4, 1)

These are the D.R.'s of the normal to the required plane. Now the equation of any plane perpendicular to PQ is $-4x + 4y + z + k = 0$ (Fig. 2.12)

If this passes through (4, 5, -6), we get, -4(4) + 4(5)-6 + k = 0   i.e., k = 2.

Hence the equation of the required plane is

$$-4x + 4y + z + 2 = 0$$

i.e., $4x - 4y - z - 2 = 0$

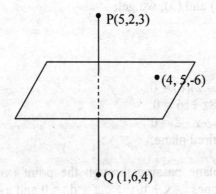

Fig. 2.12

Example 2.2.8 Find the projection of the point P (-2, 7, -1) on the plane 2x − y + z = 0.

Solution: Let N be the projection. Then PN is normal to the plane.

Equation of the plane is 2x − y + z = 0.

∴ The D.R.'s of the normal PN to the plane are (2, -1, 1).

If N is (x_1, y_1, z_1), D.R.'s of the normal PN are also equal to $(x_1 + 2, y_1 - 7, z_1 + 1)$.

Hence $\dfrac{x_1 + 2}{2} = \dfrac{y_1 - 7}{-1} = \dfrac{z_1 + 1}{1} = k$ (say)

i.e., $x_1 = 2k - 2, y_1 = -k + 7, z_1 = k - 1$ (1)

$N(x_1, y_1, z_1)$ lies on the plane 2x − y + z = 0.

∴ We get $2x_1 - y_1 + z_1 = 0$

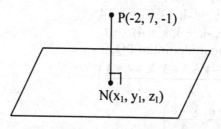

Fig. 2.13

i.e., $2(2k - 2)-(-k + 7) + (k -1) = 0$
i.e., $4k + k + k - 4 - 7 - 1 = 0$
i.e., $6k = 12$
i.e., $k = 2$.
From (1) $x_1 = 2$, $y_1 = 5$, $z_1 = 1$
i.e., the coordinates of N are $(2, 5, 1)$

Example 2.2.9 Find the image of the points $(1, 3, 4)$ in the plane $2x - y + z + 3 = 0$.

Solution:
Let $Q(x_1, y_1, z_1)$ be the image of P $(1, 3, 4)$. Let PQ meet the plane at M. Then M is the midpoint of PQ. Equation of the plane is $2x - y + z + 3 = 0$.
$\therefore$ D.R.'s of the normal to the plane are $(2, -1, 1)$.
Since PQ is perpendicular to the plane, the D.R.'s of the normal to the plane are also equal to $(x_1 - 1, y_1 - 3, z_1 - 4)$.

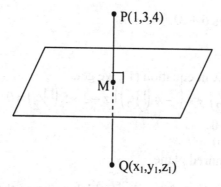

Fig. 2.14

$$\therefore \qquad \frac{x_1-1}{2} = \frac{y_1-3}{-1} = \frac{z_1-4}{1} = k \text{ (say)}$$

i.e., $\qquad x_1 = 2k+1, y_1 = -k+3, z_1 = k+4$ (1)

i.e., $\qquad$ Q is $(2k+1, -k+3, k+4)$

Coordinates of M, being the midpoint of PQ, is

$$\left(\frac{2k+1+1}{2}, \frac{-k+3+3}{2}, \frac{k+4+4}{2} \right)$$

i.e., $\qquad \left(k+1, \frac{-k+6}{2}, \frac{k+8}{2} \right)$

M lies on the plane $2x - y + z + 3 = 0$

$$\therefore \qquad 2(k+1) - \left(\frac{-k+6}{2} \right) + \frac{k+8}{2} + 3 = 0$$

i.e., $\qquad 6k+12 = 0$

i.e., $\qquad k = -2$

$\therefore$ From (1) $\quad x_1 = -3, y_1 = 5, z_1 = 2$

i.e., the image of P is Q $(-3, 5, 2)$

Example 2.2.10 Find the equation of the plane passing through the line of intersection of the planes $2x - 5y + z = 3$ and $x + y + 4z = 5$ and parallel to the plane $x + 3y + 6z = 1$.

Solution:

Equation of any plane through the line of intersection of the planes $2x - 5y + z = 3$ and $x + y + 4z = 5$

is $\qquad (2x - 5y + z - 3) + \lambda(x + y + 4z - 5) = 0$.

i.e., $\quad (2 + \lambda)x + (-5 + \lambda)y + (1 + 4\lambda)z - 3 - 5\lambda = 0$ (1)

This plane is parallel to $x + 3y + 6z - 1 = 0$

$$\therefore \frac{2+\lambda}{1} = \frac{-5+\lambda}{3} = \frac{1+4\lambda}{6}$$

$$\frac{2+\lambda}{1} = \frac{-5+\lambda}{3} \text{ implies } 6 + 3\lambda = -5 + \lambda$$

i.e., $\quad \lambda = -11/2$.

$\therefore \quad$ Substituting the value of λ in equation (1), we get,

$$\left(2 - 11/2 \right) x + \left(-5 - 11/2 \right) y + \left(1 - 4 \left(11/2 \right) \right) z - 3 + 5 \left(11/2 \right) = 0$$

i.e., $\quad -7x - 21y - 42z + 49 = 0$

i.e., $\qquad x + 3y + 6z - 7 = 0$

This is the equation of the required plane.

Example 2.2.11 Find the equation of the plane which contains the line of intersection of the planes $6x + 4y - 5z = 2$ and $x - 2y + 3z = 0$ and is perpendicular to the plane $3x - 2y + z = 5$.

Solution:

Equation of any plane through the line of intersection of the planes $6x + 4y - 5z = 2$ and $x - 2y + 3z = 0$ is

$$(6x + 4y - 5z - 2) + \lambda(x - 2y + 3z) = 0$$

i.e., $(6 + \lambda)x + (4 - 2\lambda)y + (-5 + 3\lambda)z - 2 = 0$ (1)

Plane (1) is perpendicular to the plane $3x - 2y + z = 5$

$\therefore$ Using the condition $a_1 a_2 + b_1 b_2 + c_1 c_2 = 0$, we get,

$$3(6 + \lambda) - 2(4 - 2\lambda) + 1(-5 + 3\lambda) = 0$$

i.e., $10\lambda + 5 = 0$

$$\lambda = -\tfrac{1}{2}$$

Equation (1) gives

$$\left(6 - \tfrac{1}{2}\right)x + \left(4 + 2\left(\tfrac{1}{2}\right)\right)y + \left(-5 - 3\left(\tfrac{1}{2}\right)\right)z - 2 = 0$$

i.e., $11x + 10y - 13z - 4 = 0$

This is the equation of the required plane.

Example 2.2.12 The plane $4x + 7y + 4z + 81 = 0$ is rotated through an angle 90° about its line of intersection with the plane $5x + 3y + 10z = 35$. Find the equation of the plane in its new position and the distance between the feet of the perpendiculars drawn from the origin on to the plane in its two positions.

Solution:

The equation of the plane in its new position (Fig. 2.15) after rotating 90° is

$$4x + 7y + 4z + 81 + \lambda(5x + 3y + 10z - 25) = 0$$

i.e., $(4 + 5\lambda)x + (7 + 3\lambda)y + (4 + 10\lambda)z + 81 - 25\lambda = 0$ (1)

This plane (1) is perpendicular to its original position $4x + 7y + 4z + 81 = 0$.

$\therefore$ $4(4 + 5\lambda) + 7(7 + 3\lambda) + 4(4 + 10\lambda) = 0$

i.e., $81\lambda + 81 = 0$

i.e., $\lambda = -1$.

From (1), substituting for λ, we get,

$$-x + 4y - 6z + 106 = 0$$

i.e., $x - 4y + 6z - 106 = 0$ is the equation of the plane in its new position.

The normal form of this equation is $\dfrac{x}{\sqrt{53}} - \dfrac{4y}{\sqrt{53}} + \dfrac{6z}{\sqrt{53}} = \dfrac{106}{\sqrt{53}}$

The foot of the perpendicular from the origin on to this plane is (pl, pm, pn) where

$$l = \frac{1}{\sqrt{53}}, \quad m = \frac{-4}{\sqrt{53}}, \quad n = \frac{6}{\sqrt{53}}, \quad p = \frac{106}{\sqrt{53}}$$

i.e., $(2, -8, 12)$

Similarly normal form of the equation of the plane in its original position is

$$\frac{4}{9}x + \frac{7}{9}y + \frac{4}{9}z = -9$$

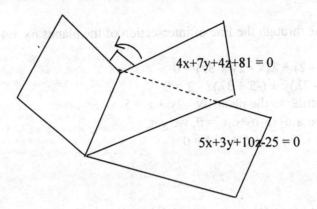

Fig. 2.15

and the foot of the perpendicular from the origin is (-4, -7, -4).
The distance between the feet of the perpendiculars is

$$\sqrt{(2+4)^2 + (-8+7)^2 + (12+4)^2} = \sqrt{293}$$

Example 2.2.13 A variable plane which remains at a constant distance 3p from the origin cuts the coordinate axes at A, B, C. Show that the locus of the centroid of the triangle ABC is $x^{-2} + z^{-2} + z^{-2} = p^{-2}$.

Solution:
Let OA = a, OB = b and OC = c.

Then the equation of the plane ABC is $\dfrac{x}{a} + \dfrac{y}{b} + \dfrac{z}{c} = 1$ (1)

The distance of the origin from this plane is $= \dfrac{1}{\sqrt{\dfrac{1}{a^2} + \dfrac{1}{b^2} + \dfrac{1}{c^2}}}$

Given this distance = 3p

$\therefore \quad \dfrac{1}{\sqrt{\dfrac{1}{a^2} + \dfrac{1}{b^2} + \dfrac{1}{c^2}}} = 3p$

i.e., $\quad \dfrac{1}{a^2} + \dfrac{1}{b^2} + \dfrac{1}{c^2} = \dfrac{1}{9p^2}$ (2)

The coordinates of the points A, B, C are (a, 0, 0), (0, b, 0), (0, 0, c) respectively.

$\therefore$ The centroid of $\triangle ABC$ is (x_1, y_1, z_1) where $x_1 = \dfrac{a}{3}$, $y_1 = \dfrac{b}{3}$, $z_1 = \dfrac{c}{3}$

i.e.,   a = $3x_1$, b = $3y_1$, c = $3z_1$.

Substituting in (2) we have $\dfrac{1}{9x_1^2} + \dfrac{1}{9y_1^2} + \dfrac{1}{9z_1^2} = \dfrac{1}{9p^2}$

i.e.,
$$\dfrac{1}{x_1^2} + \dfrac{1}{y_1^2} + \dfrac{1}{z_1^2} = \dfrac{1}{p^2}$$

$\therefore$ Locus of (x_1, y_1, z_1) is $x^{-2} + z^{-2} + z^{-2} = p^{-2}$.

Example 2.2.14 A plane meets the coordinate axes at A, B, C and the foot of the perpendicular from the origin to the plane at P. If OA = a, OB = b, OC = c, find the coordinates of P. If the plane varies such that $\dfrac{1}{a^2} + \dfrac{1}{b^2} + \dfrac{1}{c^2} = \dfrac{1}{R^2}$ where R is a constant, show that P describes the sphere $x^2 + y^2 + z^2 = R^2$.

Solution:
Since OA = a, OB = b, OC = c the equation of the plane ABC is
$$\frac{x}{a} + \frac{y}{b} + \frac{z}{c} = 1 \tag{1}$$

D.R.'s of the normal to this plane are $\left(\dfrac{1}{a}, \dfrac{1}{b}, \dfrac{1}{c}\right)$.

If $P(x_1, y_1, z_1)$ is the foot of the perpendicular from O, the D.R.'s of OP are
$(x_1 - 0, y_1 - 0, z_1 - 0)$ i.e., (x_1, y_1, z_1).
These are again the D.R.'s of the normal to the plane ABC

$\therefore$ $\dfrac{x_1}{1/a} = \dfrac{y_1}{1/b} = \dfrac{z_1}{1/c} = \lambda$ (say)

i.e., $x_1 = \dfrac{\lambda}{a}$, $y_1 = \dfrac{\lambda}{b}$, $z_1 = \dfrac{\lambda}{c}$ $\tag{2}$

$P(x_1, y_1, z_1)$ lies on the plane ABC.

$\therefore$ From (1) $\dfrac{x_1}{a} + \dfrac{y_1}{b} + \dfrac{z_1}{c} = 1$ $\tag{3}$

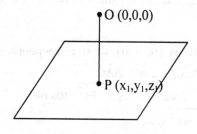

Fig. 2.16

i.e., $\dfrac{\lambda}{a^2} + \dfrac{\lambda}{b^2} + \dfrac{\lambda}{c^2} = 1$

i.e., $\lambda\left(\dfrac{1}{a^2} + \dfrac{1}{b^2} + \dfrac{1}{c^2}\right) = 1$

i.e., $\lambda = \dfrac{1}{\sum \frac{1}{a^2}}$

From (2) $x_1 = \dfrac{1}{a\sum \frac{1}{a^2}}$, $\quad y_1 = \dfrac{1}{b\sum \frac{1}{a^2}}$, $\quad z_1 = \dfrac{1}{c\sum \frac{1}{a^2}}$

$\therefore$ The Coordinates of P are $\left(\dfrac{1}{a\sum \frac{1}{a^2}}, \dfrac{1}{b\sum \frac{1}{a^2}}, \dfrac{1}{c\sum \frac{1}{a^2}}\right)$

Given $\sum \dfrac{1}{a^2} = \dfrac{1}{R^2}$

$\therefore$ The Coordinates of P are $\left(\dfrac{R^2}{a}, \dfrac{R^2}{b}, \dfrac{R^2}{c}\right)$

i.e., $x_1 = \dfrac{R^2}{a}, y_1 = \dfrac{R^2}{b}, z_1 = \dfrac{R^2}{c}$

i.e., $\dfrac{1}{a} = \dfrac{x_1}{R^2}, \dfrac{1}{b} = \dfrac{y_1}{R^2}, \dfrac{1}{c} = \dfrac{z_1}{R^2}$

Substituting for 1/a, 1/b, 1/c in (3) we get,

$$\dfrac{x_1^2}{R^2} + \dfrac{y_1^2}{R^2} + \dfrac{z_1^2}{R^2} = 1$$

i.e., $x_1^2 + y_1^2 + z_1^2 = R^2$

$\therefore$ Locus of $P(x_1, y_1, z_1)$ is $x^2 + y^2 + z^2 = R^2$, which is a sphere with radius R.

Example 2.2.15 The plane $\dfrac{x}{a} + \dfrac{y}{b} + \dfrac{z}{c} = 1$ meets the coordinate axes at A, B, C respectively. If the angle BAC is equal to $60°$, show that $a^2b^2 + b^2c^2 + c^2a^2 = 3a^4$.

Solution:
The coordinate of A, B, C are (a, 0, 0), (0, b, 0), (0, 0, c) respectively.
In $\triangle$ ABC, $BC^2 = AB^2 + AC^2 - 2AB.AC.\cos \hat{BAC}$

i.e., $b^2 + c^2 = a^2 + b^2 + a^2 + c^2 - 2\sqrt{a^2 + b^2}.\sqrt{a^2 + c^2} \cos 60$

i.e., $0 = 2a^2 - 2\sqrt{a^2 + b^2}.\sqrt{a^2 + c^2} . \tfrac{1}{2}$

i.e., $\sqrt{a^2 + b^2}.\sqrt{a^2 + c^2} = 2a^2$

i.e., $(a^2 + b^2)(a^2 + c^2) = 4a^4$

i.e., $a^2b^2 + b^2c^2 + c^2a^2 = 3a^4$.

EXERCISE 2.2

PART – A

1. Find the intercepts made by the plane $x + 2y + 3z + 4 = 0$ on the coordinate axes.
2. Find the equation of the plane passing through the points $(3, 0, 0)$, $(0, -2, 0)$ and $(0, 0, 5)$.
3. A plane meets the coordinate axes at A, B, C such that the centroid of the triangle ABC is $(1, \frac{1}{2}, -1/3)$. Find the equation of the plane.
4. The D.C.'s of the perpendicular to a plane from the origin are proportional to 1, 3, 1 and the length of the perpendicular is 2. Find the equation of the plane.
5. Find the equation of the plane through $(3, -3, 1)$ and normal to the line joining the points $(3, 4, -1)$ and $(2, -1, 5)$.
6. Find the equation of the plane which bisects at right angles the line joining the points $(3, 4, 8)$ and $(5, -2, 4)$.
7. Find the equation of the plane through the point $P(2, 3, -1)$ and perpendicular to OP.
8. Find the angle between the planes $2x - y + z = 6$ and $x + y + 2z = 3$.
9. Find the D.C.'s of the normal to the plane $2x + 3y + 6z = 1$.
10. Find the distance between the parallel planes $2x - 2y + z - 6 = 0$ and $4x - 4y + 2z - 7 = 0$.
11. Find the distance between the parallel planes $2x - 3y + 6z + 12 = 0$ and $2x - 3y + 6z - 2 = 0$.
12. Find the distance of the origin from the plane $2x + 3y - 4z = 5$.
13. Find the distance of the point $(1, -2, 3)$ from the plane $4x - 3y + 5z + 2 = 0$.
14. Find the equation of the plane through the point $(1, 2, 1)$ and the line of intersection of the planes $x + 2y - z + 1 = 0$, $3x - y + 4z + 5 = 0$.
15. Find whether the points $(2, 1, 4)$ and $(-1, 3, -2)$ lie on the same size of the plane $2x + y - 3z = 2$.

PART – B

16. Find the equation of the plane through the points $(1, 2, 2)$, $(3, -2, 1)$ and parallel to the line joining the points $(-3, 1, 2)$ and $(-8, 5, -1)$.
17. Find the equation of the plane through the points $(1, -2, 2)$ and $(-3, 1, -2)$ and perpendicular to the plane $2x + y - z + 6 = 0$.
18. Find the equation of the plane passing through the points $(3, 1, 2)$, $(3, 4, 4)$ and perpendicular to the plane $5x + y + 4z = 0$.
19. Find the equation of the plane which passes through the point $(-1, 3, 2)$ and perpendicular to the two planes $x + 2y + 2z = 5$, $3x + 3y + 2z = 8$.
20. Find the equation of the plane through the points $(2, 2 - 1)$, $(3, 4, 2)$ and $(7, 0, 6)$.
21. Show that the points $(0, 2, -4)$, $(-1, 1, -2)$, $(-2, 3, 3)$ and $(-3, -2, 1)$ are coplanar and find the equation of the plane on which they lie.

22. Find the equation of the plane which passes through the point (4, 1, 1) and is perpendicular to the planes x - 3y + 5z + 1 = 0, 3x − y + 7z - 3 = 0.

23. Find the foot of the normal from the origin to the plane 12x - 4y + 3z = 169.

24. Find the image of the point (1, 0,1) in the plane x + 2y + 3z = 6.

25. Find the image of the point (5, 3, 2) in the plane x + y -z = 5.

26. Find the equation of the plane through the line of intersection of the planes 5x + 2y + 2z + 4 = 0 and 3x + 3y + 2z + 8 = 0 and perpendicular to the plane x −y + 4z = 3.

27. Find the equation of the plane passing through the line of intersection of the planes 2x − y + 5z - 3 = 0 and 4x + 2y − z + 7 = 0 and parallel to the z-axis.

28. Prove that the planes 5x - 3y + 4z = 1, 8x + 3y + 5z = 4 and 18x - 3y + 13z = 6 contain a common line.

29. Prove that the planes 12x - 15y + 16z - 28 = 0, 6x + 6y - 7z - 8 = 0 and 2x + 35y - 39z + 12 = 0 have a common line of intersection.

30. Find the locus of the point whose distance from the plane 3x - 2y + 6z - 3 = 0 is equal to 3 times its distance from the plane 12x + 4y - 3z + 4 = 0.

31. A variable plane is at a constant distance p from the origin and meets the coordinate axes in A, B, C. Through A, B, C planes are drawn parallel to the coordinate planes. Prove that the locus of their point of intersection is given by $x^{-2} + y^{-2} + z^{-2} = p^{-2}$.

32. A variable plane is at a constant distance p from the origin and meets the axes in A, B, C. Show that the locus of the centroid of the tetrahedron OABC is $x^{-2} + y^{-2} + z^{-2} = 16p^{-2}$.

33. Prove that a variable plane which moves so that the sum of the reciprocals of its intercepts on the three coordinate axes is constant, passes through a fixed point.

34. A variable plane passes through a fixed point (a, b, c) and meets the coordinate axes in A, B, C. Show that the locus of the points common to the planes through A, B, C respectively parallel to the coordinate planes is $\dfrac{a}{x} + \dfrac{b}{y} + \dfrac{c}{z} = 1$

35. Two systems of rectangular axes have the same origin, if a plane cuts then at distances a, b, c and a_1, b_1, c_1 from the origin, show that $\dfrac{1}{a^2} + \dfrac{1}{b^2} + \dfrac{1}{c^2} = \dfrac{1}{a_1^2} + \dfrac{1}{b_1^2} + \dfrac{1}{c_1^2}$

36. The point P on the fixed plane lx + my + nz = p. The plane through P perpendicular to OP meets the axes in A, B, C. The planes through A, B, C parallel to coordinate planes intersect in Q. Show that the locus of Q is $p\left(\dfrac{1}{x^2} + \dfrac{1}{y^2} + \dfrac{1}{z^2}\right) = \dfrac{l}{x} + \dfrac{m}{y} + \dfrac{n}{z}$.

37. A variable plane is such that the sum of the intercepts made by it on the coordinate axes is a constant k. Find the locus of the foot of the perpendicular drawn to it from the origin.

38. A variable plane passes through the fixed point (a, b, c) and meets the coordinate axes at A, B, C. Show that the locus of the centroid of the triangle ABC is $\dfrac{a}{x} + \dfrac{b}{y} + \dfrac{c}{z} = 3$

39. Find the equation of the plane which passes through the line of intersection of the planes $x + 3y + 6 = 0$ and $3x - y - 4z = 0$, whose distance from the origin is unity.
40. The plane $x - y - z = 2$ is rotated through a right angle about its line of intersection with the plane $x + 2y + z = 2$. Find the equation to the plane in its new position.

2.3 THE STRAIGHT LINE

The intersection of two planes is a straight line in space. Thus two planes jointly represent a straight line. If the equations of the two planes are $ax + by + cz + d = 0$ and $a'x + b'y + c'z + d' = 0$, then the equation of the straight line which is the line of intersection of the two planes is given as $ax + by + cz + d = 0$, $a'x + b'y + c'z + d' = 0$ or $ax + by + cz + d = 0 = a'x + b'y + c'z + d$.

The equation of the x-axis, being the line of intersection of the xy and zx planes is $y = 0$, $z = 0$. Similarly the equation of the y-axis is $x = 0$, $z = 0$ and that of the z-axis is $x = 0$, $y = 0$.

2.3.1 SYMMETRIC FORM OF THE EQUATION OF A LINE

Let (l, m, n) be the D.C.'s of the line passing through a given point $A(x_1, y_1, z_1)$.
Let $P(x, y, z)$ be any point on the line and $AP = r$.

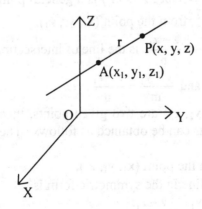

Fig. 2.17

The D.R.'s of line AP are $(x-x_1, y-y_1, z-z_1)$. But given (l, m, n) are the D.C.'s of the line AP.

$$\therefore \quad \frac{x - x_1}{l} = \frac{y - y_1}{m} = \frac{z - z_1}{n} \qquad (1)$$

This is the symmetric form of the equation of the line through $A(x_1, y_1, z_1)$ and having D.C.'s (l, m, n). Instead of the D.C.'s, the D.R.'s (a, b, c) can also be used in the symmetric form.

Note:

(i)   Each ratio in the equation (1) is equal to $\dfrac{\sqrt{(x-x_1)^2 + (y-y_1)^2 + (z-z_1)^2}}{\sqrt{l^2 + m^2 + n^2}}$

$= \sqrt{(x-x_1)^2 + (y-y_1)^2 + (z-z_1)^2}$ Since $l^2 + m^2 + n^2 = 1$.

$= AP$

$= r$

$\therefore \quad \dfrac{x-x_1}{l} = \dfrac{y-y_1}{m} = \dfrac{z-z_1}{n} = r$

i.e., $x = x_1 + lr$, $y = y_1 + mr$, $z = z_1 + nr$.

i.e., $(x_1 + lr, y_1 + mr, z_1 + nr)$ represents a general point on the line and it is at a distance r from the given point (x_1, y_1, z_1).

(ii)   If (a, b, c) are the D. R.'s of the line through (x_1, y_1, z_1), then the equation of the line is $\dfrac{x-x_1}{a} = \dfrac{y-y_1}{b} = \dfrac{z-z_1}{c} = r'$

In this case each ratio is equal to $r' = \dfrac{r}{\sqrt{a^2 + b^2 + c^2}}$.

$\therefore$ The point $(x_1 + ar', y_1 + br', z_1 + cr')$ is a general point on the line and it is at a distance $r'.\sqrt{a^2 + b^2 + c^2}$ from the point (x_1, y_1, z_1).

(iii) The line $\dfrac{x-x_1}{l} = \dfrac{y-y_1}{m} = \dfrac{z-z_1}{n}$ is the line of intersection of the planes

$\dfrac{x-x_1}{l} = \dfrac{y-y_1}{m}$ and $\dfrac{y-y_1}{m} = \dfrac{z-z_1}{n}$

(iv)   If (x_1, y_1, z_1) and (x_2, y_2, z_2) are two given points, then the equation of the line through these two points can be obtained as follows. The D.R.'s of the line are $(x_2 - x_1, y_2 - y_1, z_2 - z_1)$.

The line passes through the point (x_1, y_1, z_1).

$\therefore$ The equation of the line in the symmetric form is

$\dfrac{x-x_1}{x_2-x_1} = \dfrac{y-y_1}{y_2-y_1} = \dfrac{z-z_1}{z_2-z_1}$

(v)   **To put the equation $ax + by + cz + d = 0$, $a_1x + b_1y + c_1z + d = 0$ of a line in the symmetric form.**

Let the D.C.'s of the line be (l, m, n). The line is perpendicular to the normals of both the planes, $ax + by + cz + d = 0$ and $a_1x + b_1y + c_1z + d_1 = 0$

$\therefore \qquad al + bm + cn = 0 \qquad\qquad\qquad\qquad\qquad\qquad (1)$

$\qquad\qquad a_1l + b_1m + c_1n = 0 \qquad\qquad\qquad\qquad\qquad (2)$

From (1) and (2) $\dfrac{l}{bc_1 - b_1c} = \dfrac{m}{ca_1 - c_1a} = \dfrac{n}{ab_1 - a_1b}$

$\therefore$ D.C.'s are proportional to $(bc_1 - b_1c, \, ca_1 - c_1a, \, ab_1 - a_1b)$.

The coordinates of the point where the line meets the xy plane (if the line is parallel to xy plane choose the xz plane or the yz plane) are obtained by solving $ax + by + cz + d = 0$, $a_1x + b_1y + c_1z + d_1 = 0$ and $z = 0$.

Since $z = 0$, $\quad ax + by + d = 0$

$$a_1x + b_1y + d_1 = 0$$

$\therefore \quad \dfrac{x}{bd_1 - b_1d} = \dfrac{y}{da_1 - d_1a} = \dfrac{1}{ab_1 - a_1b}$

(We assume the given planes determining the line, as their line of intersection, are not parallel. i.e., $\quad ab_1 - a_1b \neq 0$)

The required point is $\left(\dfrac{bd_1 - b_1d}{ab_1 - a_1b}, \, \dfrac{da_1 - d_1a}{ab_1 - a_1b}, \, 0 \right)$

$\therefore$ The equation of the line in symmetric form is

$$\dfrac{x - \dfrac{bd_1 - b_1d}{ab_1 - a_1b}}{bc_1 - cb_1} = \dfrac{y - \dfrac{da_1 - d_1a}{ab_1 - a_1b}}{ca_1 - c_1a} = \dfrac{z}{ab_1 - a_1b}$$

(vi) **The length of the perpendicular from a given point $P(x_1, y_1, z_1)$ to a given line $\dfrac{x - a}{l} = \dfrac{y - b}{m} = \dfrac{z - c}{n}$**

Let M be the foot of the perpendicular from P on the given line with D.C.'s (l, m, n).

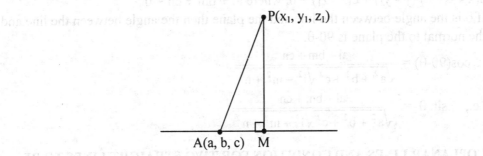

Fig. 2.18

The point $A(a, b, c)$ lies on the line.

$\therefore$ AM = Projection of AP on the line

$\qquad = l(x_1 - a) + m(y_1 - b) + n(z_1 - c)$

Also $AP = \sqrt{(x_1 - a)^2 + (y_1 - b)^2 + (z_1 - c)^2}$

From the right angled triangle APM,

$PM^2 = AP^2 - AM^2$

$= (x_1 - a)^2 + (y_1 - b)^2 + (z_1 - c)^2 - [\, l(x_1 - a) + m(y_1 - b) + n(z_1 - c)\,]^2$

This gives PM, the length of the perpendicular from P on the given line.

2.3.2   THE PLANE AND THE STRAIGHT LINE

Consider the plane $ax + by + cz + d = 0$ (1)

and the straight line $\dfrac{x - x_1}{l} = \dfrac{y - y_1}{m} = \dfrac{z - z_1}{n}$ (2).

The D.R.'s of any normal to the plane are (a, b, c) and those of the straight line are (l, m, n).

Note:

(i)   The condition for the straight line to be perpendicular to the plane is $\dfrac{a}{l} = \dfrac{b}{m} = \dfrac{c}{n}$
       (since the line is parallel to any normal to the plane).

(ii)   The condition for the straight line to be parallel to the plane is
       $al + bm + cn = 0$ (since the line is perpendicular to any normal to the plane).

(iii) If the straight line lies on the plane, then (x_1, y_1, z_1) is a point on the plane.
       $\therefore ax_1 + by_1 + cz_1 + d = 0$.
       Also the line is perpendicular to the normal to the plane.
       $\therefore al + bm + cn = 0$.
       $\therefore$ The conditions for the straight line to lie on the plane are
       $ax_1 + by_1 + cz_1 + d = 0$ and $al + bm + cn = 0$.
       Since the point (x_1, y_1, z_1) lies on the plane, the equation of the plane is
       $a(x - x_1) + b(y - y_1) + c(z - z_1) = 0$, where $al + bm + cn = 0$.

(iv)   If θ is the angle between the line and the plane then the angle between the line and
       the normal to the plane is $90 - \theta$.

$$\therefore \cos(90 - \theta) = \frac{al + bm + cn}{\sqrt{a^2 + b^2 + c^2}\,\sqrt{l^2 + m^2 + n^2}}$$

i.e., $$\sin \theta = \frac{al + bm + cn}{\sqrt{a^2 + b^2 + c^2}\,\sqrt{l^2 + m^2 + n^2}}$$

2.3.3   COPLANAR LINES AND CONDITION FOR TWO STRAIGHT LINES TO BE COPLANAR

Two lines are said to be coplanar if they lie on the same plane.

Let the equations of the straight lines be

$$\frac{x - x_1}{l_1} = \frac{y - y_1}{m_1} = \frac{z - z_1}{n_1} \tag{1}$$

and $\dfrac{x - x_2}{l_2} = \dfrac{y - y_2}{m_2} = \dfrac{z - z_2}{n_2}$ (2)

The equation of the plane containing the line (1) is

$\qquad a(x - x_1) + b(y - y_1) + c(z - z_1) = 0$ (3)

where $\qquad al_1 + bm_1 + cn_1 = 0$ (4)

When the lines (1) and (2) are coplanar, the line (2) lies on the plane (3)

∴ Plane (3) contains the point (x_2, y_2, z_2) and its normal is perpendicular to line (2).

∴ $a(x_2 - x_1) + b(y_2 - y_1) + c(z_2 - z_1) = 0$ (5)

and $\qquad al_2 + bm_2 + cn_2 = 0$ (6)

Eliminating a, b, c from (4), (5) and (6)

$$\begin{vmatrix} x_2 - x_1 & y_2 - y_1 & z_2 - z_1 \\ l_1 & m_1 & n_1 \\ l_2 & m_2 & n_2 \end{vmatrix} = 0$$

This is the condition for the lines (1) and (2) to be coplanar. The equation of the plane containing the two lines is obtained by eliminating a, b, c from (3), (4) and (6).

i.e., $$\begin{vmatrix} x - x_1 & y - y_1 & z - z_1 \\ l_1 & m_1 & n_1 \\ l_2 & m_2 & n_2 \end{vmatrix} = 0$$

Expanding this determinant, we get the equation of the required plane.

2.3.4 SHORTEST DISTANCE BETWEEN TWO SKEW LINES

If two given lines are skew lines then they are non-coplanar and hence they are neither intersecting nor parallel. Such lines have a common perpendicular. Let AB and CD be two skew lines. EF is the common perpendicular line meeting AB at E and CD at F. Then the length EF is called the shortest distance (S.D) between the two skew lines. The common perpendicular line is called the shortest distance (S.D) line.

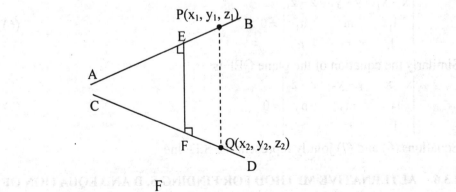

Fig. 2.19

We shall find the S.D and the equation of the S.D line between the skew lines.

$$\frac{x - x_1}{l_1} = \frac{y - y_1}{m_1} = \frac{z - z_1}{n_1} \qquad (1)$$

$$\frac{x - x_2}{l_2} = \frac{y - y_2}{m_2} = \frac{z - z_2}{n_2} \qquad (2)$$

These skew lines pass through the points $P(x_1, y_1, z_1)$ and $Q(x_2, y_2, z_2)$. If EF is the S.D between the two lines, then EF = Projection of PQ on line EF. Let (l, m, n) be the D.C.'s of the S.D line (i.e., EF). Since EF is perpendicular to both the skew lines,

$$ll_1 + mm_1 + nn_1 = 0 \qquad (3)$$

$$ll_2 + mm_2 + nn_2 = 0 \qquad (4)$$

Solving (3) and (4) we get

$$\frac{l}{m_1 n_2 - m_2 n_1} = \frac{m}{n_1 l_2 - n_2 l_1} = \frac{n}{l_1 m_2 - l_2 m_1}$$

$$\therefore l = \frac{m_1 n_2 - m_2 n_1}{\sqrt{\sum (m_1 n_2 - m_2 n_1)^2}},$$

$$m = \frac{n_1 l_2 - n_2 l_1}{\sqrt{\sum (m_1 n_2 - m_2 n_1)^2}} \quad \text{and}$$

$$n = \frac{l_1 m_2 - l_2 m_1}{\sqrt{\sum (m_1 n_2 - m_2 n_1)^2}} \qquad (5)$$

Now EF = Projection of PQ on the line EF.

i.e., S.D = $l(x_2 - x_1) + m(y_2 - y_1) + n(z_2 - z_1)$ where l, m, n are given by (5).

2.3.5 TO FIND THE EQUATION OF S. D LINE

S.D line is the intersection of the planes PEF and QEF. PEF is the plane containing the lines AB and EF. Hence its equation is

$$\begin{vmatrix} x - x_1 & y - y_1 & z - z_1 \\ l_1 & m_1 & n_1 \\ l & m & n \end{vmatrix} = 0 \qquad (6)$$

Similarly the equation of the plane QEF is

$$\begin{vmatrix} x - x_2 & y - y_2 & z - z_2 \\ l_2 & m_2 & n_2 \\ l & m & n \end{vmatrix} = 0 \qquad (7)$$

Equations (6) and (7) jointly represent the S.D line.

2.3.6 ALTERNATIVE METHOD FOR FINDING S. D AND EQUATION OF S. D LINE

The coordinates of the point E on AB can be written as

$(x_1 + l_1r_1, y_1 + m_1r_1, z_1 + n_1r_1)$.

Similarly coordinates of F on CD can be written as

$(x_2 + l_2r_2, y_2 + m_2r_2, z_2 + n_2r_2)$

∴ D.R.'s of EF are

$(x_1 - x_2 + l_1r_1 - l_2r_2, y_1 - y_2 + m_1r_1 - m_2r_2, z_1 - z_2 + n_1r_1 - n_2r_2)$

As EF is perpendicular to AB, we get,

$l_1(x_1 - x_2 + l_1r_1 - l_2r_2) + m_1(y_1 - y_2 + m_1r_1 - m_2r_2) + n_1(z_1 - z_2 + n_1r_1 - n_2r_2) = 0$ (1)

As EF is perpendicular to CD, we get,

$l_2(x_1 - x_2 + l_1r_1 - l_2r_2) + m_2(y_1 - y_2 + m_1r_1 - m_2r_2) + n_2(z_1 - z_2 + n_1r_1 - n_2r_2) = 0$ (2)

Solving (1) and (2) we get r_1 and r_2 which in turn gives the coordinates of E and F. Hence the S.D (length of EF) and the equations of the S.D line (line EF) can be found.

Example 2.3.1 Find the equation of the straight line joining points (3, 2, -4) and (1, 4, 7).

Solution:

The D.R..'s of the line are (1 - 3, 4 - 2, 7 + 4)

i.e.,   (-2, 2, 11)

The line passes through (3, 2, -4)

∴ The equation of the line in the symmetric form is $\dfrac{x-3}{-2} = \dfrac{y-2}{2} = \dfrac{z+4}{11}$

Example 2.3.2 Find the symmetric form of the equation of the line $3x - 2y + z - 1 = 0$, $5x + 4y - 6z - 2 = 0$.

Solution:

Method 1.

Let the D.C.'s of the line be (l, m, n). Since the line is perpendicular to the normals of the two planes, we get $3l - 2m + n = 0$

$$5l + 4m - 6n = 0$$

∴ $\dfrac{l}{12-4} = \dfrac{m}{5+18} = \dfrac{n}{12+10}$

∴ D.R.'s are (8, 23, 22).

The point where the line meets the yz plane are obtained by solving

$3x - 2y + z - 1 = 0$

$5x + 4y - 6z - 2 = 0$ and

$\qquad\qquad x = 0$

$-2y + z - 1 = 0$ and

$4y - 6z - 2 = 0$

$$\dfrac{y}{-2-6} = \dfrac{z}{-4-4} = \dfrac{1}{12-4}$$

i.e., $\dfrac{y}{-8} = \dfrac{z}{-8} = \dfrac{1}{8}$

$\therefore$ The required point is (0, -1, -1)

Hence the symmetric form of the equation of the line is $\dfrac{x}{8} = \dfrac{y+1}{23} = \dfrac{z+1}{22}$

Method 2.

$$3x - 2y + z - 1 = 0 \tag{1}$$
$$5x + 4y - 6z - 2 = 0 \tag{2}$$

$2*(1) + (2)$ gives, $11x - 4z - 4 = 0$

i.e., $\quad x = \dfrac{4z+4}{11}$ $\tag{3}$

Again $6*(1) + (2)$ gives, $23x - 8y - 8 = 0$

i.e., $\quad x = \dfrac{8y+8}{23}$ $\tag{4}$

From (3) and (4)

$$x = \dfrac{8y+8}{23} = \dfrac{4z+4}{11}$$

i.e., $\quad \dfrac{x}{8} = \dfrac{y+1}{23} = \dfrac{z+1}{22}$

Example 2.3.3 Find the coordinates of the foot, the length and equation of the perpendicular from the point P(-1, 3, 9) to the line $\dfrac{x-13}{5} = \dfrac{y+8}{-8} = \dfrac{z-31}{1}$

Solution:

Let M be the foot of the perpendicular from P to the line. Then the coordinates of M are
$\quad (5r + 13, -8r - 8, r + 31)$.

D.R's of line PM are $(5r + 13 +1, -8r-8 - 3, r +31-9)$

i.e., $\quad (5r + 14, -8r - 11, r + 22)$

D.R.'s of the given line are (5, -8, 1)

Since PM is perpendicular to the given line, $5(5r + 14) + (-8)(-8r - 11) + 1(r + 22) = 0$

i.e., $\quad 90r + 180 = 0$

i.e., $\quad\quad\quad r = -2.$

$\therefore$ The coordinates of M are $(5(-2) + 13, -8(-2) - 8, -2 + 31)$

i.e., $\quad (3, 8, 29)$

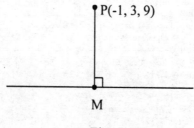

Fig. 2.20

Length of the perpendicular $= PM$
$$= \sqrt{(3+1)^2 + (8-3)^2 + (29-9)^2} = 21$$

D.R.'s of PM are $\quad (5(-2) + 14, -8(-2)-11, -2 + 22) = (4, 5, 20)$

Also PM passes through $(-1, 3, 9)$

$\therefore$ Equation of the line PM is
$$\frac{x+1}{4} = \frac{y-3}{5} = \frac{z-9}{20}$$

Example 2.3.4 Find the image of the line $\dfrac{x-1}{9} = \dfrac{y-2}{-1} = \dfrac{z+3}{-3}$ in the plane

$3x-3y + 10z - 26 = 0$

Solution:

The D.R.'s of the given line
$$\frac{x-1}{9} = \frac{y-2}{-1} = \frac{z+3}{-3} \tag{1}$$

are $(9, -1,-3)$. D.R.'s of the normal to the plane
$$3x - 3y + 10z - 26 = 0 \tag{2}$$

are $(3, -3, 10)$.

Now, $9(3) + (-1)(-3) + (-3)10 = 27 + 3 - 30 = 0$

$\therefore$ The line is parallel to the plane.

Let $Q(x_1, y_1, z_1)$ be the image of the point $P(1, 2, -3)$ on the plane. Let PQ meet the plane at M. Then PM = MQ and the image of the given line on the plane is the line through Q having D.R.'s equal to $(9, -1, -3)$, the image is parallel to the given line.

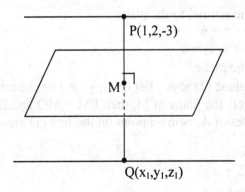

Fig. 2.21

Now D.R.'s of PQ are (x_1-1, y_1-2, z_1+3). As PQ is perpendicular to the plane, D.R.'s of PQ are also equal to $(3, -3, 10)$.

$\therefore \quad \dfrac{x_1-1}{3} = \dfrac{y_1-2}{-3} = \dfrac{z_1+3}{10} = k(\text{say})$

i.e., $x_1 = 3k + 1$, $y_1 = -3k + 2$, $z_1 = 10k - 3$

i.e.,   Q is $(3k + 1, -3k + 2, 10k - 3)$

Coordinates of M being the mid point of PQ is $\left(\dfrac{3k+2}{2}, \dfrac{-3k+4}{2}, \dfrac{10k-6}{2} \right)$

M lies on the plane $3x - 3y + 10z - 26 = 0$.

$\therefore$ $3\left(\dfrac{3k+2}{2} \right) - 3\left(\dfrac{-3k+4}{2} \right) + 10\left(\dfrac{10k-6}{2} \right) - 26 = 0$

$$118k - 118 = 0$$
$$k = 1$$

$\therefore$ $x_1 = 3(1) + 1 = 4$,

        $y_1 = -3(1) + 2 = -1$,

        $z_1 = 10(1) - 3 = 7$.

Hence Q is $(4, -1, 7)$

$\therefore$ The equation of the image of the line is

$$\frac{x-4}{9} = \frac{y+1}{-1} = \frac{z-7}{-3}$$

Example 2.3.5 Find the image of the line $\dfrac{x-1}{2} = \dfrac{y-2}{1} = \dfrac{z-3}{4}$ in the plane

$2x + y + z = 6$.

Solution:

D.R.'s of the line $\dfrac{x-1}{2} = \dfrac{y-2}{1} = \dfrac{z-3}{4}$ (1)

are $(2, 1, 4)$. The D.R.'s of the normal to the plane

                $2x + y + z = 6$ (2)

are $(2, 1, 1)$. Also $2(2) + 1(1) + 4(1) \neq 0$

$\therefore$ The line is not parallel to the plane.

$\therefore$ Let the line (1) meets the plane (2) at A. Let $Q(x_1, y_1, z_1)$ be the image of the point $P(1, 2, 3)$ on the plane. If PQ meets the plane at M, then $PM = MQ$ and the image of the line (1) is the line AQ. Coordinates of A, being a point on the line (1) are

$(2r + 1, r + 2, 4r + 3)$

A lies on plane (2)

$\therefore$ $2(2r + 1) + r + 2 + 4r + 3 = 6$

i.e., $9r + 1 = 0$

i.e., $r = -1/9$

$\therefore$ Coordinates of A are $\left(\dfrac{7}{9}, \dfrac{17}{9}, \dfrac{23}{9} \right)$

Now D.R.'s of PQ are (x_1-1, y_1-2, z_1-3). PQ being perpendicular to the plane, its D.R.'s are $(2, 1, 1)$.

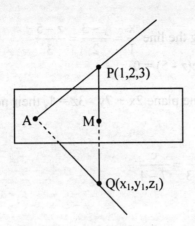

Fig. 2.22

$$\therefore \qquad \frac{x_1 - 1}{2} = \frac{y_1 - 2}{1} = \frac{z_1 - 3}{1} = k \text{ (say)}$$

i.e., $\quad x_1 = 2k + 1, y_1 = k + 2, z_1 = k + 3$

i.e., $\quad$ Q is $(2k + 1, k + 2, k + 3)$.

M, being the midpoint of PQ, has coordinates

$$\left(\frac{2k + 2}{2}, \frac{k + 4}{2}, \frac{k + 6}{2} \right)$$

Also, M lies on the plane (2).

$$\therefore \qquad 2\left(\frac{2k + 2}{2} \right) + \frac{k + 4}{2} + \frac{k + 6}{2} = 6$$

$$6k + 2 = 0$$

i.e., $k = -1/3$.

$$\therefore \qquad \text{Q is } (2(-1/3) + 1, (-1/3) + 2, (-1/3) + 3)$$

i.e., $\quad$ Q is $(1/3, 5/3, 8/3)$.

D.R.'s of AQ are $\left(\dfrac{7}{9} - \dfrac{1}{3}, \dfrac{17}{9} - \dfrac{5}{3}, \dfrac{23}{9} - \dfrac{8}{3} \right)$

i.e., $\quad (4/9, 2/9, -1/9)$

$\therefore$ Equation of line AQ is $\quad \dfrac{x - \frac{1}{3}}{\frac{4}{9}} = \dfrac{y - \frac{5}{3}}{\frac{2}{9}} = \dfrac{z - \frac{8}{3}}{-\frac{1}{9}}$

Example 2.3.6 Find the equation of the plane which contains the line $x = \dfrac{y - 3}{2} = \dfrac{z - 5}{3}$
and which is perpendicular to the plane $2x + 7y - 3z = 1$.

Solution:

Equation of any plane containing the line $\dfrac{x}{1} = \dfrac{y-3}{2} = \dfrac{z-5}{3}$ (1)

is $a(x) + b(y - 3) + c(z - 5) = 0$ (2)

where $a + 2b + 3c = 0$ (3)

If plane (2) is perpendicular to the plane $2x + 7y - 3z = 1$, their normals are perpendicular to each other.

$\therefore$ $2a + 7b - 3c = 0$ (4)

From (3) and (4) $\dfrac{a}{-6-21} = \dfrac{b}{6+3} = \dfrac{c}{7-4}$

i.e., $\dfrac{a}{-27} = \dfrac{b}{9} = \dfrac{c}{3}$

i.e., $\dfrac{a}{9} = \dfrac{b}{-3} = \dfrac{c}{-1}$

i.e.,   a, b, c are proportional to 9, -3, -1

$\therefore$ From (2) $9x - 3(y - 3) - (z - 5) = 0$

i.e., $9x - 3y - z + 14 = 0$

This is the equation of the required plane.

Example 2.3.7 Find the equation of the plane which contains the two parallel lines.

$$\dfrac{x-3}{1} = \dfrac{y-2}{-4} = \dfrac{z-1}{5}, \quad \dfrac{x-1}{1} = \dfrac{y+1}{-4} = \dfrac{z-2}{5}$$

Solution:

Any plane containing the first line is $a(x - 3) + b(y - 2) + c(z - 1) = 0$ (1)

where $a.1 + b(-4) + c(5) = 0$

i.e., $a - 4b + 5c = 0$ (2)

If plane (1) contains the second line, it contains the point $(1, -1, 2)$.

$\therefore$ $a(1-3) + b(-1-2) + c(2-1) = 0$

$\therefore$ $-2a - 3b + c = 0$ (3)

From (2) and (3) we have $\dfrac{a}{-4+15} = \dfrac{b}{-10-1} = \dfrac{c}{-3-8}$

i.e., $\dfrac{a}{11} = \dfrac{b}{-11} = \dfrac{c}{-11}$

i.e., $\dfrac{a}{1} = \dfrac{b}{-1} = \dfrac{c}{-1}$

$\therefore$ From (1) $1(x - 3) - 1(y - 2) + 1(z - 1) = 0$

i.e., $x - y - z = 0$

This is the equation of the required plane.

Example 2.3.8 Prove that the lines $\dfrac{x+1}{-3} = \dfrac{y+10}{8} = \dfrac{z-1}{2}$ and $\dfrac{x+3}{-4} = \dfrac{y+1}{7} = \dfrac{z-4}{1}$

are coplanar. Find their point of intersection and the plane through them.

Solution:

The two lines are coplanar if

$$\begin{vmatrix} x_2 - x_1 & y_2 - y_1 & z_2 - z_1 \\ l_1 & m_1 & n_1 \\ l_2 & m_2 & n_2 \end{vmatrix} = 0$$

Now LHS $= \begin{vmatrix} -3+1 & -1+10 & 4-1 \\ -3 & 8 & 2 \\ -4 & 7 & 1 \end{vmatrix}$

$= -2(8 - 14) - 9(-3 + 8) + 3(-21 + 32)$

$= -2(-6) - 9(5) + 3(11)$

$= 12 - 45 + 33$

$= 0$

∴ The two lines are coplanar. As the lines are not parallel, they intersect.

Any point on the first line is $(-3r - 1, 8r - 10, 2r + 1)$

Any point on the second line is $(-4r_1 - 3, 7r_1 - 1, r_1 + 4)$

If they are the coordinates of the point of intersection,

$-3r - 1 = -4r_1 - 3$ (1)

$8r - 10 = 7r_1 - 1$ (2)

$2r + 1 = r_1 + 4$ (3)

Solving (1) and (2) $4r_1 - 3r + 2 = 0$

$7r_1 - 8r + 9 = 0$

$$\dfrac{r_1}{-27+16} = \dfrac{r}{14-36} = \dfrac{1}{-32+21}$$

i.e., $r_1 = 1, r = 2$, which satisfy (3) also.

Substituting $r_1 = 1$ in $(-4r_1 - 3, 7r_1 - 1, r_1 + 4)$, we get the point of intersection of the two given lines. ∴ The point of intersection is $(-4 - 3, 7 - 1, 1 + 4)$ i.e., $(-7, 6, 5)$

The equation of the plane containing the two lines is

$$\begin{vmatrix} x - x_1 & y - y_1 & z - z_1 \\ l_1 & m_1 & n_1 \\ l_2 & m_2 & n_2 \end{vmatrix} = 0$$

i.e., $\begin{vmatrix} x+1 & y+10 & z-1 \\ -3 & 8 & 2 \\ -4 & 7 & 1 \end{vmatrix} = 0$

i.e., (x + 1) (8 - 14) - (y + 10) (-3 + 8) + (z - 1) (-21 + 32) = 0

i.e.,   (x + 1) (-6) - (7 + 10) 5 + (z - 1)11 = 0

i.e.,   -6x - 5y + 11z - 67 = 0

i.e.,   6x + 5y - 11z + 67 = 0

∴ The equation of the plane containing the two lines is 6x + 5y - 11z + 67 = 0

Example 2.3.9 Show that the lines $\dfrac{x-1}{1} = \dfrac{y+1}{1} = \dfrac{z-1}{-2}$ and

3x + 2y - 3z - 5 = 0, 5x + 4y − z - 7 = 0 are coplanar. Find their point of intersection and the equation of the plane containing them.

Note: The equation of the second line may be put in the symmetric form and the method in the last problem (Example 2.3.8) may be used. But a direct method is given below.

Solution:

Any point on the first line is (r + 1, r - 1, -2r + 1). If this is a point on the second line also, it will satisfy the equations of the two planes determining the line.

i.e.,   3(r + 1) + 2(r - 1) - 3(-2r + 1) - 5 = 0   (1)

and   5(r + 1) + 4(r - 1) - (-2r + 1) - 7 = 0   (2)

From (1)   3r + 2r + 6r + 3 − 2 − 3 − 5 = 0

i.e.,   11r -7 = 0

i.e.,   r = 7/11

From (2)   5r + 4r + 2r + 5 − 4 − 1 - 7 = 0

i.e.,   11r - 7 = 0

i.e.,   r = 7/11.

(1) and (2) are satisfied by the same value of r.

∴ The two lines have a common point of intersection and hence the two lines are coplanar.

Their point of intersection is $\left(\dfrac{7}{11}+1,\ \dfrac{7}{11}-1,\ (-2)\dfrac{7}{11}+1\right)$

i.e., $\left(\dfrac{18}{11}, \dfrac{-4}{11}, \dfrac{-3}{11}\right)$

Any plane containing the second line is 3x + 2y - 3z − 5 + λ(5x + 4y − z - 7) = 0

i.e.,   (3 + 5λ) x + (2 + 4λ) y + (-3 - λ) z − 5 - 7λ = 0

If this plane contains the first line, then the point (1, -1, 1) lies on this plane.

∴   (3 + 5λ) (1) + (2 + 4λ) (-1) + (-3-λ) (1) − 5 - 7λ = 0

i.e.,   -7λ-7 = 0

i.e.,   λ = -1

∴ The equation of the plane containing the two lines is

        (3 - 5) x + (2 - 4) y + (-3 + 1) z − 5 + 7 = 0

i.e.,   -2x - 2y - 2z + 2 = 0

i.e.,   x + y + z - 1 = 0

Example 2.3.10 Find the condition for the two lines
$ax + by + cz + d = 0 = a_1x + b_1y + c_1z + d_1$ and
$a_2x + b_2y + c_2z + d_2 = 0 = a_3x + b_3y + c_3z + d_3$ to be coplanar.

Solution:
Let the lines intersect at the points (x_1, y_1, z_1).
Then (x_1, y_1, z_1) lies on all the four planes determining the two lines

$$\therefore \quad a x + by + c z + d = 0$$
$$a_1x_1 + b_1y_1 + c_1z_1 + d_1 = 0$$
$$a_2x_1 + b_2y_1 + c_2z_1 + d_2 = 0$$
$$a_3x_1 + b_3y_1 + c_3z_1 + d_3 = 0$$

Eliminating x_1, y_1, z_1 from these four equations, we get,

$$\begin{vmatrix} a & b & c & d \\ a_1 & b_1 & c_1 & d_1 \\ a_2 & b_2 & c_2 & d_2 \\ a_3 & b_3 & c_3 & d_3 \end{vmatrix} = 0$$

This is the condition for the two lines to be coplanar.

Example 2.3.11 Prove that the lines $x + 2y - z - 3 = 0 = 3x - y + 2z - 1$ and
$2x - 2y + 3z - 2 = 0 = x - y + z + 1$ are coplanar and find the equation of the plane
containing them.

Solution:

The lines are coplanar if $\begin{vmatrix} 1 & 2 & -1 & -3 \\ 3 & -1 & 2 & -1 \\ 2 & -2 & 3 & -2 \\ 1 & -1 & 1 & 1 \end{vmatrix} = 0$ (By Example 2.3.10)

$$\text{LHS} = \begin{vmatrix} 3 & 2 & 2 & -3 \\ 2 & -1 & 3 & -1 \\ 0 & -2 & 5 & -2 \\ 0 & -1 & 0 & 1 \end{vmatrix} = \begin{vmatrix} 3 & -1 & 2 & -3 \\ 2 & -2 & 3 & -1 \\ 0 & -4 & 5 & -2 \\ 0 & 0 & 0 & 1 \end{vmatrix}$$

$$= \begin{vmatrix} 3 & -1 & 2 \\ 2 & -2 & 3 \\ 0 & -4 & 5 \end{vmatrix} = 3(-10 + 12) - 2(-5 + 8) = 6 - 6 = 0$$

$\therefore$ The lines are coplanar.
Any plane containing the first line is
$$x + 2y - z - 3 + \lambda(3x - y + 2z - 1) = 0 \quad\quad (1)$$

Any plane containing the second line is

$$2x - 2y + 3z - 2 + \lambda_1(x - y + z + 1) = 0 \qquad (2)$$

If (1) and (2) represent the same plane

$$\frac{1+3\lambda}{2+\lambda_1} = \frac{2-\lambda}{-2-\lambda_1} = \frac{-1+2\lambda}{3+\lambda_1} = \frac{-3-\lambda}{-2+\lambda_1}$$

Solving for λ and substituting in (1) we get the equation of the required plane.

$$\frac{1+3\lambda}{2+\lambda_1} = \frac{2-\lambda}{-2-\lambda_1} \text{ gives } 2\lambda\lambda_1 + 4\lambda + 3\lambda_1 + 6 = 0 \qquad (3).$$

$$\frac{1+3\lambda}{2+\lambda_1} = \frac{-1+2\lambda}{3+\lambda_1} \text{ gives } \lambda\lambda_1 + 5\lambda + 2\lambda_1 + 5 = 0 \qquad (4)$$

(3) – 2*(4) gives $-6\lambda - 1\lambda_1 - 4 = 0$

i.e., $\lambda_1 = -6\lambda - 4$

Substituting in (4), we get,

$$\lambda(-6\lambda - 4) + 5\lambda + 2(-6\lambda - 4) + 5 = 0$$
$$-6\lambda^2 - 11\lambda - 3 = 0$$
$$6\lambda^2 + 11\lambda + 3 = 0$$
$$(3\lambda + 1)(2\lambda + 3) = 0$$

$\lambda = -1/3$ or $-3/2$.

When $\lambda = -1/3$, we get, $\lambda_1 = -2$

When $\lambda = -3/2$, we get, $\lambda_1 = 5$

Also all the four ratios are equal (to -1/2) when $\lambda = -3/2$, $\lambda_1 = 5$.

The values $\lambda = -1/3$, $\lambda_1 = -2$ are inadmissible as the four ratios are not equal.

Substituting $\lambda = -3/2$ in equation (1) (or $\lambda_1 = 5$ in equation (2)) we get the equation of the required plane as

$$x + 2y - z - 3 - (3/2)(3x - y + 2z - 1) = 0$$

i.e., $-7x + 7y - 8z - 3 = 0$

i.e., $7x - 7y + 8z + 3 = 0$

Example 2.3.12 A line with D.R.'s (7, -5, 2) is drawn to intersect the lines $\dfrac{x-7}{-1} = \dfrac{y+2}{1} = \dfrac{z-5}{3}$, $\dfrac{x-3}{2} = \dfrac{y-6}{4} = \dfrac{z+3}{-3}$. Find the coordinates of the points of intersection and length intercepted on it.

Solution:

Let the line with D.R.'s (7, -5, 2) cut the first line at P and the second line at Q.

Coordinates of P are $(-r + 7, r - 2, 3r + 5)$

Coordinates of Q are $(2r_1 + 3, 4r_1 + 6, -3r_1 - 3)$

∴ D.R.'s of PQ are $(-r + 7 - 2r_1 - 3, r - 2 - 4r_1 - 6, 3r + 5 + 3r_1 + 3)$

i.e., $(-r - 2r_1 + 4, r - 4r_1 - 8, 3r + 3r_1 + 8)$

But, given that the D.R.'s of PQ are (7, -5, 2)

$$\therefore \frac{-r-2r_1+4}{7} = \frac{r-4r_1-8}{-5} = \frac{3r+3r_1+8}{2}$$

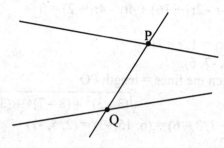

Fig. 2.23

From the first two ratios, we get,

$$5r + 10r_1 - 20 = 7r - 28r_1 - 56$$

i.e., $2r - 38r_1 - 36 = 0$ (1)

From the last two ratios, we get,

$$2r - 8r_1 - 16 = -15r - 15r_1 - 40$$

i.e., $17r + 7r_1 + 24 = 0$ (2)

Solving (1) and (2) we get, $r = -1$, $r_1 = -1$.

$\therefore$ P is $(8, -3, 2)$ and Q is $(1, 2, 0)$

Length PQ = $\sqrt{(8-1)^2 + (-3-2)^2 + (2-0)^2} = \sqrt{78}$

Example 2.3.13 Find the magnitude and the equation of the shortest distance between the lines

$$\frac{x-6}{3} = \frac{y-7}{-1} = \frac{z-4}{1} \text{ and } \frac{x}{-3} = \frac{y+9}{2} = \frac{z-2}{4}$$

Solution:

Let the shortest distance line meet the two lines at P and Q.

The coordinates of P, being a point on the line

$$\frac{x-6}{3} = \frac{y-7}{-1} = \frac{z-4}{1} = r$$ (1)

are $(3r + 6, -r + 7, r + 4)$

The Coordinates of Q, being a point on the line

$$\frac{x}{-3} = \frac{y+9}{2} = \frac{z-2}{4} = r_1$$ (2)

are $(-3r_1, 2r_1 - 9, 4r_1 + 2)$

D.R.'s of PQ are $(3r + 3r_1 + 6, -r - 2r_1 + 16, r - 4r_1 + 2)$

Since PQ is perpendicular to line (1)

$$3(3r + 3r_1 + 6) - 1(-r - 2r_1 + 16) + 1(r - 4r_1 + 2) = 0$$

i.e., $11r + 7r_1 + 4 = 0$ (3)

Since PQ is perpendicular to line (2)

$$-3(3r + 3r_1 + 6) + 2(-r - 2r_1 + 16) + 4(r - 4r_1 + 2) = 0$$

i.e., $-7r - 29r_1 + 22 = 0$ (4)

Solving (3) and (4) we get, $r_1 = 1, r = -1$.

$\therefore$ P is (3, 8, 3) and Q is (-3, -7, 6)

The shortest distance between the lines = length PQ

$$= \sqrt{(3+3)^2 + (8+7)^2 + (3-6)^2} = \sqrt{270}$$

D.R.'s of PQ are $(3 + 3, 8 + 7, 3 - 6) = (6, 15, -3) = (2, 5, -1)$

PQ passes through P (3, 8, 3).

$\therefore$ Equation of the S.D line is $\dfrac{x-3}{2} = \dfrac{y-8}{5} = \dfrac{z-3}{-1}$

Example 2.3.14 Find the shortest distance between the lines $\dfrac{x-7}{2} = \dfrac{y+4}{3} = \dfrac{z-2}{1}$ and

$2x + 5y - 8z - 52 = 0 = 3x - 3y + 2z + 27$.

Solution:

Let AB and CD be the two given lines. Through CD draw a plane parallel to AB. Then the shortest distance between the lines AB and CD will be equal to the length of the perpendicular from any point P on AB to the plane through CD. P can be taken as the point (7, -4, 2).

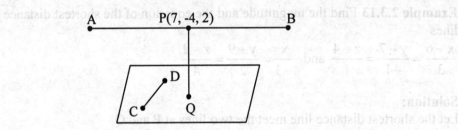

Fig. 2.24

Then PQ is the shortest distance. Now any plane containing the line CD is

$$2x + 5y - 8z - 52 + \lambda(3x - 3y + 2z + 27) = 0 \qquad (1)$$

i.e., $(2 + 3\lambda) x + (5 - 3\lambda) y + (-8 + 2\lambda) z - 52 + 27\lambda = 0.$

As this plane is parallel to line AB, the normal to the plane is perpendicular to AB.

$\therefore$ $2(2 + 3\lambda) + 3(5 - 3\lambda) + 1(-8 + 2\lambda) = 0$

i.e., $-\lambda + 11 = 0.$

i.e., $\lambda = 11.$

$\therefore$ From (1)

$$2x + 5y - 8z - 52 + 11(3x - 3y + 2z + 27) = 0$$
$$35x - 28y + 142z + 245 = 0$$

i.e.,
$$5x - 4y + 2z + 35 = 0$$

The shortest distance PQ $= \dfrac{5(7) - 4(-4) + 2(2) + 35}{\sqrt{5^2 + (-4)^2 + 2^2}}$

$$= \frac{90}{\sqrt{45}}$$

$$= 6\sqrt{5}$$

Example 2.3.15 The equations of two lines are $x = y + 2a = 6z - 6a$ and $x + a = 2y = -12z$. Show that the shortest distance between the two lines is 2a and find the equations of the line along which it lies.

Solution:

The two lines are $\quad \dfrac{x - 0}{6} = \dfrac{y + 2a}{6} = \dfrac{z - a}{1}$ \hfill (1)

and $\quad \dfrac{x + a}{12} = \dfrac{y - 0}{6} = \dfrac{z - 0}{-1}$ \hfill (2)

Let (l, m, n) be the D.C.'s of the shortest distance line. As the S.D line is perpendicular to both the lines (1) and (2), we get

$$6l + 6m + n = 0$$
$$12l + 6m - n = 0$$

$\therefore \quad \dfrac{l}{-12} = \dfrac{m}{18} = \dfrac{n}{-36}$

i.e., $\quad \dfrac{l}{2} = \dfrac{m}{-3} = \dfrac{n}{6}$

$\therefore \quad l = \dfrac{2}{7}, \quad m = \dfrac{-3}{7}, \quad n = \dfrac{6}{7}$

$\therefore$ Shortest distance = Projection of the line segment joining the points (0, -2a, a) and (-a, 0, 0) on the line with D.C.'s (2/7, -3/7, 6/7).

$$= (0 + a)(2/7) + (-2a - 0)(-3/7) + (a - 0)(6/7) = 2a$$

The equation of the S.D. line is given by

$$\begin{vmatrix} x & y+2a & z-a \\ 6 & 6 & 1 \\ 2 & -3 & 6 \end{vmatrix} = 0 = \begin{vmatrix} x+a & y & z \\ 12 & 6 & -1 \\ 2 & -3 & 6 \end{vmatrix}$$

i.e., $\quad 39x - 34y - 30z - 38a = 0 = 33x - 74y - 48z + 33a$

Example 2.3.16 Show that the equation to the plane containing the line $\dfrac{y}{b}+\dfrac{z}{c}=1$, $x=0$

and parallel to the line $\dfrac{x}{a}-\dfrac{z}{c}=1$, $y=0$ is $\dfrac{x}{a}-\dfrac{y}{b}-\dfrac{z}{c}+1=0$, and if 2d is the shortest

distance, prove that $\dfrac{1}{d^2}=\dfrac{1}{a^2}+\dfrac{1}{b^2}+\dfrac{1}{c^2}$

Solution:

Any plane containing the line AB whose equation is $\dfrac{y}{b}+\dfrac{z}{c}=1$, $x=0$ \qquad (1)

is \qquad $\dfrac{y}{b}+\dfrac{z}{c}-1+\lambda\,x=0$ \hfill (2)

Let (l, m, n) be the D.C.'s of the line CD whose equation is

$$\frac{x}{a}-\frac{z}{c}=1,\ \ y=0 \qquad\qquad (3)$$

Then \quad $\dfrac{l}{a}-\dfrac{n}{c}=0$ and $m=0$

i.e., \quad $\dfrac{l}{a}=\dfrac{n}{c}$ and $m=0$

∴ \qquad l, m, n are proportional to a, o, c.

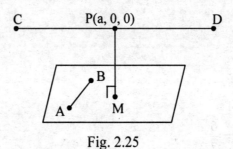

Fig. 2.25

CD is parallel to plane (2).

∴ \qquad $a.\lambda+0.1/b+c.(1/c)=0$

i.e., \qquad $a\lambda+1=0$

i.e., \qquad $\lambda=-1/a$

∴ From (2) \quad $\dfrac{y}{b}+\dfrac{z}{c}-1+\dfrac{1}{a}x=0$

i.e., \qquad $\dfrac{x}{a}-\dfrac{y}{b}-\dfrac{z}{c}+1=0$ \hfill (4)

is the required plane.

From (3), P(a, 0, 0) is a point on CD.

∴ The shortest distance between the lines AB and CD is same as the perpendicular distance PM of P(a, 0, 0) from the plane (4).

i.e., $\quad 2d = \dfrac{\dfrac{a}{a} - \dfrac{0}{b} - \dfrac{0}{c} + 1}{\sqrt{\dfrac{1}{a^2} + \dfrac{1}{b^2} + \dfrac{1}{c^2}}}$

i.e., $\quad \dfrac{2}{2d} = \sqrt{\dfrac{1}{a^2} + \dfrac{1}{b^2} + \dfrac{1}{c^2}}$

i.e., $\quad \dfrac{1}{d^2} = \dfrac{1}{a^2} + \dfrac{1}{b^2} + \dfrac{1}{c^2}$

EXERCISE 2.3

PART-A

1. Find the equation of the straight line joining the points (2, 7, 3) and (-2, 5, 6).
2. Find the equation of the line joining the origin and (5, -2, 3).
3. Prove that the line $\dfrac{x-4}{2} = \dfrac{y-2}{3} = \dfrac{z-3}{6}$ lies on the plane 3x - 4y + z = 7.
4. Find the angle between the lines $\dfrac{x}{1} = \dfrac{y}{0} = \dfrac{z}{-1}$ and $\dfrac{x}{3} = \dfrac{y}{4} = \dfrac{z}{5}$
5. Find the angle between the plane 2x + y + 2z = 0 and $\dfrac{x-2}{1} = \dfrac{y-1}{-2} = \dfrac{z+1}{2}$
6. Find the D.C.'s of the straight line x = 2y = 3z.
7. Find the points on the line $\dfrac{x-1}{2} = \dfrac{y-2}{3} = \dfrac{z-3}{6}$ at a distance 7 from the point (1, 2, 3).
8. Prove that the line $\dfrac{x-1}{2} = \dfrac{y-3}{3} = \dfrac{z-4}{-1}$ is parallel to the plane x - 2y - 4z + 7 = 0.
9. Find the equation of the normal to the plane 2x - 3y + 6z = 0 passing through (2, -3, 4).
10. Find the equation of the line through (1, 0, -2) and parallel to the line joining (4, 6, 7) and (-2, 8, 1).
11. Find the angle between the line $\dfrac{x+1}{2} = \dfrac{y-2}{3} = \dfrac{z-1}{6}$ and the plane

    3x + y + z - 1 = 0
12. Show that the lines $\dfrac{x-5}{4} = \dfrac{y-7}{4} = \dfrac{z+3}{-5}$ and $\dfrac{x-8}{7} = \dfrac{y-4}{1} = \dfrac{z-5}{3}$ are coplanar.
13. Find the equation of the plane containing the coplanar lines

    $\dfrac{x-2}{3} = \dfrac{y-3}{4} = \dfrac{z-4}{5}$ and $\dfrac{x-1}{2} = \dfrac{y-2}{3} = \dfrac{z-3}{4}$.

PART – B

14. Find the equation of the straight line $2x - 3y + 3z = 4$, $x + 2y - z = -3$ in symmetric form.

15. Find the symmetric form of the equation of the straight line $x + y + z = 1$, $2x - y - 3z + 1 = 0$.

16. Find the angle between the lines $x - 2y + z = 0 = x + 2y - 2z$ and $x + 2y + z = 0 = 3x + 9y + 5z$.

17. Find the distance of the point $(1, -2, 3)$ from the plane $x - y + z = 5$ measured parallel to the line $\dfrac{x}{2} = \dfrac{y}{3} = \dfrac{z}{-6}$.

18. Find the foot, length and equation of the perpendicular drawn from $(1, 0, -3)$ to the line $\dfrac{x-2}{3} = \dfrac{y-3}{4} = \dfrac{z-4}{5}$

19. How far is the point$(1, 1, 4)$ from the line of intersection of the planes $x + y + z = 4$, $2x + y - z + 4 = 0$?

20. Find the distance of $(-5, -10, -1)$ from the point of intersection of the line $\dfrac{x+1}{4} = \dfrac{y-2}{12} = \dfrac{z-2}{12}$ and the plane $x - y - z + 5 = 0$.

21. Find the coordinates of the foot of the perpendicular from the origin to the line $x + 2y + 3z + 4 = 0 = 2x + 3y + 4z + 5$.

22. Find the equation of the plane through $(1, 2, 3)$ and perpendicular to the line $x - y + 2z + 1 = 0 = 3x + 4y + 7z$.

23. Find the equation to the perpendicular from the origin to the line $x + 2y + 3z + 4 = 0$, $2x + 3y + 4z + 5 = 0$

24. Assuming the plane $4x - 3y + 7z = 0$ to be horizontal, find the equation of the line of greatest slope through the point $(2, 1, 1)$ in the plane $2x + y - 3z = 0$.

25. Find the projection of the line $3x - y + 2z = 1$, $x + 2y - z = 2$ on the plane $3x + 2y + z = 0$.

26. Find the equation of the orthogonal projection of the line $\dfrac{x-4}{3} = \dfrac{y-2}{4} = \dfrac{z-1}{2}$ on the plane $9x + 8y + 2z - 7 = 0$.

27. Find the equation of the image of the line $\dfrac{x-1}{9} = \dfrac{y-2}{-1} = \dfrac{z+3}{-3}$ in the plane $3x - y + 10z - 26 = 0$.

28. Find the equation of the image of the line $\dfrac{x-1}{3} = \dfrac{y-3}{5} = \dfrac{z-4}{2}$ in the plane $2x - y + z + 3 = 0$.

29. Show that the lines $\dfrac{x-4}{2} = \dfrac{y-5}{3} = \dfrac{z-6}{4}$ and $\dfrac{x-2}{3} = \dfrac{y-3}{4} = \dfrac{z-4}{5}$ are coplanar and find the equation of the plane in which they lie.

30. Show that the lines $\dfrac{x-2}{2} = \dfrac{y-3}{-1} = \dfrac{z+4}{3}$ and $\dfrac{x-3}{1} = \dfrac{y+1}{3} = \dfrac{z-1}{-2}$ are coplanar.
    Find their common point and the equation of the plane containing them.

31. Show that the lines $\dfrac{x-1}{1} = \dfrac{y+1}{1} = \dfrac{z-1}{-2}$ and
    $3x + 2y - 3z - 5 = 0 = 5x + 4y - z - 7$ are coplanar. Find their point of intersection and the equation of the plane containing them.

32. Prove that the lines $\dfrac{x-a}{1} = \dfrac{y-b}{m} = \dfrac{z-c}{n}$ and $\dfrac{x-1}{a} = \dfrac{y-m}{b} = \dfrac{z-n}{c}$ intersect. Find the point of intersection and also the plane containing them.

33. Show that the lines $\dfrac{ax}{\alpha} = \dfrac{by}{\beta} = \dfrac{cz}{\gamma}$; $\dfrac{x}{\alpha} = \dfrac{y}{\beta} = \dfrac{z}{\gamma}$; $\dfrac{x}{a\alpha} = \dfrac{y}{b\beta} = \dfrac{z}{c\gamma}$ are coplanar if
    $(a - b) (b - c) (c - a) = 0$.

34. Show that the lines $x + 2y - 5z + 9 = 0 = 3x - y + 2z - 5$ and $2x + 3y - z - 3 = 0 = 4x - 5y + z + 3$ are coplanar. Find the coordinates of their point of intersection and the equation of the plane containing them.

35. Show that the lines $x + 2y - 5z + 9 = 0 = 3x - y + 2z - 5$ and $2x + 3y - z - 3 = 0 = 4x - 5y + z + 3$ are coplanar and find the equation of the plane containing them.

36. Find the length and the equation of the shortest distance between the lines
    $\dfrac{x-1}{1} = \dfrac{y-2}{-2} = \dfrac{z-3}{3}$; $\dfrac{x+1}{2} = \dfrac{y}{-1} = \dfrac{z-1}{3}$

37. Find the length and the equation of the shortest distance between the lines
    $\dfrac{x+1}{2} = \dfrac{y+1}{3} = \dfrac{z+1}{4}$; $\dfrac{x+1}{3} = \dfrac{y}{4} = \dfrac{z}{5}$.

38. Find the shortest distance between the lines
    $\dfrac{x-3}{-3} = \dfrac{y-8}{1} = \dfrac{z-3}{-1}$; $\dfrac{x+3}{3} = \dfrac{y+7}{-2} = \dfrac{z-6}{-4}$ and find the equation of the line of shortest distance

39. A cube has edges of length a. Find the distance between a diagonal and an edge skew to it.

40. Find the length and equation of the S.D between the lines
    $\dfrac{x-3}{1} = \dfrac{y-5}{-2} = z - 7$ and $\dfrac{x+1}{7} = \dfrac{y+1}{-6} = z+1$

41. Find the length and equations of the S.D between $3x - 9y + 5z = 0 = x + y - z$ and $6x + 8y - 3z - 13 = 0 = x + 2y + z - 3$.

42. Prove that the shortest distance between any two opposite edges of the tetrahedron formed by the planes $y + z = 0$, $z + x = 0$, $x + y = 0$, $x + y + z = a$ is $\dfrac{2a}{\sqrt{6}}$ and that the three lines of S.D intersect at the point $x = y = z = -a$.

43. Find the length and equation of the S.D between the lines $\dfrac{x+1}{3} = \dfrac{y-2}{2} = \dfrac{z}{4}$ and

    $3x + 2y - 5z - 6 = 0 = 2x - 3y + z - 3$.

2.4   THE SPHERE

Definition 2.4.1 A **sphere** is the locus of a point which moves in space such that its distance from a fixed point is a constant.

The fixed point is called the **centre of the sphere** and the constant distance is called the radius of the sphere.

2.4.1 THE EQUATION OF THE SPHERE WITH CENTRE (a, b, c) AND RADIUS r

Let $P(x_1, y_1, z_1)$ be any point on the sphere.

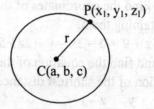

Fig. 2.26

Then its distance from the centre $C(a, b, c)$ is r.

i.e., $CP = r$

i.e., $CP^2 = r^2$

i.e., $(x_1 - a)^2 + (y_1 - b)^2 + (z_1 - c)^2 = r^2$

The locus of (x_1, y_1, z_1) is

$(x - a)^2 + (y - b)^2 + (z - c)^2 = r^2$

This is the equation of the sphere with centre (a, b, c) and radius r.

Note:

The equation of the sphere with centre at the origin and radius r is $x^2 + y^2 + z^2 = r^2$.

2.4.2   THE EQUATION $x^2 + y^2 + z^2 + 2ux + 2vy + 2wz + d = 0$ REPRESENTS A SPHERE

Consider the given equation

$x^2 + y^2 + z^2 + 2ux + 2vy + 2wz + d = 0$ (1)

i.e., $(x + u)^2 + (y + v)^2 + (z + w)^2 = u^2 + v^2 + w^2 - d$.

This shows that the equation is the locus of a point which moves such that its distance from the point (-u, -v, -w) is a constant $\sqrt{u^2 + v^2 + w^2 - d}$.

∴ Equation (1) represents a sphere with centre at the point (-u, -v, -w) and radius

$$\sqrt{u^2 + v^2 + w^2 - d}.$$

Note:

(i)   If $u^2 + v^2 + w^2 - d > 0$, then the radius $\sqrt{u^2 + v^2 + w^2 - d}$ has a positive real value and hence we get a real sphere. If $u^2 + v^2 + w^2 - d = 0$, then the radius of the sphere is zero and hence we get a sphere that reduces to a single point (-u, -v, -w). If $u^2 + v^2 + w^2 - d < 0$, then the radius is imaginary and hence we get only an imaginary sphere

(ii)   The equation $ax^2 + ay^2 + az^2 + 2ux + 2vy + 2wz + d = 0$ represents a sphere with centre at (-u/a, -v/a, -w/a) and

radius $\sqrt{\dfrac{u^2}{a^2} + \dfrac{v^2}{a^2} + \dfrac{w^2}{a^2} - \dfrac{d}{a}}$

(iii)   The general equation of the second degree in x, y, z is
$ax^2 + by^2 + cz^2 + 2fyz + 2gzx + 2hxy + 2ux + 2vy + 2wz + d = 0$
and this equation represents a sphere if a = b = c and f = g = h = 0.

2.4.3   THE EQUATION OF THE SPHERE ON THE LINE JOINING THE POINTS A(x_1, y_1, z_1) and B(x_2, y_2, z_2) AS DIAMETER

Let P(x, y, z) be any point on the sphere. AB is a diameter of the circle through the points A, B and P.

∴ AP is perpendicular to BP.

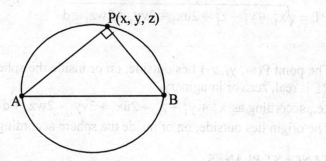

Fig. 2.27

D.R.'s of AP are   (x - x_1, y - y_1, z - z_1).
D.R.'s of BP are   (x - x_2, y - y_2, z - z_2).

Since AP is perpendicular to BP, we get,

$(x - x_1) (x - x_2) + (y - y_1) (y - y_2) + (z - z_1) (z - z_2) = 0$ (1)

Equation (1) being a second degree equation in x, y and z with a = b = c and

f = g = h = 0, represents a sphere.

Thus (1) is the required equation of the sphere on AB as diameter.

2.4.5   LENGTH OF TANGENT FROM THE POINT $P(x_1, y_1, z_1)$ TO THE SPHERE

Let the equation of the given sphere be $x^2 + y^2 + z^2 + 2ux + 2vy + 2wz + d = 0$

Let PT be a tangent from $P(x_1, y_1, z_1)$ to the sphere. If C is the centre of the sphere, then, CT is perpendicular to PT.

∴ From triangle PTC,

$$PT^2 = PC^2 - CT^2$$ (1)

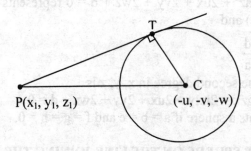

Fig. 2.28

The coordinates of C are (-u, -v, -w) and CT = $\sqrt{u^2 + v^2 + w^2 - d}$

∴ From (1), $PT^2 = (x_1 + u)^2 + (y_1 + v)^2 + (z_1 + w)^2 - (u^2 + v^2 + w^2 - d)$

i.e., $PT^2 = x_1^2 + y_1^2 + z_1^2 + 2ux_1 + 2vy_1 + 2wz_1 + d$

∴ $PT = \sqrt{x_1^2 + y_1^2 + z_1^2 + 2ux_1 + 2vy_1 + 2wz_1 + d}$

Note:

(i)   The point $P(x_1, y_1, z_1)$ lies outside, on or inside the sphere according as the length PT is real, zero or imaginary.

i.e., according as $x_1^2 + y_1^2 + z_1^2 + 2ux_1 + 2vy_1 + 2wz_1 + d$ is > = or < 0.

(ii)   The origin lies outside, on or inside the sphere according as d > = or < 0.

2.4.6   TANGENT PLANES

When a plane touches a sphere and P is the point of contact, then the normal to the plane at P passes through the centre of the sphere. i.e., the radius of the sphere through P is perpendicular to the plane. This plane is called the **tangent plane** of the sphere at P.

2.4.7 THE EQUATION OF THE TANGENT PLANE TO THE SPHERE AT A GIVEN POINT ON IT

Let $P(x_1, y_1, z_1)$ be a point on the sphere
$$x^2 + y^2 + z^2 + 2ux + 2vy + 2wz + d = 0 \qquad (1)$$
The centre C of the sphere is $(-u, -v, -w)$.

D.R.'s of CP are $(x_1 + u, y_1 + v, z_1 + w)$.

The tangent plane at P passes through P and is perpendicular to CP.

Hence the equation of the tangent plane at P is
$$(x - x_1)(x_1 + u) + (y - y_1)(y_1 + v) + (z - z_1)(z_1 + w) = 0.$$
i.e., $\quad xx_1 + yy_1 + zz_1 + ux + vy + wz = x_1^2 + y_1^2 + z_1^2 + ux_1 + vy_1 + wz_1 \qquad (2)$

Since $P(x_1, y_1, z_1)$ lies on the sphere, it satisfies the equation (1).

$\therefore \qquad x_1^2 + y_1^2 + z_1^2 + 2ux_1 + 2vy_1 + 2wz_1 + d = 0$

i.e., $\quad x_1^2 + y_1^2 + z_1^2 + ux_1 + vy_1 + wz_1 = -ux_1 - vy_1 - wz_1 - d$

Substituting this in (2) we get the equation of the tangent plane at P as
$$xx_1 + yy_1 + zz_1 + u(x + x_1) + v(y + y_1) + w(z + z_1) + d = 0.$$

2.4.8 PLANE SECTION OF A SPHERE

Consider a sphere with centre at C and radius R. Let P be any point common to the sphere and a plane intersecting the sphere (Fig. 2.29). Let N be the foot of the perpendicular from C on the plane.

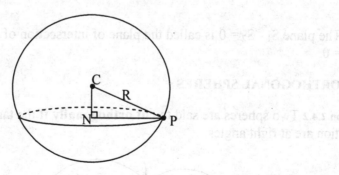

Fig. 2.29

Since the plane is a given fixed plane, N is a fixed point on the plane and the length CN is a constant. Also CN is perpendicular to the plane implies CN is perpendicular to NP and hence
$$NP^2 = CP^2 - CN^2 = R^2 - CN^2 = \text{a constant}.$$
This shows that the locus of P is a circle with centre at N and radius NP. Thus the plane section of the sphere is a circle.

Note:

(i)   If the plane passes through the centre of the sphere, the circle of intersection is of radius R and is called a great circle.

(ii)   The equations $x^2 + y^2 + z^2 + 2ux + 2vy + 2wz + d = 0$, $px + qy + rz + s = 0$ together represents a circle, which is the intersection of the sphere and the plane.

(iii)   The equation $x^2 + y^2 + z^2 + 2ux + 2vy + 2wz + d + \lambda(px + qy + rz + s) = 0$ (1) where λ is a constant represents a sphere. Also the coordinates of any point common to the sphere $x^2 + y^2 + z^2 + 2ux + 2vy + 2wz + d = 0$ and the plane $px + qy + rz + s = 0$ satisfy (1). Thus the equation (1) represents any sphere passing through the circle of intersection of the sphere given by $x^2 + y^2 + z^2 + 2ux + 2vy + 2wz + d = 0$ and the plane $px + qy + rz + s = 0$.

2.4.9   INTERSECTION OF TWO SPHERES

Let the equations of the two spheres be
$$S_1 \equiv x^2 + y^2 + z^2 + 2u_1x + 2v_1y + 2w_1z + d_1 = 0$$
$$S_2 \equiv x^2 + y^2 + z^2 + 2u_2x + 2v_2y + 2w_2z + d_2 = 0$$
The coordinates of points common to these two spheres satisfy both the equations and hence they also satisfy the equation

$S_1 - S_2 \equiv 2(u_1 - u_2)\,x + 2(v_1 - v_2)\,y + 2(w_1 - w_2)\,z + d_1 - d_2 = 0$, which represents a plane.

Thus the intersection of the two spheres is same as that of any one of them with the plane $S_1 - S_2 = 0$, which is a circle. **Hence the intersection of two spheres is a circle.** Note that, when the two spheres touch each other internally or externally, the intersection is a point circle. When the two spheres do not intersect, the circle of intersection is imaginary.

Note:

The plane $S_1 - S_2 = 0$ is called the plane of intersection of the spheres $S_1 = 0$ and $S_2 = 0$.

2.4.10   ORTHOGONAL SPHERES

Definition 2.4.2 Two spheres are said to **cut orthogonally** if the tangent planes at a point of intersection are at right angles.

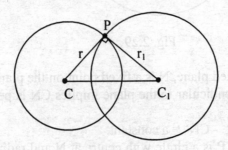

Fig. 2.30

If two spheres cut orthogonally, then the radii of such spheres through their point of intersection P, being perpendicular to the tangent planes at P are also at right angles. Thus two spheres with centers at C and C_1 and radii r and r_1 cut orthogonally, if $(CC_1)^2 = r^2 + r_1^2$.

2.4.11 THE CONDITION FOR TWO SPHERES TO CUT ORTHOGONALLY

Let the equations of the two spheres be

$$x^2 + y^2 + z^2 + 2ux + 2vy + 2wz + d = 0 \qquad (1)$$
and $\qquad x^2 + y^2 + z^2 + 2u_1x + 2v_1y + 2w_1z + d_1 = 0 \qquad (2)$

The centre and radius of sphere (1) are C (-u, -v, -w) and $r = \sqrt{u^2 + v^2 + w^2 - d}$

The centre and radius of sphere (2) are C_1 (-u_1, -v_1, -w_1) and $r_1 = \sqrt{u_1^2 + v_1^2 + w_1^2 - d_1}$

The two spheres cut orthogonally at P if $C\hat{P}C_1 = 90°$

i.e., if $\qquad (CC_1)^2 = r^2 + r_1^2$

i.e., if $\qquad (u - u_1)^2 + (v - v_1)^2 + (w - w_1)^2 = u^2 + v^2 + w^2 - d + u_1^2 + v_1^2 + w_1^2 - d_1$

i.e., if $\qquad 2uu_1 + 2vv_1 + 2ww_1 = d + d_1$

Hence the condition for the two spheres (1) and (2) to cut orthogonally is

$$\mathbf{2uu_1 + 2vv_1 + 2ww_1 = d + d_1}$$

2.4.12 RADICAL PLANE OF TWO SPHERES

Definition 2.4.3 The locus of points from which the lengths of tangents to two given spheres are equal is a plane, called the **radical plane** of the two spheres.

The equations of the two spheres be

$$S \equiv x^2 + y^2 + z^2 + 2ux + 2vy + 2wz + d = 0 \text{ and }$$
$$S_1 \equiv x^2 + y^2 + z^2 + 2u_1x + 2v_1y + 2w_1z + d_1 = 0$$

Let (x_1, y_1, z_1) be any point such that the lengths of the tangent to the two spheres are equal.

i.e., $\sqrt{x_1^2 + y_1^2 + z_1^2 + 2ux_1 + 2vy_1 + 2wz_1 + d} = \sqrt{x_1^2 + y_1^2 + z_1^2 + 2u_1x_1 + 2v_1y_1 + 2w_1z_1 + d_1}$

i.e., $\qquad 2(u - u_1) x_1 + 2(v - v_1) y_1 + 2(w - w_1) z_1 + d - d_1 = 0$.

The radical plane is the locus of (x_1, y_1, z_1).

i.e., $\qquad 2(u - u_1) x + 2(v - v_1) y + 2(w - w_1) z + d - d_1 = 0$.

i.e., $\qquad \mathbf{S - S_1 = 0}$

Note:

(i) The radical plane is perpendicular to the line of centres of the two spheres.

(ii) If two spheres intersect, then the radical plane is the common plane of intersection of the two spheres.

(iii) If the two spheres touch, then the radical plane is the common tangent plane at

the point of contact.
(iv)   The radical planes of three spheres $S_1 = 0$, $S_2 = 0$, $S_3 = 0$ taken two by two
         intersect along a line called the **radical line**, given by the equations
         $S_1 = S_2 = S_3$.
(v)   The radical planes of four spheres taken two by two pass through one point,
         called the **radical point** of the four spheres.

Example 2.4.1 Find the centre and radius of the sphere
$$3x^2 + 3y^2 + 3z^2 + 12x - 8y - 10z + 10 = 0.$$

Solution:

The equation of the given sphere is $x^2 + y^2 + z^2 + 4x - \dfrac{8}{3}y - \dfrac{10}{3}z + \dfrac{10}{3} = 0$

Comparing with $x^2 + y^2 + z^2 + 2ux + 2vy + 2wz + d = 0$, we get,
         $u = 2$, $v = -4/3$, $w = -5/3$ and $d = 10/3$.
∴   Centre is $(-u, -v, -w)$
i.e., $(-2, 4/3, 5/3)$

and radius is $\sqrt{u^2 + v^2 + w^2 - d}$ $= \sqrt{4 + \dfrac{16}{9} + \dfrac{25}{9} - \dfrac{10}{3}} = \sqrt{\dfrac{47}{9}} = \sqrt{47}\Big/3$

Example 2.4.2 Find the equation of the sphere passing through the origin and having
centre at $(2, 4, 3)$.

Solution:
If the radius of the sphere is a, its equation is
$$(x - 2)^2 + (y - 4)^2 + (z - 3)^2 = a^2 \qquad\qquad (1)$$
As this passes through the origin, $(0, 0, 0)$ satisfies (1).
∴ $(0 - 2)^2 + (0 - 4)^2 + (0 - 3)^2 = a^2$
i.e., $a^2 = 29.$
∴ (1) becomes, $(x - 2)^2 + (y - 4)^2 + (z - 3)^2 = 29$
Expanding, $x^2 + y^2 + z^2 - 4x - 8y - 6z = 0.$

Example 2.4.3 Find the equation of the sphere having the points $(-4, 5, 1)$ and $(4, 1, 7)$ as
ends of a diameter.

Solution:
The equation of the required sphere is $(x + 4)(x - 4) + (y - 5)(y - 1) + (z - 1)(z - 7) = 0.$
i.e., $x^2 + y^2 + z^2 - 6y - 8z - 4 = 0$

Example 2.4.4 Find the equation of the sphere passing through the points $(4, -1, 2)$,
$(0, -2, 3)$, $(1, 5, -1)$ and $(2, 0, 1)$.

Solution:

Let the equation of the sphere be $x^2 + y^2 + z^2 + 2ux + 2vy + 2wz + d = 0$ (1)

(1)   passes through (4, -1, 2)

∴ $8u - 2v + 4w + d = -21$ (2)

(1)   passes through (0, -2, 3)

∴ $-4v + 6w + d = -13$ (3)

(1) passes through (1, 5, -1)

∴ $2u + 10v - 2w + d = -27$ (4)

(1)   passes through (2, 0, 1)

∴ $4u + 2w + d = -5$ (5)

(2) - (3) gives $8u + 2v - 2w = -8$ (6)

(4) - (3) gives $2u + 14v - 8w = -14$ (7)

(5) - (3) gives $4u + 4v - 4w = 8$ (8)

$2 \times$(8) - (7) gives $6u - 6v = 30$ (9)

(8) - $2 \times$(6) gives $-12u = 24$ (10)

From (9) and (10) we get $u = -2$, $v = -7$.

Substituting the values of u and v in (8), we get $w = -11$

Substituting the values of v and w in (3), we get $d = 25$.

∴   From (1) we get the equation of the required sphere as

$x^2 + y^2 + z^2 - 4x - 14y - 22z + 25 = 0$

Exampe 2.4.5 Find the equation of the sphere whose centre is (1, 2, 3) and which touches the plane $2x + 2y - z = 2$.

Solution:

Radius of the sphere = Perpendicular distance of (1, 2, 3) from the plane $2x+2y-z = 2$.

$$= \frac{|2(1) + 2(2) - 3 - 2|}{\sqrt{2^2 + 2^2 + (-1)^2}}$$

$$= 1/3$$

The centre of the sphere is (1, 2, 3)

∴ The equation of the sphere is $(x - 1)^2 + (y - 2)^2 + (z - 3)^2 = 1/9$

i.e., $9x^2 + 9y^2 + 9z^2 - 18x - 36y - 54z + 125 = 0$

Example 2.4.6 Show that the spheres $x^2 + y^2 + z^2 = 25$ and $x^2 + y^2 + z^2 - 18x - 24y - 40z + 225 = 0$ touch and find the point of contact.

Solution:

The centre and radius of the sphere

$$x^2 + y^2 + z^2 = 25 \qquad\qquad\qquad (1)$$

are   A(0, 0, 0) and 5.

The centre and radius of the sphere

$$x^2 + y^2 + z^2 - 18x - 24y - 40z + 225 = 0 \qquad\qquad (2)$$

are   B(9, 12, 20) and $\sqrt{9^2 + 12^2 + 20^2 - 225} = 20$

Distance between the centres of the spheres = AB

$$= \sqrt{(9-0)^2 + (12-0)^2 + (20-0)^2}$$
$$= 25$$

Sum of the radii = 5 + 20 = 25

∴ Distance between the centres of the two spheres is equal to the sum of their radii.

∴ The two spheres touch externally (Fig. 2.31).

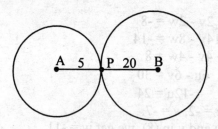

Fig. 2.31

The point of contact P divides AB internally in the ratio 5 : 20. i.e., 1 : 4

∴   P is $\left(\dfrac{4 \cdot 0 + 1 \cdot 9}{4+1}, \dfrac{4 \cdot 0 + 1 \cdot 12}{4+1}, \dfrac{4 \cdot 0 + 1 \cdot 20}{4+1} \right)$

i.e.,   P is $\left(\dfrac{9}{5}, \dfrac{12}{5}, 4 \right)$

Example 2.4.7 Show that the plane $2x - y - 2z = 16$ touches the sphere $x^2 + y^2 + z^2 - 4x + 2y + 2z - 3 = 0$ and find the point of contact.

Solution:

The centre and radius of the sphere are A(2, -1, 1) and $\sqrt{(-2)^2 + (+1)^2 + (+1)^2 + 3} = 3$

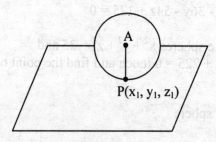

Fig. 2.32

Perpendicular distance of $(2, -1, -1)$ from the plane $2x - y - 2z - 16 = 0$ is

$$= \frac{|2(2) - (-1) - 2(-1) - 16|}{\sqrt{2^2 + (-1)^2 + (-2)^2}} = 3 = \text{radius of the sphere}$$

$\therefore$ The plane touches the sphere.

Let $P(x_1, y_1, z_1)$ be the point of contact.

Then the equation of the tangent plane at (x_1, y_1, z_1) is

$$xx_1 + yy_1 + zz_1 - 2(x + x_1) + (y + y_1) + (z + z_1) - 3 = 0$$

i.e., $(x_1 - 2) x + (y_1 + 1) y + (z_1 + 1) z - 2x_1 + y_1 + z_1 - 3 = 0$

This plane is same as $2x - y - 2z - 16 = 0$

$$\therefore \frac{x_1 - 2}{2} = \frac{y_1 + 1}{-1} = \frac{z_1 + 1}{-2} = \frac{2x_1 + y_1 + z_1 - 3}{-16} = k \ (\text{say})$$

Then $x_1 = 2k + 2$, $y_1 = -k - 1$, $z_1 = -2k - 1$ (1)

Also $-2x_1 + y_1 + z_1 - 3 = -16k$

Substituting for x_1, y_1, z_1 from (1), we get,

$-2(2k + 2) + (-k - 1) + (-2k - 1) - 3 = -16k$

i.e., $-7k - 9 = 16k$

i.e., $9k - 9 = 0$ i.e., $k = 1$

$\therefore$ From (1) $x_1 = 4$, $y_1 = -2$, $z_1 = -3$, and the point of contact P is $(4, -2, -3)$

Example 2.4.8 Show that the line whose equation is $\dfrac{x - 6}{3} = \dfrac{y - 7}{4} = \dfrac{z - 3}{5}$ touches the sphere $x^2 + y^2 + z^2 - 2x - 4y - 4 = 0$ and find the point of contact.

Solution:

The centre and the radius of the sphere are $A(1, 2, 0)$ and $\sqrt{(-1)^2 + (-2)^2 + 0^2 + 4} = 3$

Let P be the foot of the perpendicular from A to the line

$$\frac{x - 6}{3} = \frac{y - 7}{4} = \frac{z - 3}{5} \qquad (1)$$

Then the coordinates of P are $(3r + 6, 4r + 7, 5r + 3)$

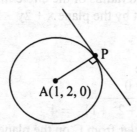

Fig. 2.33

The D.R's of AP are $(3r + 6 - 1, 4r + 7 - 2, 5r + 3 - 0)$

i.e., $(3r + 5, 4r + 5, 5r + 3)$

The D.R's of the line (1) are $(3, 4, 5)$ and as AP is perpendicular to the line (1)

$3(3r + 5) + 4(4r + 5) + 5(5r + 3) = 0$

i.e., $50r + 50 = 0$

i.e., $r = -1$

∴ The coordinates of P are $(3(-1) + 6, 4(-1) + 7, 5(-1) + 3)$ i.e., $(3, 3, -2)$

Now $AP = \sqrt{(3-1)^2 + (3-2)^2 + (-2-0)^2} = 3$

$\quad\quad$ = radius of the sphere

∴ The line (1) touches the sphere and the point of contact is P(3, 3, -2)

Example 2.4.9 Find the equation of the sphere for which the circle
$2x + 3y + 4z = 8$, $x^2 + y^2 + z^2 + 7y - 2z + 2 = 0$ is a great circle

Solution:

Equation of any sphere containing the given circle is

$$x^2 + y^2 + z^2 + 7y - 2z + 2 + \lambda(2x + 3y + 4z - 8) = 0$$

i.e., $x^2 + y^2 + z^2 - 2\lambda x + (7 + 3\lambda)y + (-2 + 4\lambda)z + 2 - 8\lambda = 0$ $\hspace{2cm}$ (1)

Centre of sphere (1) is $\left(-\lambda, \dfrac{-(7+3\lambda)}{2}, 1 - 2\lambda \right)$

Since the given circle is a great circle of sphere (1), the centre of the sphere lies on the plane $2x + 3y + 4z = 8$.

∴ $\quad 2(-\lambda) - \dfrac{3(7+3\lambda)}{2} + 4(1 - 2\lambda) = 8$

i.e., $\quad -4\lambda - 21 - 9\lambda + 8 - 16\lambda = 16$

i.e., $\quad\quad\quad\quad\quad -29\lambda = 29$ i.e.,

i.e., $\quad\quad\quad\quad\quad \lambda = -1$

∴ The equation (1) becomes $x^2 + y^2 + z^2 - 2x + 4y - 6z + 10 = 0$.

This is the equation of the required sphere.

Example 2.4.10 Find the centre and radius of the circle in which the sphere
$x^2 + y^2 + z^2 + 2y + 4z - 11 = 0$ is cut by the plane $x + 2y + 2z + 15 = 0$.

Solution:

The centre and radius of the sphere

$$x^2 + y^2 + z^2 + 2y + 4z - 11 = 0 \hspace{3cm} (1)$$

are C(0, -1, -2) and $CP = \sqrt{0^2 + 1^2 + 2^2 + 11} = 4$

Let N be the foot of the perpendicular from C on the plane

$$x + 2y + 2z + 15 = 0 \hspace{3cm} (2)$$

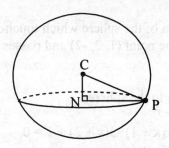

Fig. 2.34

Then $CN = \dfrac{0 + 2(-1) + 2(-2) + 15}{\sqrt{1^2 + 2^2 + 2^2}} = 3$

$\therefore$ The radius of the circle $= \sqrt{CP^2 - CN^2}$

$\qquad\qquad\qquad\qquad\quad = \sqrt{4^2 - 3^2} = \sqrt{7}$

The centre of the circle is N

Now CN is perpendicular to plane (2).

$\therefore$ Its D.R.'s are (1, 2, 2). Also CN passes through C (0, -1, -2).

$\therefore$ The equation of line CN is $\dfrac{x-0}{1} = \dfrac{y+1}{2} = \dfrac{z+2}{2}$

Any point on this line is (r, 2r - 1, 2r - 2).

If this point is N, it satisfies plane (2).

$\therefore \quad$ r + 2(2r - 1) + 2(2r - 2) + 15 = 0

i.e., $\qquad\qquad\qquad\qquad 9r + 9 = 0$

i.e., $\qquad\qquad\qquad\qquad\quad r = -1.$

$\therefore$ The coordinates of N are (-1, 2(-1)-1, 2(-1)-2)

i.e., $\quad$ (-1, -3, -4)

Thus the centre and radius of the circle are (-1, -3, -4) and $\sqrt{7}$

Example 2.4.11 Find the equation of the sphere through the circle $x^2 + y^2 + z^2 = 9$, $2x + 3y + 4z = 5$ and the point (1, 2, 3).

Solution:
Equation of any sphere through the given circle is
$\qquad x^2 + y^2 + z^2 - 9 + \lambda(2x + 3y + 4z - 5) = 0.$
When this passes through (1, 2, 3), we get $1^2 + 2^2 + 3^2 - 9 + \lambda(2.1 + 3.2 + 4.3 - 5) = 0$
$\qquad 5 + \lambda.15 = 0$
$\qquad \lambda = -1/3.$
$\therefore$ The required sphere is $x^2 + y^2 + z^2 - 9 + (-1/3)(2x + 3y + 4z - 5) = 0.$

i.e., $\quad 3x^2 + 3y^2 + 3z^2 - 2x - 3y - 4z - 22 = 0$

Example 2.4.12 Find the equation of the sphere which touches the sphere $x^2 + y^2 + z^2 + 2x - 6y + 1 = 0$ at the point $(1, 2, -2)$ and passes through the origin.

Solution:
The equation of the tangent plane to the sphere
$$x^2 + y^2 + z^2 + 2x - 6y + 1 = 0 \tag{1}$$
at $(1, 2, -2)$ is $x.1 + y.2 + z(-2) + (x + 1) - 3(y + 2) + 1 = 0$
i.e., $\quad 2x - y - 2z - 4 = 0 \tag{2}$
Since the required sphere touches the sphere (1), it contains the point circle of intersection of the sphere (1) and the tangent plane at $(1, 2, -2)$.

$\therefore$ Its equation is $x^2 + y^2 + z^2 + 2x - 6y + 1 + \lambda(2x - y - 2z - 4) = 0$.
This sphere passes through origin.
$\therefore \quad 1 + \lambda(-4) = 0$
i.e., $\quad \lambda = 1/4$

$\therefore$ The required sphere is $x^2 + y^2 + z^2 + 2x - 6y + 1 + \dfrac{1}{4}(2x - y - 2z - 4) = 0$

i.e., $\quad 4(x^2 + y^2 + z^2) + 10x - 25y - 2z = 0$.

Example 2.4.13 A sphere touches the plane $x - 2y - 2z - 7 = 0$ in the point $(3, -1, -1)$ and passes through the point $(1, 1, -3)$. Find its equation.

Solution:
The equation of the point sphere with centre at $(3, -1, -1)$ is
$$(x - 3)^2 + (y + 1)^2 + (z + 1)^2 = 0$$
i.e., $\quad x^2 + y^2 + z^2 - 6x + 2y + 2z + 11 = 0 \tag{1}$
The required sphere touches the plane
$$x - 2y - 2z - 7 = 0 \tag{2}$$
at $(3, -1, -1)$
$\therefore$ It contains the point circle of intersection of sphere (1) and plane (2).
Hence the equation of the required sphere is of the form
$$x^2 + y^2 + z^2 - 6x + 2y + 2z + 11 + \lambda(x - 2y - 2z - 7) = 0$$
and passes through $(1, 1, -3)$
$\therefore \quad 1^2 + 1^2 + (-3)^2 - 6.1 + 2.1 + 2(-3) + 11 + \lambda(1 - 2.1 - 2(-3) - 7) = 0$.
$$12 + \lambda(-2) = 0$$
$$\lambda = 6$$
$\therefore$ The required sphere is
$$x^2 + y^2 + z^2 - 6x + 2y + 2z + 11 + 6(x - 2y - 2z - 7) = 0$$
i.e., $\quad x^2 + y^2 + z^2 - 10y - 10z - 31 = 0$

Example 2.4.14 Show that the circles $x^2 + y^2 + z^2 - y + 2z = 0$, $x - y + z = 2$ and $x^2 + y^2 + z^2 + x - 3y + z - 5 = 0$, $2x - y + 4z = 1$ lie on the same sphere and find its equation.

Solution:
Any sphere containing the first circle is
$$x^2 + y^2 + z^2 - y + 2z + \lambda(x - y + z - 2) = 0 \qquad (1)$$
Any sphere containing the second circle is
$$x^2 + y^2 + z^2 + x - 3y + z - 5 + k(2x - y + 4z - 1) = 0 \qquad (2)$$
If the two circles lie on the same sphere, equations (1) and (2) are same for some values of λ and k. Comparing equations (1) and (2)

$$\lambda = 1 + 2k, \qquad -1 - \lambda = -3 - k, \qquad 2 + \lambda = 1 + 4k, \qquad -2\lambda = -5 - k$$

$\lambda = 3$, $k = 1$ satisfy all these equations.

∴ The two given circles lie on the same sphere and the equation of the sphere is
$$x^2 + y^2 + z^2 - y + 2z + 3(x - y + z - 2) = 0$$
i.e., $\qquad x^2 + y^2 + z^2 + 3x - 4y + 5z - 6 = 0$

Example 2.4.15 A sphere of constant radius 2k passes through the origin and meets the axes in A, B, C. Show that the locus of the centroid of the tetrahedron OABC is the sphere $x^2 + y^2 + z^2 = k^2$.

Solution:
Let the equation of the sphere be $x^2 + y^2 + z^2 + 2ux + 2vy + 2wz + d = 0$
It passes through the origin. ∴ $d = 0$.
At the points where the sphere meets the x-axis, $y = z = 0$.
∴ $\qquad x^2 + 2ux = 0$.
i.e., $\qquad x = 0$ or $x = -2u$.
$\qquad x = 0$ corresponds to the origin.
∴ The coordinates of A are (-2u, 0, 0).
Similarly coordinates of B and C are (0, -2v, 0) and (0, 0, -2w) respectively.
The radius of the sphere is 2k.
∴ $\qquad u^2 + v^2 + w^2 = 4k^2 \qquad\qquad (1)$
Let the coordinates of the tetrahedron OABC be (x_1, y_1, z_1).

Then $\quad x_1 = \dfrac{-2u + 0 + 0}{4}, \quad y_1 = \dfrac{0 - 2v + 0 + 0}{4}, \quad z_1 = \dfrac{0 + 0 - 2w + 0}{4}$

i.e., $\quad x_1 = -u/2, \qquad y_1 = -v/2, \qquad z_1 = -w/2$
i.e., $\quad u = -2x_1, \qquad v = -2y_1, \qquad w = -2z_1$.
Substituting the values of u, v, w in (1), we get

$$4x_1^2 + 4y_1^2 + 4z_1^2 = 4k^2$$

i.e., $\quad x_1^2 + y_1^2 + z_1^2 = k^2$

$\therefore$ Locus of the centroid (x_1, y_1, z_1) is $x^2 + y^2 + z^2 = k^2$.

Example 2.4.16 A plane passes through a fixed point (a, b, c). Show that the locus of the foot of the perpendicular from the origin on the plane is the sphere
$x^2 + y^2 + z^2 - ax - by - cz = 0$.

Solution:
Let the equation of the plane be
$$lx + my + nz = p \qquad (1)$$
This plane passes through (a, b, c)
$$\therefore \quad la + mb + nc = p \qquad (2)$$
The equation of the perpendicular from the origin on the plane is
$$\frac{x-0}{l} = \frac{y-0}{m} = \frac{z-0}{n} = k \qquad (3)$$
The coordinates (x_1, y_1, z_1) of the foot of the perpendicular will satisfy the equations (1) and (3).
$$\therefore \quad lx_1 + my_1 + nz_1 = p \qquad (4)$$
and
$$\frac{x_1}{l} = \frac{y_1}{m} = \frac{z_1}{n} = k \qquad (5)$$
Eliminate l, m, n from (2), (4) and (5)

From (5), $l = \dfrac{x_1}{k}$, $m = \dfrac{y_1}{k}$, $n = \dfrac{z_1}{k}$

Substituting the values of l, m, n in (2) and (4)
$$\frac{ax_1}{k} + \frac{by_1}{k} + \frac{cz_1}{k} = p$$
and
$$\frac{x_1^2}{k} + \frac{y_1^2}{k} + \frac{z_1^2}{k} = p$$
$$\therefore \quad \frac{1}{k}(x_1^2 + y_1^2 + z_1^2) = p$$
$$= \frac{1}{k}(ax_1 + by_1 + cz_1)$$
i.e., $\quad x_1^2 + y_1^2 + z_1^2 - ax_1 - by_1 - cz_1 = 0$
$\therefore$ Locus of the point (x_1, y_1, z_1) is the sphere $x^2 + y^2 + z^2 - ax - by - cz = 0$.

Example 2.4.17 A sphere of constant radius r passes through the origin O and cuts the axes in A, B, C. Prove that the locus of the foot of the perpendicular from O on the plane ABC is $(x^2 + y^2 + z^2)^2 (x^{-2} + y^{-2} + z^{-2}) = 4r^2$

Solution:

Let the equation of the plane ABC be $\dfrac{x}{a} + \dfrac{y}{b} + \dfrac{z}{c} = 1 \qquad (1)$

The coordinates of A, B and C are (a, 0, 0), (0, b, 0) and (0, 0, c) respectively.
Equation of the sphere through O, A, B, C is

$$x^2 + y^2 + z^2 - ax - by - cz = 0 \qquad (2)$$

Radius of this sphere $= \sqrt{u^2 + v^2 + w^2 - d}$

i.e., $\qquad r = \sqrt{\dfrac{a^2}{4} + \dfrac{b^2}{4} + \dfrac{c^2}{4}}$

i.e., $\qquad 4r^2 = a^2 + b^2 + c^2 \qquad (3)$

The equation of the perpendicular from the origin on the plane ABC is

$$\frac{x-0}{1/a} = \frac{y-0}{1/b} = \frac{z-0}{1/c} = k$$

i.e., $\qquad ax = by = cz = k \qquad (4).$

The coordinates (x_1, y_1, z_1) of the foot of the perpendicular will satisfy equations (1) and (4)

$$\therefore \qquad \frac{x_1}{a} + \frac{y_1}{b} + \frac{z_1}{c} = 1 \qquad (5)$$

and $\qquad ax_1 = by_1 = cz_1 = k \qquad (6)$

Eliminate a, b, c from (3), (5) and (6).

From (6), $a = \dfrac{k}{x_1}, \ b = \dfrac{k}{y_1}, \ c = \dfrac{k}{z_1}$

Substituting for a, b, c in (3) we get

$$4r^2 = \frac{k^2}{x_1^2} + \frac{k^2}{y_1^2} + \frac{k^2}{z_1^2}$$

i.e., $\qquad k^2 \left(\dfrac{1}{x_1^2} + \dfrac{1}{y_1^2} + \dfrac{1}{z_1^2} \right) = 4r^2 \qquad (7)$

Also from (5) $\dfrac{x_1^2}{k} + \dfrac{y_1^2}{k} + \dfrac{z_1^2}{k} = 1$

i.e., $\qquad x_1^2 + y_1^2 + z_1^2 = k \qquad (8)$

Using (8) in (7)

$$(x_1^2 + y_1^2 + z_1^2)^2 (x_1^{-2} + y_1^{-2} + z_1^{-2}) = 4r^2$$

$\therefore$ Locus of (x_1, y_1, z_1) is

$$(x^2 + y^2 + z^2)^2 (x^{-2} + y^{-2} + z^{-2}) = 4r^2$$

Example 2.4.18 Show that the spheres $x^2 + y^2 + z^2 + 6y + 2z + 8 = 0$ and $x^2 + y^2 + z^2 + 6x + 8y + 4z + 20 = 0$ intersect at right angles. Find the equation of their plane of intersection.

Solution:
The condition for two spheres to intersect at right angles is

$$2uu_1 + 2vv_1 + 2ww_1 = d + d_1 \qquad\qquad (1)$$

From $x^2 + y^2 + z^2 + 6y + 2z + 8 = 0$, we have $u = 0$, $v = 3$, $w = 1$, $d = 8$

From $x^2 + y^2 + z^2 + 6x + 8y + 4z + 20 = 0$, we have $u_1 = 3$, $v_1 = 4$, $w_1 = 2$, $d_1 = 20$.

$\therefore \qquad 2uu_1 + 2vv_1 + 2ww_1 = 2.0.3 + 2.3.4 + 2.1.2 = 28$

and $\quad d + d_1 = 8 + 20 = 28$

Condition (1) is satisfied.

$\therefore$ The two spheres intersect at right angles.

The equation of the plane of intersection is $S_1 - S_2 = 0$

i.e., $\quad (x^2 + y^2 + z^2 + 6y + 2z + 8) - (x^2 + y^2 + z^2 + 6x + 8y + 4z + 20) = 0$

i.e., $\quad -6x - 2y - 2z - 12 = 0$

i.e., $\qquad 3x + y + z + 6 = 0$

This is the required equation of the plane of intersection.

Example 2.4.19 Find the equation of the sphere that passes through the circle $x^2 + y^2 + z^2 + x - 3y + 2z - 1 = 0$, $2x + 5y - z + 7 = 0$ and cuts orthogonally the sphere whose equation $x^2 + y^2 + z^2 - 3x + 5y - 7z - 6 = 0$

Solution:

The equation of any sphere passing through the given circle is

$$x^2 + y^2 + z^2 + x - 3y + 2z - 1 + \lambda(2x + 5y - z + 7) = 0$$

i.e., $\quad x^2 + y^2 + z^2 + (1 + 2\lambda)x + (-3 + 5\lambda)y + (2 - \lambda)z - 1 + 7\lambda = 0 \qquad (1)$

Sphere (1) cuts the sphere $x^2 + y^2 + z^2 - 3x + 5y - 7z - 6 = 0$ orthogonally.

$\therefore \qquad 2uu_1 + 2vv_1 + 2ww_1 = d + d_1$

i.e., $\quad 2\left(\dfrac{1+2\lambda}{2}\right)\left(\dfrac{-3}{2}\right) + 2\left(\dfrac{-3+5\lambda}{2}\right)\left(\dfrac{5}{2}\right) + 2\left(\dfrac{2-\lambda}{2}\right)\left(\dfrac{-7}{2}\right) = -1 + 7\lambda - 6$

i.e., $\quad -3 - 6\lambda - 15 + 25\lambda - 14 + 7\lambda = -14 + 14\lambda$

i.e., $\qquad\qquad\qquad 12\lambda - 18 = 0 \qquad$ i.e., $\lambda = 3/2$

$\therefore$ Equation of the required sphere is

$$x^2 + y^2 + z^2 + x - 3y + 2z - 1 + (3/2)(2x + 5y - z + 7) = 0$$

i.e., $\quad 2(x^2 + y^2 + z^2) + 8x + 9y + z + 19 = 0$

Example 2.4.20 Find the equations to the straight line from any point of which the tangents to the spheres $x^2 + y^2 + z^2 = 1$, $x^2 + y^2 + z^2 + 2x + y + z = 4$, $x^2 + y^2 + z^2 + x + 2y + z = 4$ are equal. Find also the coordinates of the point on the line from which the length of the tangents is least.

Solution:

The equations to the straight line (**radical line**) are given by

$$x^2 + y^2 + z^2 - 1 = x^2 + y^2 + z^2 + 2x + y + z - 4 = x^2 + y^2 + z^2 + x + 2y + z - 4$$

i.e., $\quad -1 = 2x + y + z - 4 = x + 2y + z - 4$

i.e., $\quad -z + 3 = 2x + y = x + 2y$

$$2x + y = x + 2y \Rightarrow x = y$$
$$-z + 3 = 2x + y \Rightarrow -z + 3 = 3x$$

i.e., $\dfrac{-z+3}{3} = x$

$\therefore \qquad x = y = \dfrac{-z+3}{3}$ \hfill (1)

Let (x_1, y_1, z_1) be any point on the line (1).

Then $\quad x_1 = y_1 = \dfrac{-z_1+3}{3}$ \hfill (2)

The length of the tangent from the point (x_1, y_1, z_1) to the spheres is given by

$$L = \sqrt{x_1^2 + y_1^2 + z_1^2 - 1}$$

i.e., $\quad L^2 = x_1^2 + x_1^2 + (-3x_1 + 3)^2 - 1 \qquad$ (Using (2))

i.e., $\quad L^2 = 11x_1^2 - 18x_1 + 8$

L is minimum when $\dfrac{dL^2}{dx_1} = 0$ and $\dfrac{d^2L^2}{dx_1^2} > 0$

Now $\quad \dfrac{dL^2}{dx_1} = 22x_1 - 18 = 0 \Rightarrow x_1 = \dfrac{9}{11}$

Also $\quad \dfrac{d^2L^2}{dx_1^2} = 22 > 0$

$\therefore$ When L is least, $x_1 = 9/11$, $y_1 = 9/11$ and $z_1 = -3x_1+3 = -3(9/11) + 3 = 6/11$

The required point is $(9/11, 9/11, 6/11)$.

EXERCISE 2.4

PART – A

1. What are the conditions for a general second degree equation in x, y, z to represent a sphere?

2. Find the centre and radius of the sphere $a(x^2 + y^2 + z^2) + 2ux + 2vy + 2wz + d = 0$

3. Find the centre and radius of the sphere $2x^2 + 2y^2 + 2z^2 + 6x - 6y + 8z + 9 = 0$.

4. Find the radius of the sphere whose centre is at $(6, -1, 2)$ and touches the plane $2x - y + 2z - 2 = 0$

5. Find the equation of the sphere on the join of $(1, -1, -1)$ and $(-3, 4, 5)$ as diameter.

6. Find the equation of the sphere which passes through the points $(0, 0, 0)$, $(-a, b, c)$, $(a, -b, c)$, $(a, b, -c)$.

7. Find the equation of the sphere through the points $(0, 0, 0)$, $(a, 0, 0)$, $(0, b, 0)$, $(0, 0, c)$.

8. Find the equation of the sphere through the points $(1, 2, 3)$, $(2, 3, 4)$ and has its centre on the line $x = y = z$.

9. Find the equation of the sphere which touches the XOY – plane and has centre at (2, 3, 5)

10. Find the condition for the plane $lx + my + nz = p$ to be a tangent plane to the sphere $x^2 + y^2 + z^2 = r^2$.

11. Find the equation of the sphere having the circle $x^2 + y^2 + z^2 = 9$, $x - 2y + 2z = 5$ for a great circle.

12. Find the equation of the tangent plane to the sphere $3x^2 + 3y^2 + 3z^2 - 2x - 3y - 4z - 22 = 0$ at the point (1, 2, 3)

13. Find the equation of the tangent plane to the sphere $x^2 + y^2 + z^2 - 4x - 6y - 8z - 9 = 0$ at the point (1, 2, -2).

14. Find the length of the tangent drawn from the point (1, 2, 3) to the sphere $x^2 + y^2 + z^2 - 3x + 4y + 5z - 7 = 0$.

15. Find the position of the point (-1, 3, 2) relative to sphere $x^2 + y^2 + z^2 - 2x + 6y + 4z - 35 = 0$.

16. A point moves such that the lengths of the tangents from the point to two given spheres $S_1 = 0$ and $S_2 = 0$ are equal. Find the locus of the moving point. (The locus is a plane called the radical plane of the two spheres).

17. Find the equation to the sphere which passes through the point (a, b, c) and the circle $x^2 + y^2 + z^2 = r^2$, $z = 0$.

18. Find the equation of the sphere which passes through the point (1, 3, -2) and the circle $x^2 + y^2 + z^2 = 25$; $x = 0$.

19. Show that the spheres $x^2 + y^2 + z^2 - 2ay - 2az + a^2 = 0$ and $x^2 + y^2 + z^2 - 2ax - 2az + a^2 = 0$ intersect orthogonally.

20. Test whether the plane $2x - y + 2z = 1$ cuts the sphere $x^2 + y^2 + z^2 - 4x + 2y - 4 = 0$ or not.

21. Find the equation to that section of the sphere $x^2 + y^2 + z^2 = r^2$ of which a given internal point (a, b, c) is the centre.

PART – B

22. Find the equation of the sphere passing through the points (1, 2, 3), (0, -2, 4), (4, -4, 2) and (3, 1, 4).

23. Find the equation of the sphere passing through the three points (0, 2, 3), (1, 1, -1), (-5, 4, 2) and having its centre on the plane $3x + 4y + 2z = 6$.

24. Find the equation of the sphere passing through the points (1, 1, -2) and (-1, 1, 2) and having its centre on the line $x + y - z - 1 = 0 = 2x - y + z - 2$

25. Find the equation of the sphere which passes through the points (1, -4, 3), (1, -5, 2) and has its centre on the line $\dfrac{x+4}{-4} = \dfrac{y+2}{1} = \dfrac{z-6}{3}$

26. Find the condition that $lx + my + nz = p$ is a tangent plane to the sphere $x^2 + y^2 + z^2 + 2ux + 2vy + 2wz + d = 0$.

27. Prove that the two spheres $x^2 + y^2 + z^2 - 2x + 4y - 4z = 0$ and $x^2 + y^2 + z^2 + 10x + 2z + 10 = 0$ touch each other and find the coordinates of

the point of contact.

28. Show that the spheres $x^2 + y^2 + z^2 = 64$ and $x^2 + y^2 + z^2 - 12x + 4y - 6z + 48 = 0$ touch internally. Find their point of contact.

29. Show that the intersection of the sphere $x^2 + y^2 + z^2 - 2x - 4y - 6z - 2 = 0$ and the plane $x + 2y + 2z - 20 = 0$ is a circle of radius $\sqrt{7}$ with its centre at the point $(2, 4, 5)$.

30. Show that the centre of all sections of the sphere $x^2 + y^2 + z^2 = r^2$ by planes through a point (a, b, c) lie on the sphere $x(x - a) + y(y - b) + z(z - c) = 0$.

31. A sphere of constant radius k passes through the origin and meets the axes in A, B, C. Prove that the centroid of the triangle ABC lies on the sphere $9(x^2 + y^2 + z^2) = 4k^2$.

32. A plane passes through a fixed point (a, b, c) and cuts the axes in A, B, C. Show that the locus of the centre of the sphere OABC is $\dfrac{a}{x} + \dfrac{b}{y} + \dfrac{c}{z} = 2$

33. Prove that the equation of the sphere which lies in the octant OXYZ and touches the coordinate planes is of the form $x^2 + y^2 + z^2 - 2k(x + y + z) + 2k^2 = 0$.

34. Find the centre and radius of the circle $x^2 + y^2 + z^2 - 2y - 4z = 11$, $x + 2y + 2z = 15$.

35. Find the equation of the sphere for which the circle $x^2 + y^2 + z^2 - 3x + 4y - 2z - 5 = 0$, $5x - 2y + 4z + 7 = 0$ is a great circle.

36. Find the equation of the sphere which passes through the circle $x^2 + y^2 + z^2 - 2x - 4y = 0$, $x + 2y + 3z = 8$ and touch the plane $4x + 3y = 25$.

37. Find the equation of the sphere having its centre on the plane $4x - 5y - z = 3$ and passing through the circle $x^2 + y^2 + z^2 - 2x - 3y + 4z + 8 = 0$, $x - 2y + z = 8$.

38. Find the equation of the sphere through the circle $x^2 + y^2 + z^2 - 2x + 3y + 4z - 5 = 0$ and $x^2 + y^2 + z^2 - 3x - 4y + 5z - 6 = 0$ passing through the point $(1, 1, 2)$.

39. Prove that the circles $x^2 + y^2 + z^2 - 2x + 3y + 4z - 5 = 0$, $5y + 6z + 1 = 0$ and $x^2 + y^2 + z^2 - 3x - 4y + 5z + 6 = 0$, $x + 2y - 7z = 0$ lie on the same sphere and find its equation.

40. Show that the centres of the spheres which cut both the spheres $x^2 + y^2 + z^2 + 2ax + c = 0$, $x^2 + y^2 + z^2 + 2bx + c = 0$ in great circles lie on the plane $x + a + b = 0$.

41. Show that the plane $2x - y - 2z = 16$ touches the sphere $x^2 + y^2 + z^2 - 4x + 2y + 2z - 3 = 0$ and find the point of contact.

42. Find the equations of the two spheres which passes through the circle $x^2 + y^2 + z^2 - 2x + 2y + 4z - 3 = 0$, $2x + y + z = 4$ and touch the plane $3x + 4y = 14$.

43. Find the equation of the tangent line to the circle $x^2 + y^2 + z^2 = 3$, $3x - 2y + 4z + 3 = 0$ at the point $(1, 1, -1)$.

44. Find the equation of the tangent line to the circle $x^2 + y^2 + z^2 - x + 4z = 0$, $3x - 2y + 4z + 1 = 0$ at the point $(1, -2, -2)$.

45. Find the points on the sphere $x^2 + y^2 + z^2 + 2x - 4z - 4 = 0$ the tangent planes at which are parallel to the plane $x - 2y - 2z + 1 = 0$.

46. Find the equation of the sphere which touches the sphere $4(x^2 + y^2 + z^2) - 25x - 2y + 10z = 0$ at $(2, -2, 1)$ and passes through $(0, 0, -1)$.

47. Find the coordinates of the point on the sphere $x^2 + y^2 + z^2 + 2x - 4z - 4 = 0$ which is nearest to the plane $4x + y + z - 7 = 0$.

48. Find the equation of the sphere which has its centre at $(5, -2, 3)$ and which touches the line $\dfrac{x-1}{6} = \dfrac{y+1}{2} = \dfrac{z-12}{-3}$

49. Find the equation of two tangent planes to the sphere $x^2 + y^2 + z^2 = 9$ which pass through the line $x + y - 6 = 0 = x - 2z - 3$.

50. Find the values of k for which the plane $x + y + z = k\sqrt{3}$ touches the sphere $x^2 + y^2 + z^2 - 2x - 2y - 2z - 6 = 0$

51. Find the equation of the sphere which touches the plane $3x + 2y - z + 2 = 0$ at $(1, -2, 1)$ and cuts orthogonally the sphere $x^2 + y^2 + z^2 - 4x + 6y + 4 = 0$.

52. Find the equation of the sphere that passes through the circle $x^2 + y^2 + z^2 + 3 + y + 2z - 2 = 0$, $x + 3y - 2z + 1 = 0$ and cuts orthogonally the sphere $x^2 + y^2 + z^2 + x - 3z - 2 = 0$.

53. Show that the sphere through the circle $x^2 + y^2 + z^2 - 2x + 3y - 4z + 6 = 0$, $3x - 4y + 5z - 15 = 0$ and cutting orthogonally the sphere $x^2 + y^2 + z^2 + 2x + 4y - 6z + 11 = 0$ is $5(x^2 + y^2 + z^2) - 13x + 19y - 25z + 45 = 0$.

54. Show that the sphere passing through the points $(0, 3, 0)$, $(-2, -1, -4)$ and cutting orthogonally the two spheres $x^2 + y^2 + z^2 + x - 3z - 2 = 0$ and $2x^2 + 2y^2 + 2z^2 + x + 3y + 4 = 0$ is $x^2 + y^2 + z^2 + 2x - 2y + 4z - 3 = 0$.

55. Show that the locus of the points from which the tangents to the three spheres $(x - 2)^2 + y^2 + z^2 = 1$, $x^2 + (y - 3)^2 + z^2 = 6$, $(x + 2)^2 + (y + 1)^2 + (z - 2)^2 = 6$ are all equal is the line $\dfrac{x}{3} = \dfrac{y}{2} = \dfrac{z}{7}$. Find the coordinates of the point on this line from which the length of the tangents to the three spheres is also equal to that of the tangent to the sphere $(2x + 1)^2 + 4y^2 + (2z - 1)^2 = 6$.

56. Show that the locus of the point from which equal tangents may be drawn to the three spheres $x^2 + y^2 + z^2 = 1$, $x^2 + y^2 + z^2 + 2x - 2y + 2z - 1 = 0$ and $x^2 + y^2 + z^2 - x + 4y - 6z - 2 = 0$ is the straight line $\dfrac{x-1}{2} = \dfrac{y-2}{5} = \dfrac{z-1}{3}$.

ANSWERS

EXERCISE 2.1

(6) $(2/7, 3/7, 6/7)$

(7) $\left(\pm \dfrac{1}{\sqrt{3}}, \pm \dfrac{1}{\sqrt{3}}, \pm \dfrac{1}{\sqrt{3}} \right); 4$

(8) $(2/3, -2/3, 1/3)$

(10) $13; (12/13, 4/13, 3/13)$

(11) 7

(12) (2/3, -1/3, 2/3)

(13) $\left(\frac{1}{\sqrt{6}}, \frac{-2}{\sqrt{6}}, \frac{1}{\sqrt{6}} \right)$

(14) 90°

(15) 60°

(20) k = 2

(25) $\cos^{-1}(1/3)$

(26) $\dfrac{10}{\sqrt{33}}$

(27) (-2, 4, 2)

(28) (8, -7, 8)

(29) (-26, 3, -59)

(33) 60° or 120°

(34) $\cos\theta = 1/6$

(36) 60°

(40) $\dfrac{1}{2}\sqrt{a^2 b^2 + b^2 c^2 + c^2 a^2}$

EXERCISE 2.2

(1)   -4, -2, -4/3

(2) $\dfrac{x}{3} - \dfrac{y}{2} + \dfrac{z}{5} = 1$

(3)   x + 2y - 3z = 3

(4)   x + 3y + z = $\pm 2\sqrt{11}$

(5)   x + 5y - 6z + 18 = 0

(6)   x - 3y + 2z + 11 = 0

(7) 2x + 3y - z = 14

(8) $\theta = \pi/3$

(9) (2/7, 3/7, 6/7)

(10) 5/6

(11) 2

(12) $\dfrac{5}{\sqrt{29}}$

(13) $\dfrac{27}{\sqrt{50}}$

(14) x - 5y + 6z + 3 = 0

(15) lie on opposite sides.

(16)   16x + 11y - 12z + 14 = 0

(17) x - 12y - 10z - 5 = 0

(18) 2x + 2y - 3z -2 = 0

(20) 5x + 2y - 3z -17 = 0

(21) 3x − y + z + 6 = 0

(22) 2x − y - z = 6

(23) (12, -4, 3)

(24) (9/7, 4/7, 13/7)

(25) (13/3, 7/3, 8/3)

(26) 7x - 17y - 6z - 56 = 0

(27) 22x + 9y + 32 = 0

(30) 291.x + 58y + 15z + 45 = 0 or 213 x + 110y - 141z + 123 = 0

(37) $(x^2 + y^2 + z^2)(x^{-1} + y^{-1} + z^{-1}) = k.$

(39) 2x + y - 2z + 3 = 0 and x - 2y - 2z - 3 = 0

(40)   5x + 4y + z - 10 = 0

EXERCISE 2.3

(1) $\dfrac{x-2}{4} = \dfrac{y-7}{2} = \dfrac{z-3}{-3}$

(2) $\dfrac{x}{5} = \dfrac{y}{-2} = \dfrac{z}{3}$

(4) $\cos^{-1}(-1/5)$

(5) $\sin^{-1}(4/9)$

(6) (6/7, 3/7, 2/7)

(7) (3, 5, 9) and (-1, -1, -3)

(9) $\dfrac{x-2}{2} = \dfrac{y-3}{-3} = \dfrac{z-4}{6}$

(10) $\dfrac{x-1}{6} = \dfrac{y-0}{-2} = \dfrac{z+2}{6}$

(11) $\sin\theta = \dfrac{15}{\sqrt{539}}$

(14) $\dfrac{x+1}{-3} = \dfrac{y}{5} = \dfrac{z-2}{7}$

(15) $\dfrac{x}{2} = \dfrac{y-1}{-5} = \dfrac{z}{3}$

(16) $\cos^{-1} \dfrac{8}{\sqrt{406}}$

(17) 1

(18) $(-1, -1, -1)$; 3 and $\dfrac{x-1}{-2} = \dfrac{y}{-1} = \dfrac{z+3}{2}$

(19) $\sqrt{\dfrac{27}{14}}$

(20) 13

(21) $(2/3, -1/3, -4/3)$

(22) $15x + y - 7z + 4 = 0$

(23) $x - 2y + z = 0 = 3x + 2y + z$

(24) $x + 13y + 5z - 20 = 0 = 2x + y - 3z$

(25) $3x - 8y + 7z + 4 = 0 = 3x + 2y + z$

(26) $9x + 8y + 2z - 7 = 0 = 2x - 3y + 3z - 5$

(27) $\dfrac{x-4}{9} = \dfrac{y+1}{-1} = \dfrac{z-7}{-3}$

(28) $\dfrac{x+5}{1} = \dfrac{y+7}{6} = \dfrac{z}{1}$

(29) $x - 2y + z = 0$

(30) $(4, 2, -1)$; $-x + y + z + 3 = 0$

(31) $(18/11, -4/11, -3/11)$, $x + y + z = 1$

(32) $(a + l, b + m, c + n)$; $(mc - nb)x + (na - lc)y + (lb - ma)z = 0$.

(34) $(1/2, 3/2, 5/2)$, $83x - 2y - 7z - 21 = 0$

(35) $83x - 2y - 7z - 21 = 0$

(36) $\dfrac{2\sqrt{3}}{3}$; $\dfrac{x - \frac{25}{9}}{1} = \dfrac{y + \frac{14}{9}}{1} = \dfrac{z - \frac{75}{9}}{1}$

(37) $\dfrac{1}{\sqrt{6}}$; $\dfrac{x - \frac{5}{3}}{\frac{1}{6}} = \dfrac{y - 3}{-\frac{1}{3}} = \dfrac{z - \frac{13}{3}}{\frac{1}{6}}$

(38) $3\sqrt{30}$; $\dfrac{x-3}{2} = \dfrac{y-8}{5} = \dfrac{z-3}{-1}$

(39) $\dfrac{1}{2}\cdot\sqrt{2a}$

(40) $\sqrt{116}$; $\dfrac{x-3}{2} = \dfrac{y-5}{3} = \dfrac{z-7}{8}$

(41) $\dfrac{11}{\sqrt{342}}$; $10x - 29y + 16z = 0 = 13x + 82y + 55z - 109$.

(43) $\dfrac{97}{13\sqrt{6}}$; $2x + 11y - 7z - 20 = 0 = 13y - 13z - 3$.

EXERCISE 2.4

1) Coefficients of x^2, y^2, z^2 are equal and the terms xy, yz, zx are absent.

(2) Centre is $\left(-\dfrac{u}{a}, -\dfrac{v}{a}, -\dfrac{w}{a}\right)$ and radius is $\sqrt{\dfrac{u^2}{a^2} + \dfrac{v^2}{a^2} + \dfrac{w^2}{a^2} - \dfrac{d}{a}}$

(3) Centre (-3/2, 3/2, -2) radius 2.　　　　(4) 5

(5) $x^2 + y^2 + z^2 + 2x - 3y - 4z - 12 = 0$

(6) $x^2 + y^2 + z^2 - (a^2 + b^2 + c^2)\left(\dfrac{x}{a} + \dfrac{y}{b} + \dfrac{z}{c}\right) = 0$

(7) $x^2 + y^2 + z^2 - ax - by - cz = 0$

(8) $x^2 + y^2 + z^2 - 5(x + y + z) + 16 = 0$.

(9) $x^2 + y^2 + z^2 - 4x - 6y - 10z + 13 = 0$

(10) $p^2 = r^2(l^2 + m^2 + n^2)$

(11) $9(x^2 + y^2 + z^2) = 10(x - 2y + 2z) + 31$

(12) $4x + 9y + 14z - 64 = 0$

(13) $x + y + 6z + 9 = 0$

(14) $3\sqrt{3}$　　　(15) (-1, 3, 2) lies outside the sphere　　　(16) $S_1 - S_2 = 0$

(17) $c(x^2 + y^2 + z^2 - r^2) - z(a^2 + b^2 + c^2 - r^2) = 0$.

(18) $x^2 + y^2 + z^2 + 11x = 25$.

(20) The plane cuts the sphere

(21) $a(x - a) + b(y - b) + c(z - c) = 0$.

(22) $x^2 + y^2 + z^2 - 4x + 2y - 2z + 8 = 0$

(23) $x^2 + y^2 + z^2 + 4x - 6y - 1 = 0$

(24) $x^2 + y^2 + z^2 - 2x - y - z - 5 = 0$

(25) $x^2 + y^2 + z^2 - 4x + 7y - 3z + 15 = 0$

(26) $(lu + mv + nw + p)^2 = (l^2 + m^2 + n^2)(u^2 + v^2 + w^2 - d)$

(27) (-11/7, -8/7, 5/7)　　　(28) (48/7, -16/7, 24/7)　　　(34) (1, 3, 4), $\sqrt{7}$

(35) $x^2 + y^2 + z^2 + 2x + 2y + 2z + 2 = 0$

(36) $x^2 + y^2 + z^2 + 6z - 16 = 0$ and $5(x^2 + y^2 + z^2) - 14x - 28y + 32 = 0$.

(37) $13(x^2 + y^2 + z^2) - 35x - 21y + 43z + 176 = 0$

(38) $7(x^2 + y^2 + z^2) - 24x - 49y + 38z - 45 = 0$

(39) $x^2 + y^2 + z^2 - 2x - 2y - 2z - 6 = 0$

(41) (4, -2, -3)

(42) $x^2 + y^2 + z^2 - 2x + 2y + 4z - 3 = 0$,　$x^2 + y^2 + z^2 + 2x + 4y + 6z - 11 = 0$

(43)　$\dfrac{x - 0.6}{-2} = \dfrac{y - 2.4}{7} = \dfrac{z}{5}$　　　(44)　$\dfrac{x - 1}{8} = \dfrac{y + 2}{2} = \dfrac{z + 2}{-5}$

(45) (0, -2, 0), (-2, 2, 4)　　　(46) $x^2 + y^2 + z^2 - 6x + 2z + 1 = 0$

(47) $\left(-1+2\sqrt{2},\ \dfrac{1}{\sqrt{2}},\ 2+\dfrac{1}{\sqrt{2}}\right)$

(48) $x^2 + y^2 + z^2 - 10x + 4y - 6z - 11 = 0$

(49) $2x + y - 2z - 9 = 0,\ x + 2y + 2z - 9 = 0$ (50) $k = \sqrt{3} \pm 3$

(51) $x^2 + y^2 + z^2 + 7x + 10y - 5z + 12 = 0$ $\sqrt{(-2)^2 + (+1)^2 + (+1)^2 + 3} = 3$

(52) $x^2 + y^2 + z^2 + 2x - 2y + 4z - 3 = 0$ (55) $\left(\dfrac{3}{2},\ 1,\ \dfrac{7}{2}\right)$

CHAPTER 3

GEOMETRICAL APPLICATIONS OF DIFFERENTIAL CALCULUS

3.0   INTRODUCTION

Consider a function $y = f(x)$ which is continuous in the closed interval $[a, b]$ and differentiable in the open interval (a, b). If $c \in (a, b)$ then the derivative of $y = f(x)$ at c is

$$f'(c) = \underset{h \to 0}{Lt} \frac{f(c+h) - f(c)}{h} \tag{1}$$

If P is the point $(c, f(c))$ and Q is the point $(c + h, f(c + h))$ on the graph of the function, draw PM and QN perpendicular to x-axis and PR perpendicular to QN. Then PM = f(c) and QN = f(c + h). Also PR = h. Let the chord PQ meet the x-axis at S and makes an angle θ with the positive direction of the x-axis, measured in the anticlockwise direction. Let the tangent at P to the curve meet the x-axis at T and makes an angle ϕ with the positive direction of the x-axis, measured in

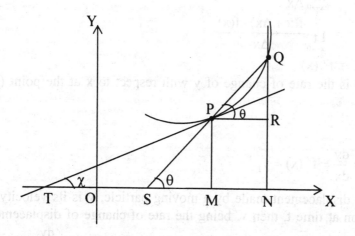

Fig.3.1

the anticlockwise direction. Now as $h \to 0$, $Q \to P$ along the curve and the chord PQ tends to the tangent at P. Also the angle θ tends to the angle χ.

Then from (1)
$$f'(c) = \underset{h \to 0}{Lt} \frac{QN - PM}{PR}$$

$$= \underset{h \to 0}{Lt} \frac{QR}{PR}$$

$$= \underset{h \to 0}{Lt} \tan \theta$$

$$= \tan \chi$$

Thus $f'(c)$ is the slope of the tangent to the curve at the point $P(c, f(c))$.

3.0.1 DERIVATIVE AS RATE OF CHANGE

Let $y = f(x)$ be a given function. Δx be a small change in x. Δy be the corresponding change in y. Then, $y + \Delta y = f(x + \Delta x)$

i.e., $\Delta y = f(x + \Delta x) - y$

i.e., $\Delta y = f(x + \Delta x) - f(x)$

i.e., $\dfrac{\Delta y}{\Delta x} = \dfrac{f(x + \Delta x) - f(x)}{\Delta x}$

Δy is the total change in y in the interval $(x, x + \Delta x)$.

Hence $\dfrac{\Delta y}{\Delta x}$ is the average rate of change of y with respect to x in the interval $(x, x + \Delta x)$.

When $\Delta x \to 0$, the interval $(x, x + \Delta x)$ becomes the point x.

$\therefore$ At the point (x, y) the rate of change of y with respect to x is

$$= \underset{\Delta x \to 0}{Lt} \frac{\Delta y}{\Delta x}$$

$$= \underset{\Delta x \to 0}{Lt} \frac{f(x + \Delta x) - f(x)}{\Delta x}$$

$$= f'(x)$$

Thus $f'(x)$ is the rate of change of y with respect to x at the point (x, y) on the curve $y = f(x)$.

Note:

(i) $\quad \underset{\Delta x \to 0}{Lt} \dfrac{\Delta y}{\Delta x} = \dfrac{dy}{dx} = f'(x) = y_1$

(ii) If s is the displacement made by a moving particle, v is its velocity and a is its acceleration at time t, then v, being the rate of change of displacement, is equal to $\dfrac{ds}{dt}$ and a, being the rate of change of velocity, is equal to $\dfrac{dv}{dt}$.

3.0.2   DERIVATIVE OF THE LENGTH OF AN ARC

Let $P(x, y)$ be any point on the curve $y = f(x)$ and $Q(x + \Delta x, y + \Delta y)$ be a neighboring point on the curve. Let A be any fixed point on the curve and s be the length of the arc AP. Let Δs be the length of the arc PQ. While $\Delta x \to 0$, $Q \to P$ and the arc PQ and chord PQ becomes almost equal.

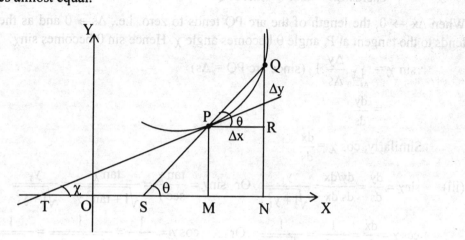

Fig.3.2

$$\therefore \underset{\Delta x \to 0}{\text{Lt}} \frac{\text{chord PQ}}{\text{arc PQ}} = 1 \tag{1}$$

Now from triangle PQR, $(\text{chord PQ})^2 = PR^2 + QR^2$

i.e., $(\text{chord PQ})^2 = \Delta x^2 + \Delta y^2$.

i.e., $\left(\dfrac{\text{chord PQ}}{\Delta x}\right)^2 = 1 + \left(\dfrac{\Delta y}{\Delta x}\right)^2$

i.e., $\left(\dfrac{\text{chord PQ}}{\text{arc PQ}}\right)^2 \left(\dfrac{\text{arc PQ}}{\Delta x}\right)^2 = 1 + \left(\dfrac{\Delta y}{\Delta x}\right)^2$

When $\Delta x \to 0$ we get,

$$\left(\underset{\Delta x \to 0}{\text{Lt}} \frac{\Delta s}{\Delta x}\right)^2 = 1 + \underset{\Delta x \to 0}{\text{Lt}} \left(\frac{\Delta y}{\Delta x}\right)^2 \quad \text{(Using (1) and arc PQ} = \Delta s)$$

i.e., $\dfrac{ds}{dx} = \sqrt{1 + \left(\dfrac{dy}{dx}\right)^2} = \sqrt{1 + y_1^2}$

i.e., $\left(\dfrac{ds}{dx}\right)^2 = 1 + \left(\dfrac{dy}{dx}\right)^2$

Note:

(i) $\dfrac{ds}{dy} = \sqrt{1 + \left(\dfrac{dx}{dy}\right)^2}$

(ii) $\sin\theta = \dfrac{\Delta y}{\text{chord PQ}} = \dfrac{\Delta y}{\text{arcPQ}} \cdot \dfrac{\text{arc PQ}}{\text{chord PQ}}$

When $\Delta x \to 0$, the length of the arc PQ tends to zero. i.e., $\Delta s \to 0$ and as the chord PQ tends to the tangent at P, angle θ becomes angle χ. Hence $\sin\theta$ becomes $\sin\chi$.

$$\therefore \sin\chi = \underset{\Delta s \to 0}{Lt} \dfrac{\Delta y}{\Delta s} \cdot 1 \quad (\text{since arc PQ} = \Delta s)$$

$$= \dfrac{dy}{ds}$$

Similarly, $\cos\chi = \dfrac{dx}{ds}$

(iii) $\sin\chi = \dfrac{dy}{ds} = \dfrac{dy/dx}{ds/dx} = \dfrac{y_1}{\sqrt{1 + y_1^2}}$ Or $\sin\chi = \dfrac{\tan\chi}{\sec\chi} = \dfrac{\tan\chi}{\sqrt{1 + \tan^2\chi}} = \dfrac{y_1}{\sqrt{1 + y_1^2}}$

$\cos\chi = \dfrac{dx}{ds} = \dfrac{1}{ds/dx} = \dfrac{1}{\sqrt{1 + y_1^2}}$ Or $\cos\chi = \dfrac{1}{\sec\chi} = \dfrac{1}{\sqrt{1 + \tan^2\chi}} = \dfrac{1}{\sqrt{1 + y_1^2}}$

3.0.3 POLAR COORDINATES

Let O be a fixed point in the plane and OA a fixed line through O. O is called the **pole** and OA is called the **initial line**. Let P be any point in the plane. Let r be the distance of P from O and θ be the angle that OP makes with OA, measured in the anticlockwise direction. The pair (r, θ) is called the **polar coordinates** of P. r is called the length of the **radius vector** OP and θ is called the **vectorial angle** of P.

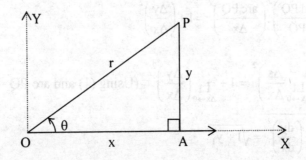

Fig.3.3

Note:

Generally we take the initial line OA as the x axis. If the cartesian coordinates of P are x and y, then $x = r \cos \theta$ and $y = r \sin \theta$. For conversion from Cartesian coordinates to polar coordinates substitute $x = r \cos \theta$ and $y = r \sin \theta$. Conversely for conversion from polar coordinates to Cartesian coordinates put $r^2 = x^2 + y^2$ and $\tan \theta = y/x$.

3.0.4 ANGLE BETWEEN RADIUS VECTOR AND TANGENT

Let $P(r, \theta)$ and $Q(r + \Delta r, \theta + \Delta \theta)$ be neighbouring points on a curve. If PN is drawn perpendicular to OQ, then we have from triangle OPN,

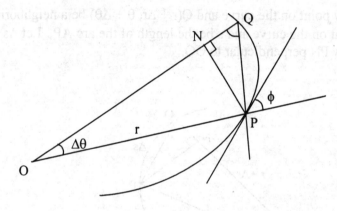

Fig. 3.4

$$PN = OP \sin \Delta\theta$$
$$= r \sin \Delta\theta$$
$$ON = OP \cos \Delta\theta$$
$$= r \cos \Delta\theta$$
$$QN = OQ - ON$$
$$= r + \Delta r - r \cos \Delta\theta$$
$$= \Delta r + r (1 - \cos \Delta\theta)$$
$$= \Delta r + 2r.\sin^2 (\Delta\theta / 2)$$

Let ϕ be the angle between the radius vector OP and the tangent at P. While $\Delta\theta \to 0$, $Q \to P$ and the chord QP becomes the tangent at P. Also the angle between the OQ and chord PQ will tend to ϕ.

Now $\tan P\hat{Q}N = \dfrac{PN}{QN} = \dfrac{r \sin \Delta\theta}{\Delta r + 2r.\sin^2 \dfrac{\Delta\theta}{2}}$

$$\frac{r.\dfrac{\sin \Delta\theta}{\Delta\theta}}{\dfrac{\Delta r}{\Delta\theta} + r.\dfrac{\sin \dfrac{\Delta\theta}{2}}{\dfrac{\Delta\theta}{2}}.\sin \dfrac{\Delta\theta}{2}} =$$

When $\Delta\theta \to 0$, we get, $\tan \varphi = \dfrac{r.1}{\dfrac{dr}{d\theta} + r.1.0} = r.\dfrac{d\theta}{dr}$

3.0.5   LENGTH OF AN ARC IN POLAR COORDINATES

Let $P(r, \theta)$ be any point on the curve and $Q(r + \Delta r, \theta + \Delta\theta)$ be a neighboring point. Let A be any fixed point on the curve and s be the length of the arc AP. Let Δs be the length of the arc PQ. Draw PN perpendicular to OQ.

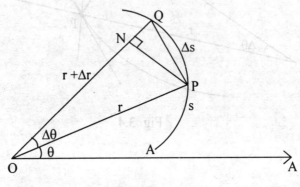

Fig 3.5

Then $PN = r \sin \Delta\theta$

$ON = r \cos \Delta\theta$

$QN = OQ - ON = r + \Delta r - r \cos \Delta\theta = \Delta r + r (1 - \cos \Delta\theta)$

$= \Delta r + 2r \sin^2 (\Delta\theta/2)$

Now, $(\text{chord } PQ)^2 = PN^2 + QN^2$

$= (r \sin \Delta\theta)^2 + (\Delta r + 2r \sin^2(\Delta\theta/2))^2$

i.e., $\left(\dfrac{\text{chord } PQ}{\Delta\theta}\right)^2 = \left(r.\dfrac{\sin \Delta\theta}{\Delta\theta}\right)^2 + \left(\dfrac{\Delta r}{\Delta\theta} + r.\dfrac{\sin \dfrac{\Delta\theta}{2}}{\dfrac{\Delta\theta}{2}}.\sin \dfrac{\Delta\theta}{2}\right)^2$ (1)

$$\frac{\text{chord PQ}}{\Delta\theta} = \frac{\text{chord PQ}}{\text{arc PQ}} \cdot \frac{\text{arc PQ}}{\Delta\theta}$$

$$= \frac{\text{chord PQ}}{\text{arc PQ}} \cdot \frac{\Delta s}{\Delta\theta}$$

$$\underset{\Delta\theta\to 0}{\text{Lt}} \frac{\text{chord PQ}}{\Delta\theta} = 1 \cdot \frac{ds}{d\theta}$$

∴ When $\Delta\theta \to 0$, from (1) we get, $\left(\dfrac{ds}{d\theta}\right)^2 = r^2 + \left(\dfrac{dr}{d\theta}\right)^2$

Note:

(i) $\left(\dfrac{ds}{dr}\right)^2 = \left(r \cdot \dfrac{d\theta}{dr}\right)^2 + 1$

(ii) $\sin P\hat{Q}N = \dfrac{PN}{\text{chord PQ}} = \dfrac{r\sin\Delta\theta}{\text{chord PQ}}$

$$= \frac{r \cdot \sin\Delta\theta}{\text{arc PQ}} \cdot \frac{\text{arc PQ}}{\text{chord PQ}}$$

$$= \frac{r \cdot \sin\Delta\theta}{\Delta\theta} \cdot \frac{\Delta\theta}{\Delta s} \cdot \frac{\text{arc PQ}}{\text{chord PQ}}$$

When $\Delta\theta \to 0$, $P\hat{Q}N \to \phi$, the angle between OP and the tangent at P

∴ $\sin\phi = r \cdot 1 \cdot \dfrac{d\theta}{ds} \cdot 1 = r \cdot \dfrac{d\theta}{ds}$

Similarly, we can derive $\cos\phi = \dfrac{dr}{ds}$

3.1 CURVATURE AND RADIUS OF CURVATURE

Definition 3.1.1 Let P be any point on a given curve $y = f(x)$. Let Q be a neighbouring point on the curve (Fig. 3.6). Let A be a fixed point on the curve and s be the length of the arc AP. Let Δs be the length of arc PQ. Let the tangents to the curve at P and Q make angles χ and $\chi+\Delta\chi$ with the x-axis. Then the angle between these tangents is $\Delta\chi$. Thus for a change of Δs in the arc length of the curve, the direction of the tangent to the curve changes by $\Delta\chi$. Hence $\dfrac{\Delta\chi}{\Delta s}$ is the average rate of bending of the curve on the arc PQ. $\dfrac{\Delta\chi}{\Delta s}$ is called the **average curvature** of the arc PQ. The **curvature of the curve** at the point P on the curve is defined as $\underset{\Delta s\to 0}{\text{Lt}} \dfrac{\Delta\chi}{\Delta s} = \dfrac{d\chi}{ds}$,

i.e., the curvature of the curve at the point P is the rate of bending of the curve with respect to the arc length at P.

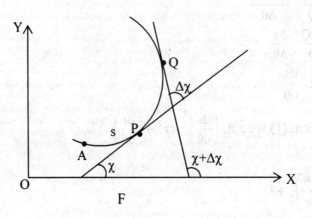

Fig.3.6

3.1.1   CURVATURE OF A CIRCLE

The curvature of a circle at any point on it will be a constant, and it is equal to the reciprocal of its radius.

Consider a circle with centre at C and radius r (Fig. 3.7). Let A be any point on the circle. Draw the tangent AT to the circle at A. Let P be any other point on the circle and the tangent to the circle at P meets AT at M. Let χ be the angle between the tangents at P and A.

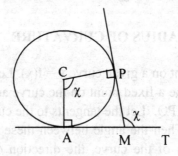

Fig.3.7

Then $P\hat{M}T = A\hat{C}P = \chi$

If arc AP = s, we have $s = r\chi$.

Differentiating with respect to χ we get, $\dfrac{ds}{d\chi} = r$. $\therefore \dfrac{d\chi}{ds} = \dfrac{1}{r}$

i.e., curvature of the circle at any point on it is the reciprocal of its radius.

Definition 3.1.2 The **radius of curvature** of a curve at any point on it is defined as the reciprocal of the curvature of the curve at that point and is denoted by ρ. i.e., $\rho = \dfrac{ds}{d\chi}$

Note:

Curvature at a point P on a given curve is the reciprocal of the radius of curvature at P.

3.1.2 FORMULA FOR RADIUS OF CURVATURE

Let $y = f(x)$ be a given curve and $P(x, y)$ be any point on it. If χ is the angle made by the tangent at P with the positive direction of x-axis, then we have

$$\tan \chi = \frac{dy}{dx} \qquad\qquad (1)$$

$$\sin \chi = \frac{dy}{ds} \qquad\qquad (2)$$

$$\cos \chi = \frac{dx}{ds} \qquad\qquad (3)$$

Differentiating (1) w. r. t. x we get,

$$\sec^2\chi \, \frac{d\chi}{dx} = \frac{d^2y}{dx^2}$$

i.e., $\sec^2\chi \, \dfrac{d\chi}{ds} \cdot \dfrac{ds}{dx} = \dfrac{d^2y}{dx^2}$

$$\therefore \quad \frac{ds}{d\chi} = \frac{\sec^3\chi}{\dfrac{d^2y}{dx^2}} = \frac{(1+\tan^2\chi)^{3/2}}{\dfrac{d^2y}{dx^2}} \qquad \text{(Using (3))}$$

i.e., $\rho = \dfrac{\left[1+\left(\dfrac{dy}{dx}\right)^2\right]^{3/2}}{\dfrac{d^2y}{dx^2}} = \dfrac{(1+y_1^2)^{3/2}}{y_2}$ $\qquad \left(\text{Since } \rho = \dfrac{ds}{d\chi}\right)$

Note:

(i)   The radius of curvature is same even if the x and y axes are interchanged.

$$\therefore \; \rho \text{ is also given by the formula } \rho = \frac{\left[1+\left(\dfrac{dx}{dy}\right)^2\right]^{3/2}}{\dfrac{d^2x}{dy^2}}$$

(ii)   When the curve is given by its parametric equations $x = f(t)$, $y = g(t)$; we have

$$\dot{x} = \frac{dx}{dt} = f'(t) \quad \text{and} \quad \dot{y} = \frac{dy}{dt} = g'(t);$$

$$\frac{dy}{dx} = \frac{dy/dt}{dx/dt} = \frac{\dot{y}}{\dot{x}}$$

$$\therefore \frac{d^2y}{dx^2} = \frac{\dot{x}.\ddot{y} - \dot{y}.\ddot{x}}{(\dot{x})^2} \cdot \frac{dt}{dx}$$

$$= \frac{\dot{x}.\ddot{y} - \dot{y}.\ddot{x}}{(\dot{x})^2} \cdot \frac{1}{\dot{x}} = \frac{\dot{x}.\ddot{y} - \dot{y}.\ddot{x}}{(\dot{x})^3}$$

$$\therefore \rho = \frac{\left[1 + \left(\dfrac{dy}{dx}\right)^2\right]^{3/2}}{\dfrac{d^2y}{dx^2}} = \frac{\left(1 + \left(\dfrac{\dot{y}}{\dot{x}}\right)^2\right)^{3/2}}{\dfrac{\dot{x}.\ddot{y} - \dot{y}.\ddot{x}}{(\dot{x})^3}} = \frac{(\dot{x}^2 + \dot{y}^2)^{3/2}}{\dot{x}.\ddot{y} - \dot{y}.\ddot{x}}$$

3.1.3 RADIUS OF CURVATURE IN POLAR COORDINATES

Method 1.

Let $r = f(\theta)$ be the equation of the curve in polar coordinates.

We have $\rho = \dfrac{(1 + y_1^2)^{3/2}}{y_2}$ in Cartesian coordinates.

To convert this into polar coordinates, the transformation is $x = r\cos\theta$, $y = r\sin\theta$

$$y_1 = \frac{dy}{dx} = \frac{dy/d\theta}{dx/d\theta} = \frac{r.\cos\theta + r_1.\sin\theta}{-r.\sin\theta + r_1\cos\theta} \qquad \text{where } r_1 = \frac{dr}{d\theta}$$

$$y_2 = \frac{d^2y}{dx^2} = \frac{d}{d\theta}\cdot\left(\frac{dy}{dx}\right)\cdot\frac{d\theta}{dx}$$

$$= \left(\frac{\dfrac{(-r\sin\theta + r_1\cos\theta)[-r.\sin\theta + r_1\cos\theta + r_2\sin\theta + r_1\cos\theta]}{(-r\sin\theta + r_1\cos\theta)^2}}{} \\ -\dfrac{(r\cos\theta + r_1\sin\theta)[-r_1\sin\theta - r\cos\theta + r_2\cos\theta - r_1\sin\theta]}{(-r\sin\theta + r_1\cos\theta)^2}\right) \cdot$$

$$\frac{1}{(-r\sin\theta + r_1\cos\theta)}$$

$$\text{where } r_2 = \frac{d^2 r}{d\theta^2}$$

$$= \frac{r^2 - r.r_2 + 2.r_1^2}{(-r\sin\theta + r_1\cos\theta)^3}$$

$$1 + y_1^2 = 1 + \left(\frac{r\cos\theta + r_1\sin\theta}{-r\sin\theta + r_1\cos\theta}\right)^2 = \frac{r^2 + r_1^2}{(-r\sin\theta + r\cos\theta)^2}$$

$$\therefore \rho = \frac{\left(1 + y_1^2\right)^{3/2}}{y_2} = \frac{(r^2 + r_1^2)^{3/2}}{r^2 - r.r_2 + 2r_1^2}$$

Method 2.

Let r = f(θ) be the equation of the curve. Let P(r, θ) be any point on the curve. If ϕ is the angle between the radius vector OP and the tangent at P and χ is the angle made by the tangent at P with the initial line OA, we have $\chi = \theta + \phi$.

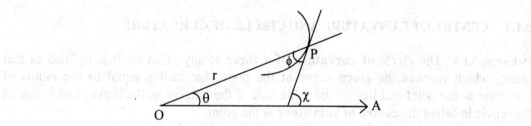

Fig.3.8

$$\therefore \frac{d\chi}{d\theta} = 1 + \frac{d\phi}{d\theta} \tag{1}$$

We have $\tan\phi = r.\dfrac{d\theta}{dr} = \dfrac{r}{(dr/d\theta)}$ \hfill (2)

Differentiating (2) w.r.t. θ, we get,

$$\sec^2\phi.\frac{d\phi}{d\theta} = \frac{\dfrac{dr}{d\theta}.\dfrac{dr}{d\theta} - r.\dfrac{d^2 r}{d\theta^2}}{\left(\dfrac{dr}{d\theta}\right)^2} = \frac{r_1^2 - r.r_2}{r_1^2}$$

$$\frac{d\phi}{d\theta} = \frac{1}{1 + \tan^2\phi}\left(\frac{r_1^2 - r.r_2}{r_1^2}\right) = \frac{1}{1 + \left(\dfrac{r}{r_1}\right)^2}\left(\frac{r_1^2 - r.r_2}{r_1^2}\right)$$

$$= \frac{r_1^2 - r.r_2}{r^2 + r_1^2}$$

$\therefore$ From (1) $\dfrac{d\chi}{d\theta} = 1 + \dfrac{d\varphi}{d\theta} = \dfrac{r^2 + 2r_1^2 - r.r_2}{r^2 + r_1^2}$ (3)

Also in polar coordinates $\left(\dfrac{ds}{d\theta}\right)^2 = r^2 + \left(\dfrac{dr}{d\theta}\right)^2 = r^2 + r_1^2$

$$\rho = \frac{ds}{d\chi} = \frac{ds}{d\theta} \cdot \frac{d\theta}{d\chi}$$

$$= \left(r^2 + r_1^2\right)^{1/2} \frac{\left(r^2 + r_1^2\right)}{r^2 + 2r_1^2 - r.r_2} \quad \text{by (3)}$$

i.e., $\rho = \dfrac{\left(r^2 + r_1^2\right)^{3/2}}{r^2 + 2r_1^2 - r.r_2}$

3.1.4   CENTRE OF CURVATURE AND CIRCLE OF CURVATURE

Definition 3.1.3   The **circle of curvature** of a curve at any point on it is defined as that circle, which touches the given curve at the point, has radius equal to the radius of curvature at the point and lies on the same side of the tangent as the curve. The centre of this circle is called the **centre of curvature** at the point.

Let P(x, y) be any point on the curve y = f(x). Draw PT, the tangent to the curve at P and PN, the inward normal to the curve at P. On PN, cut off a length PC = radius of

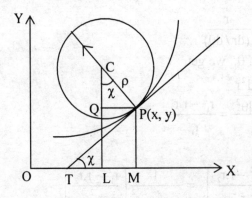

Fig.3.9

curvature of the curve at P. (i.e., PC = ρ). Then C is the **centre of curvature** of the curve at P. The circle with centre at C and radius ρ touches the given curve at P. This circle is the circle of curvature at P.

Draw PM, CL perpendicular to x-axis and PQ perpendicular to CL. Let the tangent at P meets the x-axis at T and makes an angle χ with the positive direction of x-axis. Then $P\hat{C}Q = \chi$.

Let $(\bar{x}, \bar{y})$ be the coordinates of C.

Then $\bar{x} = OL$

$$= OM - QP$$

$$= x - \rho \sin \chi$$

$$= x - \frac{\rho}{\sqrt{1 + \cot^2\chi}}$$

$$= x - \frac{\rho \tan \chi}{\sqrt{1 + \tan^2 \chi}}$$

$$= x - \frac{\rho . y_1}{\sqrt{1 + y_1^2}} \quad \text{where } y_1 = \tan \chi = \frac{dy}{dx}$$

$$= x - \frac{\left[1 + y_1^2\right]^{3/2}}{y_2} . \frac{y_1}{\sqrt{1 + y_1^2}} \qquad \text{since} \quad \rho = \frac{\left[1 + y_1^2\right]^{3/2}}{y_2}$$

i.e., $\bar{x} = x - \dfrac{y_1(1 + y_1^2)}{y_2}$

And $\bar{y} = CL = PM + CQ$

$$= y + \rho \cos \chi$$

$$= y + \frac{\rho}{\sqrt{1 + \tan^2\chi}}$$

$$= y + \frac{\left[1 + y_1^2\right]^{3/2}}{y_2\sqrt{1 + y_1^2}}$$

$$= y + \frac{(1 + y_1^2)}{y_2}$$

The circle of curvature has centre $C(\bar{x}, \bar{y})$ and radius ρ. Hence the equation of the circle of curvature is

$$(x - \bar{x})^2 + (y - \bar{y})^2 = \rho^2.$$

Example 3.1.1 Find the radius of curvature of the curve y = sin x at the point x = $\pi/2$.

Solution:

Given $y = \sin x$

$\therefore \qquad y_1 = \cos x$ and $y_2 = -\sin x$.

$\therefore \qquad$ At $x = \pi/2$, $y_1 = 0$ and $y_2 = -1$

$\therefore \qquad \rho_{(x=\pi/2)} = \dfrac{(1+y_1^2)^{3/2}}{y_2} = \dfrac{(1+0)^{3/2}}{-1} = -1$

$\therefore$ The magnitude of $\rho = 1$.

Example 3.1.2 Find the radius of curvature and curvature of the curve $\sqrt{x} + \sqrt{y} = 1$ at $(\frac{1}{4}, \frac{1}{4})$.

Solution:

Given $\sqrt{x} + \sqrt{y} = 1$

Differentiating w.r.t x, we get

$$\frac{1}{2\sqrt{x}} + \frac{1}{2\sqrt{y}}.y_1 = 0$$

$$y_1 = -\sqrt{y}\Big/\sqrt{x}$$

$$y_{1\,(1/4,1/4)} = -\sqrt{1/4}\Big/\sqrt{1/4} = -1$$

Now, $\quad y_1 = -\sqrt{y}\Big/\sqrt{x}$

Differentiating w.r.t x

$$y_2 = \frac{-\sqrt{x}.\dfrac{1}{2\sqrt{y}}y_1 + \sqrt{y}.\dfrac{1}{2\sqrt{x}}}{x}$$

$$y_{2(1/4,1/4)} = \frac{-\sqrt{1/4}.\dfrac{1}{2\sqrt{1/4}}.(-1) + \sqrt{1/4}.\dfrac{1}{2\sqrt{1/4}}}{1/4}$$

$$= \frac{1/2 + 1/2}{1/4} = 4$$

$$\rho = \frac{(1+y_1^2)^{3/2}}{y_2} = \frac{(1+1)^{3/2}}{4} = \frac{2.\sqrt{2}}{4} = \frac{1}{\sqrt{2}}$$

i.e., radius of curvature $= \dfrac{1}{\sqrt{2}}$

$\therefore$ Curvature $= \dfrac{1}{\rho} = \sqrt{2}$

Example 3.1.3 Find the radius of curvature of the curve $xy^2 = a^3 - x^3$ at $(a, 0)$.

Solution:

Given $xy^2 = a^3 - x^3$ (1)

Differentiating (1) w.r.t y

$$x.2y + \frac{dx}{dy}.y^2 = -3x^2.\frac{dx}{dy}$$

$$\therefore \quad \frac{dx}{dy} = \frac{-2xy}{y^2 + 3x^2} \qquad\qquad\qquad (2)$$

$$\frac{dx}{dy}_{(a,0)} = 0$$

From (2) $\quad \dfrac{d^2x}{dy^2} = \dfrac{(y^2 + 3x^2)(-2x - 2y\frac{dx}{dy} + 2xy(2y + 6x\frac{dx}{dy})}{(y^2 + 3x^2)^2}$

$$\therefore \quad \frac{d^2x}{dy^2}_{(a,0)} = \frac{3a^2(-2a)}{9a^4} = \frac{-2}{3a}$$

$$\therefore \quad \rho = \frac{[1+0]^{3/2}}{\frac{-2}{3a}} = \frac{-3a}{2}$$

$$\therefore \text{Magnitude of } \rho = \frac{3a}{2}$$

Example 3.1.4 Prove that the radius of curvature at any point on the common catenary $y = c \cosh x/c$ is y^2/c.

Solution:

Given $y = c \cosh x/c$

Differentiating w.r.t. x

$$y_1 = c.\sinh(x/c).(1/c)$$

$$y_1 = \sinh x/c \qquad\qquad\qquad (1)$$

Differentiating again w.r.t. x

$$y_2 = (1/c)\cosh x/c.$$

$$\therefore \quad \rho = \frac{\left(1 + y_1^2\right)^{3/2}}{y_2} = \frac{(1 + \sinh^2 x/c)^{3/2}}{\frac{1}{c}\cosh x/c}$$

$$= \frac{c.(\cosh^2 \frac{x}{c})^{3/2}}{\cosh \frac{x}{c}}$$

$$= c.(\cosh^2 \frac{x}{c})$$

$$= c. \, y^2/c^2 \quad = y^2/c.$$

Example 3.1.5 Find the radius of curvature at $(a\cos^3\theta, \, a\sin^3\theta)$ on $x^{2/3} + y^{2/3} = a^{2/3}$

Solution:
Given $x = a \cos^3\theta$

$\therefore \quad \dfrac{dx}{d\theta} = a.3\cos^2\theta \, (- \sin \theta) = -3a \sin \theta \cos^2\theta$

$y = a \sin^3\theta$

$\therefore \quad \dfrac{dy}{d\theta} = a.3.\sin^2\theta.\cos\theta = 3a \sin^2\theta\cos\theta$

$\therefore \quad y_1 = \dfrac{dy}{dx} = \dfrac{dy/d\theta}{dx/d\theta} = \dfrac{3a \sin^2\theta\cos\theta}{-3a \sin\theta \cos^2\theta} = -\tan \theta$

$$y_2 = \frac{d^2y}{dx^2} = -\sec^2\theta.\frac{d\theta}{dx} = -\sec^2\theta.\frac{1}{-3a \sin\theta \cos^2\theta}$$

$$= \frac{1}{3a\sin\theta \cos^4\theta}$$

$\therefore \quad \rho = \dfrac{(1+y_1^2)^{3/2}}{y_2} = (1+\tan^2\theta)^{3/2}.3a\sin\theta \cos^4\theta$

$$= (\sec^2\theta)^{3/2}. \, 3a \sin \theta \cos^4\theta = 3a \sin \theta \cos \theta$$

Example 3.1.6 Find ρ at $t = 1$ on the curve $x = ct, \, y = c/t$

Solution:
Given $x = ct$
Differentiating w.r.t. t
$\qquad \dot{x} = c \quad \therefore \text{At } t = 1, \; \dot{x} = c, \; \ddot{x} = 0$

Similarly, $\; y = c/t \;$ gives $\dot{y} = -\dfrac{c}{t^2}$

$$\ddot{y} = \frac{2c}{t^3}$$

$\therefore \quad$ At $t = 1, \; \dot{y} = -c, \; \ddot{y} = 2c$

Now $\; \rho = \dfrac{(\dot{x}^2 + \dot{y}^2)^{3/2}}{\dot{x}\,\ddot{y} - \dot{y}\,\ddot{x}}$

$$= \frac{(c^2 + c^2)^{3/2}}{c.2c + c.0} = \frac{(2c^2)^{3/2}}{2c^2} = \sqrt{2}.c$$

Example 3.1.7 For the ellipse $\dfrac{x^2}{a^2} + \dfrac{y^2}{b^2} = 1$, show that the radius of curvature at an end of the major axis is equal to the semi-latus rectum.

Solution: Differentiating $\dfrac{x^2}{a^2} + \dfrac{y^2}{b^2} = 1$ w.r.t. x, we get,

$$\frac{2x}{a^2} + \frac{2y}{b^2}.y_1 = 0$$

$$y_1 = \frac{-b^2 x}{a^2 y}$$

$$y_2 = \frac{a^2 y.(-b^2) + b^2 x.a^2 y_1}{a^4 y^2} = \frac{-a^2.b^2.y + b^2 x.a^2\left(\dfrac{-b^2 x}{a^2 y}\right)}{a^4 y^2} = \frac{-b^2(a^2 y^2 + b^2 x^2)}{a^4 y^3}$$

$$\therefore \rho = \frac{(1+y_1^2)^{3/2}}{y_2} = \frac{\left(1 + \dfrac{b^4 x^2}{a^4 y^2}\right)^{3/2}.a^4 y^3}{-b^2(a^2 y^2 + b^2 x^2)} = \frac{-\left(a^4 y^2 + b^4 x^2\right)^{3/2}}{a^2 b^2(a^2 y^2 + b^2 x^2)}$$

At the end of the major axis $x = \pm a$, $y = 0$

$$\therefore \rho = \frac{-(b^4 a^2)^{3/2}}{a^2 b^2 b^2 a^2} = \frac{-b^2}{a}$$

Magnitude of $\rho = \dfrac{b^2}{a}$

= semi-latus rectum of the ellipse.

Example 3.1.8 Prove that the radius of curvature of the curve $r = a(1 + \cos\theta)$ at any point on it is $\dfrac{4a}{3}\cos\theta/2$

Solution:

Given $r = a(1 + \cos\theta)$

$\therefore$ $r_1 = a(-\sin\theta)$

$r_2 = -a\cos\theta$

$$\rho = \frac{\left(r^2 + r_1^2\right)^{3/2}}{r^2 + 2r_1^2 - r.r_2}$$

$$= \frac{\left[a^2(1+\cos\theta)^2 + a^2\sin^2\theta\right]^{3/2}}{a^2(1+\cos\theta)^2 + a(1+\cos\theta).a\cos\theta + 2.a^2\sin^2\theta}$$

$$= \frac{\left(2a^2 + 2a^2\cos\theta\right)^{3/2}}{a^2 + 2a^2(\cos^2\theta + \sin^2\theta) + 2a^2\cos\theta + a^2\cos\theta}$$

$$= \frac{a^3\left[2(1+\cos\theta)\right]^{3/2}}{3a^2(1+\cos\theta)} = \frac{a.2.\sqrt{2}}{3}.(1+\cos\theta)^{1/2}$$

$$= \frac{2.\sqrt{2}.a}{3}.\sqrt{2}.\cos\theta/2 \qquad (\text{ Since } 1+\cos\theta = 2\cos^2\theta/2)$$

$$= \frac{4a}{3}\cos\theta/2$$

Example 3.1.9 Find the radius of curvature of the $r^n = a^n \cos n\theta$ at any point 'θ'

Solution:

Given $r^n = a^n \cos n\theta$

Taking log on both sides, we get

$n.\log r = n.\log a + \log\cos n\theta$

Differentiating w.r.t θ, we get,

$$\frac{n}{r}.\frac{dr}{d\theta} = \frac{-n.\sin n\theta}{\cos n\theta}$$

$$\therefore \qquad r_1 = \frac{dr}{d\theta} = -r\tan n\theta$$

$$r_2 = \frac{d^2r}{d\theta^2} = -r_1\tan n\theta - r.n.\sec^2 n\theta$$

$$= -(-r\tan n\theta).\tan n\theta - r.n.\sec^2 n\theta$$

$$= r.\tan^2 n\theta - nr.\sec^2 n\theta$$

$$\therefore \rho = \frac{(r^2 + r_1^2)^{3/2}}{r^2 + 2.r_1^2 - r.r_2}$$

$$= \frac{(r^2 + r^2\tan^2 n\theta)^{3/2}}{r^2 + 2.r^2\tan^2 n\theta - r(r\tan^2 n\theta - nr\sec^2 n\theta)}$$

$$= \frac{r^3\sec^3 n\theta}{r^2[1 + 2.\tan^2 n\theta - \tan^2 n\theta + n\sec^2 n\theta)}$$

$$= \frac{r\sec^3 n\theta}{(1+n).\sec^2 n\theta}$$

$$= \frac{r.\sec n\theta}{1+n} = \frac{r}{(1+n)\cos n\theta} = \frac{r.a^n}{(n+1).r^n} = \frac{a^n}{(n+1).r^{n-1}}$$

Example 3.1.10 Find the centre of curvature and circle of curvature of the curve $x = a(\cos t + t \sin t)$, $y = a(\sin t - t \cos t)$ at any point 't'.

Solution:

$$\frac{dx}{dt} = a(-\sin t + \sin t + t \cos t) = at \cos t$$

$$\frac{dy}{dt} = a(\cos t + t \sin t - \cos t) = at \sin t$$

$$y_1 = \frac{dy}{dx} = \frac{at \sin t}{at \cos t} = \tan t$$

$$y_2 = \frac{d^2 y}{dx^2} = \sec^2 t \cdot \frac{dt}{dx} = \sec^2 t \cdot \frac{1}{at \cos t} = \frac{1}{at \cos^3 t}$$

$$\rho = \frac{(1 + y_1^2)^{3/2}}{y_2}$$

$$= (1 + \tan^2 t)^{3/2} \cdot at \cos^3 t$$

$$= (\sec^2 t)^{3/2} \cdot at \cos^3 t$$

$$= \sec^3 t \cdot at \cdot \cos^3 t$$

$$= at$$

If $(\bar{x}, \bar{y})$ is the centre of curvature

$$\bar{x} = x - \frac{y_1(1 + y_1^2)}{y_2}$$

$$= a(\cos t + t \sin t) - \tan t(1 + \tan^2 t) at \cos^3 t$$

$$= a \cos t + at \sin t - at \tan t \cdot \sec^2 t \cos^3 t$$

$$= a \cos t + at \sin t - at \sin t$$

$$= a \cos t$$

$$\bar{y} = y + \frac{(1 + y_1^2)}{y_2}$$

$$= a(\sin t - t \cos t) + (1 + \tan^2 t)(at \cos^3 t)$$

$$= a \sin t - at \cos t + \sec^2 t \cdot at \cos^3 t$$

$$= a \sin t - at \cos t + at \cos t$$

$$= a \sin t$$

∴ The centre of curvature is $(a \cos t, a \sin t)$.

The circle of curvature is $(x - \bar{x})^2 + (y - \bar{y})^2 = \rho^2$

i.e., $(x - a \cos t)^2 + (y - a \sin t)^2 = a^2 t^2$.

Example 3.1.11 Find the centre of curvature and the circle of curvature of $y = 4x^2$ at the origin.

Solution:

We have $y = 4x^2$

$\therefore$ $y_1 = 8x$ and $y_2 = 8$

$\therefore$ At $(0, 0)$, $y_1 = 0$, $y_2 = 8$

The radius of curvature $\rho = \dfrac{(1 + y_1^2)^{3/2}}{y_2} = \dfrac{1}{8}$

If $(\bar{x}, \bar{y})$ is the centre of curvature,

$$\bar{x} = x - \frac{y_1}{y_2}(1 + y_1^2) = 0 - 0 = 0$$

$$\bar{y} = y + \frac{1 + y_1^2}{y_2} = 0 + \frac{1}{8} = \frac{1}{8}$$

$\therefore$ Centre of curvature $= (0, 1/8)$

The equation of the circle of curvature at $(0, 0)$ is

$$(x - \bar{x})^2 + (y - \bar{y})^2 = \rho^2$$

i.e., $(x - 0)^2 + (y - \tfrac{1}{8})^2 = \tfrac{1}{8}$

i.e., $x^2 + (y - \tfrac{1}{8})^2 = \tfrac{1}{8}$

EXERCISE 3.1

PART - A

1. Define curvature and radius of curvature.
2. What is the relation between curvature and radius of curvature?
3. What is the radius of curvature of the circle $x^2 + y^2 = r^2$.
4. Prove that the radius of curvature of a circle is its radius.
5. Prove that the curvature of a circle is the reciprocal of its radius.
6. Write down the formula for radius of curvature in polar coordinates.
7. Find the radius of curvature of $x^4 + y^4 = 2$ at $(1, 1)$.
8. Find the radius of curvature of $y = e^x$ at the point where it crosses the y axis.
9. Find the radius of curvature of $xy = 30$ at $(3, 10)$
10. Find the radius of curvature of $\sqrt{x} + \sqrt{y} = \sqrt{a}$ at $(a/4, a/4)$
11. Find the radius of curvature of $xy^3 = a^4$ at (a, a).
12. Find the radius of curvature of $y = c \log \sec(x/c)$ at any point on the curve.
13. Find the radius of curvature of $y = \sin x - \sin 2x$ at $x = \pi/2$.

14. Find the curvature of $y = \log \sin x$ at any point on the curve.

15. Find the curvature of $y = e^{\sqrt{3}x}$ at $x = 0$.

16. Find the radius of curvature of $r = a\theta$ at the pole.

17. Find the curvature of $r\theta = a$ at any point on it.

18. Find the curvature of $r = e^{\theta}$ at any point on it.

19. For $r = a \cos\theta$, prove $\rho = a/2$

20. For $\sqrt{r} = \sqrt{a}\cos\theta/2$, prove $\rho = \dfrac{2}{3}\sqrt{ar}$

PART – B

Find the curvature of the following curves.

21. $y = c \cos h(x/c)$

22. $xy = c^2$ at (c, c)

23. $y = x^2(x - 3)$ at the points where the tangent is parallel to the x-axis.

24. $x = e^t \cos t$, $y = e^t \sin t$ at $(1, 0)$

25. $r^2 = a^2 \cos 2\theta$.

Find the radius of curvature of the following curves.

26. $y = \dfrac{\log x}{x}$ at $x = 1$

27. $x^3 - 2x^2y + 3xy^2 - 4y^3 + 5x^2 - 6xy + 7y^2 - 8y = 0$ at $(0, 0)$.

28. $x^3 + y^3 = 3axy$ at $(3a/2, 3a/2)$

29. $xy^2 = a^3 - x^3$ at $(a, 0)$.

30. $\sqrt{\dfrac{x}{a}} + \sqrt{\dfrac{y}{b}} = 1$ at any point on the curve.

31. $x = a(\theta + \sin\theta)$, $y = a(1 - \cos\theta)$

32. $x = ae^{\theta}(\sin\theta - \cos\theta)$, $y = ae^{\theta}(\sin\theta + \cos\theta)$

33. $x = a[\log \cot \theta/2 - \cos\theta]$, $y = a \sin\theta$.

34. $x = 3a \cos\theta - a \cos 3\theta$, $y = 3a \sin\theta - a \sin 3\theta$

35. $x = a \log \sec\theta$, $y = a(\tan\theta - \theta)$ at the point 'θ'.

36. $x = at^2$, $y = 2at$ at the point t.

37. $x = 6t^2 - 3t^4$, $y = 8t^3$ at the point t.

38. $x = a \log(\sec\theta + \tan\theta)$, $y = \sec\theta$.

39. $x = a(\cos t + t \sin t)$, $y = a(\sin t - t \cos t)$

40. $r = a \sin n\theta$ at the pole.

41. $r = ae^{\theta \cot\alpha}$

42. $a^2 = r^2 \cos 2\theta$.

43. Show that for the rectangular hyperbola $xy = c^2$, $\rho = \dfrac{r^3}{2c^2}$ where r is the distance of the point from the origin.

44. Prove that ρ at any point of $x^{2/3} + y^{2/3} = a^{2/3}$ is three times the length of the perpendicular from the origin to the tangent at the point.

45. For the curve $x = a\cos\theta$, $y = b\sin\theta$ show that the magnitude of ρ at the end of the major axis is the semi-latus rectum.

46. For the ellipse $\dfrac{x^2}{a^2} + \dfrac{y^2}{b^2} = 1$, show that ρ at P is given by $\dfrac{a^2 b^2}{p^3}$ where p is the length of the perpendicular from the centre upon the tangent at the point P.

47. Show that the curves $y = \dfrac{a}{2}\left(e^{x/a} + e^{-x/a}\right)$ and $y = \dfrac{a}{2}\left(2 + \dfrac{x^2}{a^2}\right)$ have the same curvature at their crossing with the y-axis.

48. Show that the radius of curvature at any point of $y = a\cos h(x/a)$ is numerically equal to the length of the normal intercepted between the curve and its directrix (i.e., the x-axis)

49. For the parabola of projection, $x = ut\cos\alpha$, $y = ut\sin\alpha - (1/2)gt^2$, show that the radius of curvature at the point of projection is $\dfrac{-u^2}{g\cos\alpha}$.

50. Prove that the radius of curvature at any point on the cardioid $r = a(1 + \cos\theta)$ varies as the square root of the radius vector. If ρ_1 and ρ_2 are the radius of curvatures at the extrimities of any chord which passes through the pole, then prove that
$9(\rho_1^2 + \rho_2^2) = 16a^2$

51. The tangents at the points P and Q on the cycloid $x = a(\theta - \sin\theta)$;
$y = a(1 - \cos\theta)$ are at right angles. If ρ_1 and ρ_2 are the radii of curvature at these points prove that $\rho_1^2 + \rho_2^2 = 16a^2$

Find the centre of curvature of the following curves.

52. $y = x^2$ at $(0, 0)$.

53. $y = x\log x$ at the point where $y' = 0$.

54. $xy = c^2$ at (c, c).

55. $y = x^3 - 6x^2 + 3x + 1$ at $(1, -1)$.

56. Find the equation of the circle of curvature of the parabola $y^2 = 12x$ at the point $(3, 6)$.

57. Find the equation of the circle of curvature of the curve $x^3 + y^3 = 3axy$ at the point $(3a/2, 3a/2)$.

58. Find the equation of the circle of curvature of the curve $\sqrt{x} + \sqrt{y} = \sqrt{a}$ at $\left(\dfrac{a}{4}, \dfrac{a}{4}\right)$.

3.2   EVOLUTES AND INVOLUTES

Definition 3.2.1 Let P be any point on a given curve C. When P moves on the curve C, the centre of curvature of C at P traces a curve C' called the **evolute** of C. Thus the **evolute** of a curve is the locus of the centre of curvature. When C' is the evolute of C, then C is called the **involute** of C'.

3.2.1   METHOD TO FIND THE EVOLUTE OF A CURVE

Let the equation of the given curve be

$$y = f(x) \qquad\qquad (1)$$

Let $(\bar{x}, \bar{y})$ be the centre of curvature corresponding to the point (x, y) on the curve (1).

Then $\quad \bar{x} = x - \dfrac{y_1}{y_2}\left(1 + y_1^2\right) \qquad\qquad (2)$

and $\quad \bar{y} = y + \dfrac{1}{y_2}\left(1 + y_1^2\right) \qquad\qquad (3)$

Eliminating x and y from the equations (1) and (2) and (3), we get a relation $g(\bar{x}, \bar{y}) = 0$ between $\bar{x}$ and $\bar{y}$. To obtain the locus of $(\bar{x}, \bar{y})$, replace $\bar{x}$ by x and $\bar{y}$ by y in the above relation $g(\bar{x}, \bar{y}) = 0$. The resulting equation $g(x, y) = 0$ is the equation of the evolute of the given curve $y = f(x)$.

Example 3.2.1 Find the equation of the evolute of the parabola $y^2 = 4ax$

Solution:

Given $y^2 = 4ax$

∴ Differentiating w.r.t. x

$$2.y.y_1 = 4a$$
$$y_1 = 2a / y$$
$$y_2 = (-2a / y^2). y_1 = (-2a / y^2). 2a / y$$
$$= -4a^2 / 4axy = -a / xy$$

If $(\bar{x}, \bar{y})$ is the center of curvature

$$\bar{x} = x - \frac{y_1}{y_2}(1 + y_1^2) = x - \left(\frac{2a\big/y}{-a/xy}\right)\left(1 + \frac{4a^2}{y^2}\right) = x + 2x\left(1 + \frac{4a^2}{4ax}\right)$$

∴ $\quad \bar{x} = 3x + 2a \qquad\qquad (1)$

$$\bar{y} = y - \frac{(1 + y_1^2)}{y_2} = y - \left(\frac{xy}{a}\right)\left(1 + \frac{4a^2}{y^2}\right) = y - \frac{xy}{a} - \frac{4ax}{y}$$

$$= y - \frac{xy}{a} - y \qquad \text{since } y^2 = 4ax$$

$$= \frac{-x}{a}.2\sqrt{ax}$$

$$\therefore \quad \overline{y} = \frac{-2}{\sqrt{a}} x^{3/2}$$

$$\therefore \quad \overline{y}^2 = \frac{4}{a} x^3$$ (2)

$$\therefore \quad \overline{y}^2 = \frac{4}{a}\left(\frac{\overline{x}-2a}{3}\right)^3 \qquad \text{(Using (1) in (2))}$$

i.e., $27a.\overline{y}^2 = 4(\overline{x}-2a)^3$

$\therefore$ Locus of $(\overline{x},\overline{y})$ is $27a.y^2 = 4(x-2a)^3$

Example 3.2.2 Find the equation of the evolute of the ellipse $\dfrac{x^2}{a^2} + \dfrac{y^2}{b^2} = 1$

Solution:

The parametric equations of the ellipse are $x = a\cos\theta$ and $y = b\sin\theta$

$$\therefore \quad \frac{dx}{d\theta} = -a\sin\theta \quad \text{and} \quad \frac{dy}{d\theta} = b\cos\theta$$

$$y_1 = \frac{dy}{dx} = \frac{-b\cos\theta}{a\sin\theta} = \frac{-b}{a}\cot\theta$$

$$y_2 = \frac{d^2y}{dx^2} = \frac{d}{d\theta}\left(\frac{-b}{a}\cot\theta\right)\frac{d\theta}{dx}$$

$$= \left(\frac{-b}{a}\right)(-\operatorname{cosec}^2\theta)\frac{1}{-a\sin\theta}$$

$$= \left(\frac{-b}{a^2}\right)(\operatorname{cosec}^3\theta)$$

Let $(\overline{x},\overline{y})$ be the coordinates of the centre of curvature.

Then $\overline{x} = x - \dfrac{y_1}{y_2}(1+y_1^2)$

$$= a\cos\theta - \frac{\left(\dfrac{-b}{a}\cot\theta\right)}{\left(\dfrac{-b}{a^2}\operatorname{cosec}^3\theta\right)}.(1+\frac{b^2}{a^2}\cot^2\theta)$$

$$= a\cos\theta - a\cot\theta.\sin^3\theta\left(1+\frac{b^2}{a^2}\cot^2\theta\right)$$

$$= a\cos\theta - a\cos\theta.\sin^2\theta\left(1+\frac{b^2}{a^2}\cot^2\theta\right)$$

$$= a\cos\theta - a\cos\theta.\sin^2\theta - \frac{b^2}{a}\cos^3\theta$$

$$= a\cos\theta\,(1-\sin^2\theta) - \frac{b^2}{a}\cos^3\theta$$

$$= a\cos^3\theta - \frac{b^2}{a^2}\cos^3\theta$$

i.e., $\bar{x} = \dfrac{a^2-b^2}{a}.\cos^3\theta$ \hfill (1)

Now $\bar{y} = y - \dfrac{(1+y_1^2)}{y_2}$

$$= b\sin\theta + \frac{1}{\left(\dfrac{-b}{a^2}.\operatorname{cosec}^3\theta\right)}\cdot\left(1+\frac{b^2}{a^2}\cot^2\theta\right)$$

$$= b\sin\theta - \frac{a^2}{b}.\sin^3\theta\left(1+\frac{b^2}{a^2}\cot^2\theta\right)$$

$$= b\sin\theta - \frac{a^2}{b}.\sin^3\theta - b\sin\theta\cos^2\theta$$

$$= b\sin\theta\,(1-\cos^2\theta) - \frac{a^2}{b}\sin^3\theta$$

$$= b\sin^3\theta - \frac{a^2}{b}\sin^3\theta$$

i.e $\bar{y} = -\left(\dfrac{a^2-b^2}{b}\right)\sin^3\theta$ \hfill (2)

Locus of $(\bar{x},\bar{y})$ is obtained by eliminating θ from (1) and (2)

From (1) $\cos\theta = \left(\dfrac{a\bar{x}}{a^2-b^2}\right)^{1/3}$

From (2) $\sin\theta = \left(\dfrac{-b\bar{y}}{a^2-b^2}\right)^{1/3}$

$$\cos^2\theta + \sin^2\theta = 1$$

i.e., $\left(\dfrac{a\bar{x}}{a^2-b^2}\right)^{2/3} + \left(\dfrac{-b\bar{y}}{a^2-b^2}\right)^{2/3} = 1$

i.e., $(a\overline{x})^{2/3} + (b\overline{y})^{2/3} = (a^2 - b^2)^{2/3}$

∴ The equation of the evolute of the ellipse is $(ax)^{2/3} + (by)^{2/3} = (a^2 - b^2)^{2/3}$

Example 3.2.3 Show that the evolute of the cycloid $x = a(\theta - \sin\theta)$, $y = a(1 - \cos\theta)$ is again a cycloid.

Solution:

Given $x = a(\theta - \sin\theta)$ and $y = a(1 - \cos\theta)$

Differentiating w.r.t θ

∴ $\dfrac{dx}{d\theta} = a(1 - \cos\theta)$ and $\dfrac{dy}{d\theta} = a\sin\theta$

   $y_1 = \dfrac{dy}{dx} = \dfrac{a\sin\theta}{a(1 - \cos\theta)} = \dfrac{2.\sin\theta/2.\cos\theta/2}{2\sin^2\theta} = \cot\theta/2$

i.e., $y_1 = \cot\theta/2$ \hfill (1)

   $y_2 = \dfrac{d^2y}{dx^2} = \dfrac{d}{d\theta}(\cot\theta/2).\dfrac{d\theta}{dx}$

   $= \dfrac{-1}{2a}\operatorname{cosec}^2(\theta/2).\dfrac{1}{a(1 - \cos\theta)}$

   $= \dfrac{-1}{2a}\operatorname{cosec}^2(\theta/2).\dfrac{1}{2\sin^2\theta/2}$

   $= \dfrac{-1}{4a}\operatorname{cosec}^4(\theta/2)$ \hfill (2)

If $(\overline{x}, \overline{y})$ is the centre of curvature,

   $\overline{x} = x - \dfrac{y_1}{y_2}(1 + y_1^2)$

   $= a(\theta - \sin\theta) - \dfrac{\cot\theta/2}{\left(-\dfrac{1}{4a}\right)\operatorname{cosec}^4\theta/2}\left(1 + \cot^2\theta/2\right)$

   $= a(\theta - \sin\theta) + \dfrac{4a.\cot\theta/2}{\operatorname{cosec}^4\theta/2}.\cos ec^2\theta/2.$

   $= a(\theta - \sin\theta) + 4a.\sin\theta/2.\cos\theta/2.$

   $= a(\theta - \sin\theta) + 2a\sin\theta$

i.e., $\overline{x} = a(\theta + \sin\theta)$ \hfill (3)

   $\overline{y} = x - \dfrac{(1 + y_1^2)}{y_2}$

$$= a (1 - \cos \theta) + \frac{1}{\dfrac{-1}{4a} \cosec^4 \theta/2} (1 + \cot^2 \theta/2)$$

$$= a (1 - \cos\theta) - 4a \sin^2 \theta /2$$

$$= a (1 - \cos\theta) - 2a (1 - \cos \theta)$$

i.e., $\quad \bar{y} = -a(1 - \cos\theta)$ \hfill (4)

The evolute of the given cycloid is the locus of $(\bar{x}, \bar{y})$. Eliminating θ from (1) and (2) we get the equation of the evolute. Otherwise, the parametric equations of the locus of $(\bar{x}, \bar{y})$ are $x = a (\theta + \sin \theta)$ and $y = -a (1 - \cos \theta)$. These are the parametric equations of a cycloid. Thus the evolute of a cycloid is again a cycloid.

Example 3.2.4 Prove that the evolute of the tractrix $x = a (\cos t + \log \tan (t/2))$, $y = a \sin t$ is a catenary.

Solution:
Given $x = a (\cos t + \log \tan (t/2))$

$$\therefore \quad \frac{dx}{dt} = a\left(-\sin t + \frac{1}{\tan (t/2)} \sec^2 (t/2).(1/2) \right)$$

$$= a\left(-\sin t + \frac{1}{2} \cdot \frac{\cos (t/2)}{\sin (t/2) \cos^2 (t/2)} \right)$$

$$= a\left(-\sin t + \frac{1}{2 \sin (t/2) \cos (t/2)} \right)$$

$$= a\left(-\sin t + \frac{1}{\sin t} \right) = a\left(\frac{1 - \sin^2 t}{\sin t} \right)$$

i.e., $\quad \dfrac{dx}{dt} = a\left(\dfrac{\cos^2 t}{\sin t} \right)$

$\qquad y = a \sin t$

$$\therefore \quad \frac{dy}{dt} = a \cos t$$

$$y_1 = \frac{dy}{dx} = \frac{a \cos t}{a\left(\dfrac{\cos^2 t}{\sin t} \right)}$$

i.e., $\quad y_1 = \tan t$ \hfill (1)

$$y_2 = \frac{d^2y}{dx^2} = \frac{d}{dt}(\tan t).\frac{dt}{dx} = \sec^2 t.\frac{\sin t}{a\cos^2 t}$$

i.e., $\quad y_2 = \dfrac{\sin t}{a\cos^4 t}$ \hfill (2)

If $(\overline{x}, \overline{y})$ is the centre of curvature, then

$$\overline{x} = x - \frac{y_1}{y_2}(1+y_1^2)$$

$$= a\,(\cos t + \log\tan(t/2)) - \frac{\tan t}{\left(\dfrac{\sin t}{a\cos^4 t}\right)}(1+\tan^2 t) \quad \text{(Using (1) and (2))}$$

$$= a\,(\cos t + a\log\tan(t/2)) - a.\frac{\sin t}{\cos t}.\frac{\cos^4 t}{\sin t}.\sec^2 t$$

$$= a\cos t + a\log\tan(t/2) - a\cos t$$

i.e., $\quad \overline{x} = a\log\tan(t/2)$ \hfill (3)

$$\overline{y} = y - \frac{(1+y_1^2)}{y_2}$$

$$= a\sin t + \frac{1}{\left(\dfrac{\sin t}{a\cos^4 t}\right)}(1+\tan^2 t)$$

$$= a\sin t + \frac{a\cos^4 t}{\sin t}.\sec^2 t$$

$$= \frac{a}{\sin t}(\sin^2 t + \cos^2 t)$$

i.e., $\quad \overline{y} = \dfrac{a}{\sin t}$ \hfill (4)

From (3) and (4) we get the parametric equation of the evolute as

$$x = a\log\tan(t/2), \qquad y = \frac{a}{\sin t}$$

A relation between x and y is obtained by eliminating t.

Now from $\quad x = a\log\tan(t/2)$, we get $x/a = \log\tan(t/2)$,

$\therefore e^{x/a} = \tan(t/2)$ and $e^{-x/a} = \cot(t/2)$

Now $\cosh x/a = \dfrac{e^{x/a} + e^{-x/a}}{2}$

$$= \frac{\tan t/2 + \cot t/2}{2} = \frac{\dfrac{\sin t/2}{\cos t/2} + \dfrac{\cos t/2}{\sin t/2}}{2}$$

$$= \frac{\sin^2 t/2 + \cos^2 t/2}{2\sin t/2 \cos t/2}$$

i.e., $\cosh x/a = \dfrac{1}{\sin t}$

$\therefore$ $a \cosh x/a = \dfrac{a}{\sin t} = y$

i.e., $y = a \cosh x/a$, which is a catenary.

Example 3.2.5 Show that the evolute of the curve $x = a(\cos\theta + \theta\sin\theta)$;
$y = a(\sin\theta - \theta\cos\theta)$ is a circle.

Solution:
Given $x = a(\cos\theta + \theta.\sin\theta)$

$\dfrac{dx}{d\theta} = a(-\sin\theta + 1.\sin\theta + \theta.\cos\theta) = a\,\theta\cos\theta$

$y = a(\sin\theta - \theta\cos\theta)$

$\dfrac{dy}{d\theta} = a(\cos\theta - 1.\cos\theta - \theta.(-\sin\theta)) = a\,\theta\sin\theta$

$\therefore$ $y_1 = \dfrac{dy}{dx} = \dfrac{a\,\theta\sin\theta}{a\,\theta\cos\theta} = \tan\theta$

$y_2 = \dfrac{d^2y}{d\theta^2} = \dfrac{d}{d\theta}(\tan\theta).\dfrac{d\theta}{dx} = \sec^2\theta.\dfrac{1}{a\,\theta\cos\theta} = \dfrac{1}{a\,\theta\cos^3\theta}$

If $(\bar{x}, \bar{y})$ is the centre of curvature, then

$\bar{x} = x - \dfrac{y_1}{y_2}(1 + y_1^2)$

$= a(\cos\theta + \theta\sin\theta) - \tan\theta(1 + \tan^2\theta)\,a\,\theta\cos^3\theta$

$= a\cos\theta + a\,\theta\sin\theta - a\,\theta\tan\theta.\sec^2\theta\cos^3\theta$

$= a\cos\theta + a\,\theta\sin\theta - a\,\theta\sin\theta$

i.e., $\bar{x} = a\cos\theta$ (1)

$\bar{y} = y - \dfrac{(1 + y_1^2)}{y_2}$

$= a(\sin\theta - \theta\cos\theta) + (1 + \tan^2\theta).a\,\theta\cos^3\theta$

$= a\sin\theta - a\,\theta\cos\theta + a\,\theta\sec^2\theta.\cos^3\theta$

$= a\sin\theta - a\,\theta\cos\theta + a\,\theta\cos\theta$

i.e., $\bar{y} = a\sin\theta$ (2)

The locus of $(\bar{x}, \bar{y})$ is obtained by eliminating θ between (1) and (2)

$\bar{x}^2 + \bar{y}^2 = a^2\cos^2\theta + a^2\sin^2\theta$

i.e $\bar{x}^2 + \bar{y}^2 = a^2$

∴ Locus of $(\bar{x}, \bar{y})$ is $x^2 + y^2 = a^2$, which is a circle.

3.2.2   PROPERTIES OF THE EVOLUTES

Property 1

The normal at any point on a curve touches the evolute at the corresponding centre of curvature.

Proof: Let $y = f(x)$ be a given curve and $P(x, y)$ be any point on it. Let $C(\bar{x}, \bar{y})$ be the curvature at p and ρ be the radius of curvature.

The $\bar{x} = x - \rho \sin \chi, \qquad \bar{y} = y + \rho \cos \chi$

Differentiating $\bar{x}$ w.r.t s, the length of the arc AP from a fixed point A on the given curve, we get,

$$\frac{d\bar{x}}{ds} = \frac{dx}{ds} - \rho \cos\chi . \frac{d\chi}{ds} - \frac{d\rho}{ds}.\sin\chi$$

$$= \cos\chi - \rho\cos\chi . \frac{1}{\rho} - \frac{d\rho}{ds}.\sin\chi$$

i.e., $\dfrac{d\bar{x}}{ds} = -\dfrac{d\rho}{ds}.\sin\chi$ (1)

$$\frac{d\bar{y}}{ds} = \frac{dy}{ds} + \rho(-\sin\chi).\frac{d\chi}{ds} + \frac{d\rho}{ds}.\cos\chi$$

$$= \sin\chi + \rho.(-\sin\chi).\frac{1}{\rho} + \frac{d\rho}{ds}.\cos\chi$$

i.e., $\dfrac{d\bar{y}}{ds} = \dfrac{d\rho}{ds}.\cos\chi$ (2)

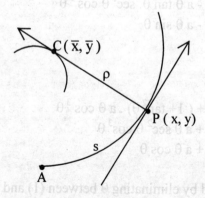

Fig.3.10

From (1) and (2)

$$\frac{dy}{dx} = \frac{dy/ds}{dx/ds} = -\cot\chi = \frac{-1}{\tan\chi} = \frac{-1}{y_1}$$

Since $\frac{dy}{dx}$ is the slope of the tangent to the evolute at $C(\bar{x}, \bar{y})$ and $\frac{-1}{y_1}$ is the slope of the normal PC to the given curve at $P(x, y)$, the normal at P to the given curve touches its evolute at C, the corresponding centre of curvature.

Property 2

The difference between the radii of curvature at two points of a curve is equal to the length of the arc of the evolute between the two corresponding points.

Proof: Let $(\bar{x}, \bar{y})$ be the centre of curvature at the point (x, y) on the given curve.

Then $\quad \dfrac{d\bar{x}}{ds} = -\sin\chi \dfrac{d\rho}{ds}$ and $\dfrac{d\bar{y}}{ds} = \cos\chi \dfrac{d\rho}{ds}$

$$d\bar{x} = -\sin\chi\, d\rho \quad \text{and} \quad d\bar{y} = \cos\chi\, d\rho$$

If $\bar{s}$ is the length of the arc of the evolute measured from a fixed point B on it.

Then $\quad d\bar{s}^2 = d\bar{x}^2 + d\bar{y}^2$

$$= (-\sin\chi . d\rho)^2 + (\cos\chi . d\rho)^2$$

$$= (\sin^2\chi + \cos^2\chi)\, d\rho^2 = d\rho^2$$

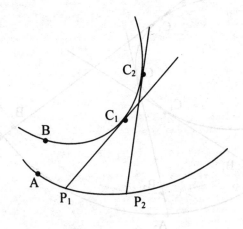

F

Fig.3.11

i.e., $\quad d\bar{s} = \pm d\rho$

Integrating, $\bar{s} = \pm\rho + k$ (1)

Let C_1 and C_2 be two points on the evolute corresponding to the points P_1 and P_2 on the curve, ρ_1 and ρ_2 the radii of curvature at P_1 and P_2 and arc $BC_1 = s_1$, arc $BC_2 = s_2$.

Then from (1) we have $s_1 = \pm \rho_1 + k$, $s_2 = \pm \rho_2 + k$

Hence $s_2 - s_1 = \rho_2 \sim \rho_1$ i.e. $C_1 C_2 = \rho_2 \sim \rho_1$

Property 3

There are infinite number of curves having the same evolute. i.e., there is one evolute, but an infinite number of involutes.

Proof: Let $A_1 A_2 A_3 \ldots$ be a given curve and $C_1 C_2 C_3 \ldots$ be its evolute (Fig. 3.12). Then the curve $A_1 A_2 A_3 \ldots$ is an involute of the curve $C_1 C_2 C_3 \ldots$ Along the normals at $A_1 A_2 A_3 \ldots$ take equal lengths $A_1 B_1$, $A_2 B_2$, $A_3 B_3 \ldots$ Then a new curve $B_1 B_2 B_3 \ldots$ is formed and this curve is 'parallel' to the original curve, having the same normals as the original curve. Further,

$$\text{arc } C_1 C_2 = A_1 C_1 \sim A_2 C_2 = B_1 C_1 \sim B_2 C_2 \quad \text{and}$$

$$\text{arc } C_2 C_3 = A_2 C_2 \sim A_3 C_3 = B_2 C_2 \sim B_3 C_3 \quad \text{etc.}$$

Hence the curve $B_1 B_2 B_3 \ldots$ also has the same evolute $C_1 C_2 C_3 \ldots$. Then $B_1 B_2 B_3 \ldots$ is another involute of the curve $C_1 C_2 C_3 \ldots$. i.e., any curve 'parallel' to the given curve is also an involute of the curve $C_1 C_2 C_3 \ldots$. Thus the curve $C_1 C_2 C_3 \ldots$, which is an evolute, can have infinite number of involutes.

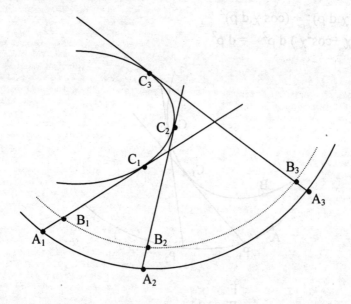

Fig. 3.12

Note:

The involutes of a given curve form a family of parallel curves.

EXERCISE 3.2

PART – A

1. Define evolute of a curve.
2. Define involute.
3. What is the connection between normal at a point on a curve and evolute?
4. How will you find the evolute of a given curve?
5. If the centre of curvature of a curve at a variable point 'θ' is (a cos θ, a sin θ), find the evolute of the curve.
6. If the centre of curvature of a curve at a variable point 'θ' is
   (a cos³θ , b sin³θ) find the evolute of the curve.
7. If the centre of curvature of a curve at a variable point 't' is $(2a + 3at^2, -2at^3)$, find the evolute of the curve.
8. If the centre of curvature of a curve at a variable point 'θ'
   is $\left(\dfrac{(a^2 - b^2)}{a} \cos^3\theta, \ \dfrac{(b^2 - a^2)}{b} \sin^3\theta \right)$, find the evolute of the curve.

PART –B

9. Find the equation in parametric form of the evolute of the curve
   x = 2 cos t + cos 2t, y = 2 sin t + sin 2t.
10. Find the evolute of the four-cusped hypocycloid $x^{2/3} + y^{2/3} = a^{2/3}$
11. Find the evolute of the rectangular hyperbola $xy = c^2$
12. Show that the evolute of the cycloid given by x = a (θ + sin θ), y = a(1- cos θ) is again a cycloid.
13. Find the evolute of the hyperbola $\dfrac{x^2}{a^2} - \dfrac{y^2}{b^2} = 1$.
14. Find the evolute of the parabola $x^2 = 4ay$.
15. Find the evolute of the curve $\sqrt{x} + \sqrt{y} = \sqrt{a}$

3.3   ENVELOPES

Consider the equation f(x, y, α) = 0   (1)

For a given value of α, equation (1) represents a particular curve. For different values of α we get a family of curves and hence the equation (1) is said to represent a **one parameter family of curves** with α as the only parameter.

3.3.1   ENVELOPS OF ONE PARAMETER FAMILY OF CURVES

The equation $y = mx + \dfrac{a}{m}$ where a is a constant and m is a parameter represents a family of straight lines, each member of the family touching the parabola $y^2 = 4ax$. Similarly the equation $x \cos \theta + y \sin \theta = p$ where p is a constant and θ is a parameter represents a family of straight lines touching the circle $x^2 + y^2 = p^2$. In these examples, we see that a family of curves touches a curve, in the first case a parabola and in the second case a circle. In other words, a curve (here a parabola or a circle) touches every member of a family of curves (straight lines). In general a curve which touches every member of a given family curves is called the envelope of that family of curves.

A formal definition of envelope of a family of curves is given below and the concepts given in the above paragraph are derived from this definition.

Definition 3.3.1 Let $f(x, y, \alpha) = 0$ be a one parameter family of curves such that any two curves corresponding to adjacent values of α intersect. Consider two adjacent curves of the family given by $f(x, y, \alpha) = 0$ and $f(x, y, \alpha + h) = 0$ where h is small. As $h \to 0$, the points of intersection of the above two curves will tend to a limiting position and the locus of all these limiting positions is called the **envelope** of the given family of curves.

3.3.2   METHOD OF FINDING THE EQUATION OF ENVELOPE

Let $f(x, y, \alpha) = 0$ be a given family of curves. Consider the adjacent members of this family given by

$$f(x, y, \alpha) = 0 \tag{1}$$
$$f(x, y, \alpha + h) = 0 \tag{2}$$

The coordinates of the points of intersection of these curves satisfy both the equations (1) and (2) and hence satisfy the equation.

$$\frac{f(x, y, \alpha + h) - f(x, y, \alpha)}{h} = 0$$

∴   The coordinates of the limiting positions of the points of intersection satisfy the equation

$$\lim_{h \to 0} \frac{f(x, y, \alpha + h) - f(x, y, \alpha)}{h} = 0$$

i.e., $\dfrac{\partial f}{\partial \alpha} = 0 \tag{3}$

∴ The equation of the envelope is found by eliminating α between (1) and (3).

Theorem 3.3.1 The envelope of a family of curves touches each member of the family.

Proof: Let the equation of the family of curves be

$$f(x, y, \alpha) = 0 \tag{1}$$

Equation of envelope is obtained by eliminating α between (1) and the equation

$$\frac{\partial f}{\partial \alpha} = 0 \tag{2}$$

Differentiating (1) w.r.t. x, we get,

$$\frac{\partial f}{\partial x} + \frac{\partial f}{\partial y} \cdot \frac{dy}{dx} = 0$$

i.e., $\quad \dfrac{dy}{dx} = - \dfrac{\frac{\partial f}{\partial x}}{\frac{\partial f}{\partial y}}$

∴ Slope of the tangent to curve (1) is $- \dfrac{\frac{\partial f}{\partial x}}{\frac{\partial f}{\partial y}}$

Now substituting in (1) the value of α computed from (2), we get the equation of the envelope as $f(x, y, \alpha) = 0$ where α is a function of x and y.

∴ Differentiating (1) w.r.t. x considering α as a function of x and y, we get

$$\frac{\partial f}{\partial x} + \frac{\partial f}{\partial y} \cdot \frac{dy}{dx} + \frac{\partial f}{\partial \alpha} \left(\frac{\partial \alpha}{\partial x} + \frac{\partial \alpha}{\partial y} \cdot \frac{dy}{dx} \right) = 0$$

i.e., $\dfrac{\partial f}{\partial x} + \dfrac{\partial f}{\partial y} \cdot \dfrac{dy}{dx} = 0 \qquad$ (Using (2))

i.e., $\quad \dfrac{dy}{dx} = - \dfrac{\frac{\partial f}{\partial x}}{\frac{\partial f}{\partial y}}$

is the slope of the tangent to the envelope.

∴ The slopes of the tangents to the curve and the envelope at the common points are equal.

Hence they touch each other.

Example 3.3.1 Find the envelope of the family of straight lines $y = mx + a/m$ where m is a parameter.

Solution:

Given $\quad y = mx + a/m \qquad\qquad\qquad\qquad$ (1)

Differentiating partially w.r.t. m

$\qquad 0 = x - a/m^2$

i.e., $m^2 = a/x \qquad\qquad\qquad\qquad\qquad\qquad$ (2)

The equation of the envelope is obtained by eliminating m between (1) and (2)

From (1) $\quad ym = m^2 x + a$

$\qquad\qquad y^2 m^2 = (m^2 x + a)^2$

$\qquad\qquad y^2 \cdot \dfrac{a}{x} = (a + a)^2 \qquad\qquad$ (Using (2))

$\qquad\qquad y^2 a = x \cdot 4a^2$

$y^2 = 4ax.$

∴ The envelope is a parabola.

Example 3.3.2 Find the envelope of the family of straight lines $y = mx - 2am - am^3$, where m is a parameter.

Solution:

Given $y = mx - 2am - am^3$ (1)

Differentiating partially w.r.t m

        $0 = x - 2a - 3am^2$ (2)

Eliminate m between (1) and (2)

From (2), $m^2 = \dfrac{x - 2a}{3a}$

From (1), $y = m(x - 2a - am^2)$

           $y = m\left[x - 2a - \dfrac{x - 2a}{3}\right]$

           $y = m.\dfrac{2}{3}.(x - 2a)$

    ∴ $m = \dfrac{3y}{2(x - 2a)}$

Now from (2) $0 = x - 2a - 3a\left[\dfrac{3y}{2(x - 2a)}\right]^2$

i.e., $4(x - 2a)^3 = 27ay^2$ is the equation of the envelope.

Example 3.3.3 Find the envelope of the family of straight lines $x \cos \theta + y \sin \theta = a$, where θ is a parameter.

Solution:

Given $x \cos \theta + y \sin \theta = a$ (1)

Differentiating partially w.r.t θ

        $x.(-\sin \theta) + y \cos \theta = 0$ (2)

The equation of the envelope is obtained by eliminating θ between (1) and (2)

Now squaring and adding (1) and (2)

$x^2 \cos^2 \theta + y^2 \sin^2 \theta + 2xy \sin \theta \cos \theta + x^2 \sin^2 \theta + y^2 \cos^2 \theta - 2xy \sin \theta \cos \theta = a^2$

∴ $x^2 + y^2 = a^2$ which is a circle.

Example 3.3.4 Find the envelope of the family of curves $\left(\dfrac{a^2}{x}\right)\cos\theta - \left(\dfrac{b^2}{y}\right)\sin\theta = c$ where

θ is a parameter.

Solution:

Given $\left(\dfrac{a^2}{x}\right)\cos\theta - \left(\dfrac{b^2}{y}\right)\sin\theta = c$ (1)

Differentiating w.r.t θ

$$\left(\frac{a^2}{x}\right)(-\sin\theta) - \left(\frac{b^2}{y}\right)\cos\theta = 0$$

$$\left(\frac{a^2}{x}\right)\sin\theta + \left(\frac{b^2}{y}\right)\cos\theta = 0 \qquad (2)$$

The equation of the envelope is obtained by eliminating θ between (1) and (2)

Now squaring and adding (1) and (2)

$$\left(\frac{a^2}{x}\right)^2(\cos^2\theta + \sin^2\theta) + \left(\frac{b^2}{y}\right)^2(\sin^2\theta + \cos^2\theta) = c^2$$

$$\frac{a^4}{x^2} + \frac{b^4}{y^2} = c^2$$

i.e., $a^4 y^2 + b^4 x^2 = c^2 x^2 y^2$.

3.3.3   TO FIND THE EQUATION OF ENVELOPE WHEN THE EQUATION OF THE FAMILY OF CURVES IS A QUADRATIC IN THE PARAMETER α

Let the equation of the family of curves be $A\alpha^2 + B\alpha + C = 0$ (1)
where A, B, C are functions of x and y and $A \neq 0$.

Differentiating (1) partially w.r.t α

$$A.2\alpha + B = 0 \qquad (2)$$

Eliminate α between (1) and (2)

From (2) $\alpha = -B/2A$

Substituting the value of α in (1)

$$A\left(\frac{-B}{2A}\right)^2 + B\left(\frac{-B}{2A}\right) + C = 0$$

$$AB^2 - 2A\,B^2 + 4A^2C = 0$$

$$-AB^2 + 4A^2C = 0$$

i.e., $B^2 = 4AC$.   (Since $A \neq 0$)

Hence the equation of the envelope is $B^2 = 4AC$.

Example 3.3.5 Find the envelope of the family of curves $y = mx + a\sqrt{1+m^2}$ where m is a parameter.

Solution:

We have $y = mx + a\sqrt{1 + m^2}$

$$y - mx = a\sqrt{1 + m^2}$$

Squaring both sides $(y - mx)^2 = a^2(1 + m^2)$

i.e., $y^2 - 2mxy + m^2x^2 = a^2(1 + m^2)$

$$(x^2 - a^2)m^2 - 2xy.m + y^2 - a^2 = 0$$

This being a quadratic equation in m, the equation of the envelope is given by

$$B^2 = 4AC, \text{ where } A = x^2 - a^2, B = -2xy, C = y^2 - a^2$$

i.e., $4x^2y^2 = 4(x^2 - a^2)(y^2 - a^2)$

$$x^2y^2 = x^2y^2 - x^2a^2 - y^2a^2 + a^4$$

$x^2 + y^2 = a^2$. The envelope is a circle.

Example 3.3.6 Find the envelope of the following of circles $(x - \alpha)^2 + (y - \alpha)^2 = 2\alpha$, where α is a parameter.

Solution:

Given $(x - \alpha)^2 + (y - \alpha)^2 = 2\alpha$

$$x^2 + y^2 - 2x\alpha - 2y\alpha + 2\alpha^2 = 2\alpha$$

$$2\alpha^2 - (x + y + 1)2\alpha + x^2 + y^2 = 0$$

Thus the equation of the family is a quadratic in the parameter α. Hence the equation of the envelope is given by $B^2 = 4AC$, where $A = 2$, $B = -(x + y + 1)$, $C = x^2 + y^2$

i.e., $4(x + y + 1)^2 = 4.2.(x^2 + y^2)$

i.e., $(x + y + 1)^2 = 2(x^2 + y^2)$

3.3.4   EVOLUTE AS THE ENVELOPE OF NORMALS

We have proved that normals to a curve are tangents to its evolute. i.e., evolute of a curve touches the family of normals to the curve. Hence the evolute of a curve is the envelope of the family of normals to the curve. Thus we have another method of finding the evolute of a given curve.

Example 3.3.7 Find the evolute of the parabola $y^2 = 4ax$ considering it as the envelope of its normals.

Solution:

The equation of normal at any point $(at^2, 2at)$ on the parabola $y^2 = 4ax$ is

$$y + xt = 2at + at^3 \tag{1}$$

Differentiating partially w.r.t. t, we get,

$$0 + x = 2a + 3at^2 \tag{2}$$

To find the envelope of the family of normals, eliminate t between (1) and (2).

From (2), $\quad t = \left(\dfrac{x - 2a}{3a}\right)^{1/2}$

Substituting in (1), we get

$$y + x\left(\frac{x - 2a}{3a}\right)^{1/2} = 2a\left(\frac{x - 2a}{3a}\right)^{1/2} + a\left(\frac{x - 2a}{3a}\right)^{3/2}$$

$$y = -\,(x - 2a)\cdot\left(\frac{x - 2a}{3a}\right)^{1/2} + a\left(\frac{x - 2a}{3a}\right)^{3/2}$$

$$y = \left(\frac{x - 2a}{3a}\right)^{3/2}(-3a + a)$$

$$y^2 = \left(\frac{x - 2a}{3a}\right)^{3}4a^2$$

$$y^2 = \frac{(x - 2a)^3}{27a^3}\cdot 4a^2$$

$$27ay^2 = 4(x - 2a)^3$$

This is the equation of the evolute of the given parabola.

Example 3.3.8 Find the evolute of the ellipse $\dfrac{x^2}{a^2} + \dfrac{y^2}{b^2} = 1$, treating it as the envelope of its normals.

Solution:

The normal at any point $(a\cos\theta, b\sin\theta)$ on the ellipse $\dfrac{x^2}{a^2} + \dfrac{y^2}{b^2} = 1$ is

$$\frac{ax}{\cos\theta} - \frac{by}{\sin\theta} = a^2 - b^2 \tag{1}$$

Differentiating (1) w.r.t θ

$$\frac{ax}{\cos^2\theta}\cdot\sin\theta + \frac{by}{\sin^2\theta}\cdot\cos\theta = 0$$

$$\frac{ax}{\cos^3\theta} + \frac{by}{\sin^3\theta} = 0 \tag{2}$$

Eliminate θ between (1) and (2)

From (2) $\quad \dfrac{ax}{\cos^3\theta} = \dfrac{-\,by}{\sin^3\theta} = k$ (say)

$$\therefore \cos\theta = \left(\frac{ax}{k}\right)^{1/3} \quad \text{and} \quad \sin\theta = \left(\frac{-by}{k}\right)^{1/3} \tag{3}$$

$$\cos^2\theta + \sin^2\theta = 1$$

$$\therefore \left(\frac{ax}{k}\right)^{2/3} + \left(\frac{-by}{k}\right)^{2/3} = 1$$

$$k^{2/3} = (ax)^{2/3} + (by)^{2/3} \qquad (4)$$

Substituting in (1) for $\sin\theta$ and $\cos\theta$ from (3)

$$\left[(ax)^{2/3} + (-by)^{2/3}\right]k^{1/3} = a^2 - b^2$$

$$\left[(ax)^{2/3} + (by)^{2/3}\right]^2 k^{2/3} = \left(a^2 - b^2\right)^2$$

$$\left[(ax)^{2/3} + (by)^{2/3}\right]^3 = \left(a^2 - b^2\right)^2 \qquad \text{(Using (4)))}$$

$$\text{i.e.,} (ax)^{2/3} + (by)^{2/3} = \left(a^2 - b^2\right)^{2/3}$$

This is the evolute of the ellipse $\dfrac{x^2}{a^2} + \dfrac{y^2}{b^2} = 1$

Example 3.3.9 Find the evolute of the hyperbola $\dfrac{x^2}{a^2} - \dfrac{y^2}{b^2} = 1$, treating it as the envelope of its normals.

Solution:

The normal at any point $(a\sec\theta, b\tan\theta)$ on the hyperbola $\dfrac{x^2}{a^2} - \dfrac{y^2}{b^2} = 1$ is

$$\frac{ax}{\sec\theta} + \frac{by}{\tan\theta} = a^2 + b^2 \qquad (1)$$

Differentiating (1) w.r.t. θ, we get,

$$ax(-\sin\theta) + by(-\operatorname{cosec}^2\theta) = 0 \qquad (2)$$

Eliminate θ between (1) and (2)

From (2) $\quad ax\sin\theta + \dfrac{by}{\sin^2\theta} = 0$

$$\sin^3\theta = \frac{-by}{ax}$$

$$\therefore \sin\theta = \frac{(-by)^{1/3}}{(ax)^{1/3}} \quad \text{and} \quad \sin^2\theta = \frac{(by)^{2/3}}{(ax)^{2/3}}$$

Now $\quad \cos\theta = \sqrt{1 - \sin^2\theta} = \sqrt{\dfrac{(ax)^{2/3} - (by)^{2/3}}{(ax)^{2/3}}}$

From (1) $\quad ax\cos\theta + \dfrac{by\cos\theta}{\sin\theta} = a^2 + b^2$

Substituting for $\sin\theta$ and $\cos\theta$, we get,

$$ax.\sqrt{\frac{(ax)^{2/3} - (by)^{2/3}}{(ax)^{2/3}}} + by.\sqrt{\frac{(ax)^{2/3} - (by)^{2/3}}{(ax)^{2/3}}}.\frac{(ax)^{1/3}}{(-by)^{1/3}} = a^2 + b^2$$

i.e., $(ax)^{2/3} \cdot \sqrt{(ax)^{2/3} - (by)^{2/3}} - (-by)^{2/3} \cdot \sqrt{(ax)^{2/3} - (by)^{2/3}} = a^2 + b^2$

i.e., $\sqrt{(ax)^{2/3} - (by)^{2/3}} \cdot \left[(ax)^{2/3} - (by)^{2/3}\right] = a^2 + b^2$

Squaring both sides, we get,

$$\left[(ax)^{2/3} - (by)^{2/3}\right]^3 = \left(a^2 + b^2\right)^2$$

i.e., $(ax)^{2/3} - (by)^{2/3} = \left(a^2 + b^2\right)^{2/3}$

3.3.5 ENVELOPE OF TWO PARAMETER FAMILY OF CURVES WHEN THE PARAMETERS ARE CONNECTED BY A RELATION

Let the two parameter family of curves be

$$f(x, y, a, b) = 0 \qquad (1)$$

Suppose the parameters a and b are connected by a relation

$$g(a, b) = 0 \qquad (2)$$

Now b is considered as a function of a given by (2).

$\therefore$ Differentiating (1) and (2) w.r.t. a, we get,

$$\frac{\partial f}{\partial a} + \frac{\partial f}{\partial b} \cdot \frac{db}{da} = 0 \qquad (3)$$

$$\frac{\partial g}{\partial a} + \frac{\partial g}{\partial b} \cdot \frac{db}{da} = 0 \qquad (4)$$

Comparing (3) and (4) we get

$$\frac{f_a}{g_a} = \frac{f_b}{g_b} = k \text{ (say)}$$

$\therefore \qquad f_a = k g_a, \; f_b = k g_b \qquad (5)$

Eliminating a, b and k from (1), (2) and (5) we get the equation of the envelope.

Example 3.3.10 Find the envelope of the straight lines $\dfrac{x}{a} + \dfrac{y}{b} = 1$ where the parameters a and b are connected by the relation $a^2 + b^2 = c^2$ and c is a constant.

Solution:

Given $\dfrac{x}{a} + \dfrac{y}{b} = 1 \qquad (1)$

Differentiating w.r.t a

$$\frac{-x}{a^2} - \frac{y}{b^2} \cdot \frac{db}{da} = 0 \qquad (2)$$

Given $a^2 + b^2 = c^2$

Differentiating w.r.t a

$$2a + 2b.\frac{db}{da} = 0 \qquad (3)$$

Eliminate a and b between (1), (2) and (3).

Comparing (1) and (2)

$$\frac{-x/a^2}{2a} = \frac{-y/b^2}{2b}$$

i.e., $\dfrac{x}{a^3} = \dfrac{y}{b^3} = \dfrac{1}{k^3}$ (say)

Then $a = k.x^{1/3}$, $b = k.y^{1/3}$

Now $a^2 + b^2 = c^2$

i.e., $k^2 x^{2/3} + k^2 y^{2/3} = c^2$

$$\therefore k^2 = \frac{c^2}{x^{2/3} + y^{2/3}} \qquad\qquad k = \pm \frac{c}{\sqrt{x^{2/3} + y^{2/3}}}$$

$\therefore$ From (1) $\dfrac{x}{kx^{1/3}} + \dfrac{y}{ky^{1/3}} = 1$

i.e., $\qquad x^{2/3} + y^{2/3} = k$

i.e., $\qquad x^{2/3} + y^{2/3} = \pm \dfrac{c}{\sqrt{x^{2/3} + y^{2/3}}}$

Cross multiplying and squaring, we get,

$$(x^{2/3} + y^{2/3})^3 = c^2$$

i.e., $\qquad x^{2/3} + y^{2/3} = c^{2/3}$

Example 3.3.11 Find the envelope of the family of ellipses $\dfrac{x^2}{a^2} + \dfrac{y^2}{b^2} = 1$ where a and b are connected by the relation $a + b = c$ and c is a constant.

Solution:

Method 1.

Given $\dfrac{x^2}{a^2} + \dfrac{y^2}{b^2} = 1 \qquad (1)$

and $\qquad a + b = c \qquad (2)$

Differentiating (1) w.r.t. a

$$- x^2 . \frac{2}{a^3} - y^2 . \frac{2}{b^3} . \frac{db}{da} = 0 \qquad (3)$$

Differentiating (2) w.r.t a

$$1 + \frac{db}{da} = 0 \tag{4}$$

Comparing (3) and (4)

$$-\frac{2x^2}{a^3} = -\frac{2y^2}{b^3} = -2k \text{ (say)}$$

$$\therefore a = \left(\frac{x^2}{k}\right)^{1/3} \quad \text{and} \quad b = \left(\frac{y^2}{k}\right)^{1/3}$$

From (2) $a + b = c$

i.e., $\left(\frac{x^2}{k}\right)^{1/3} + \left(\frac{y^2}{k}\right)^{1/3} = c$

$$\therefore x^{2/3} + y^{2/3} = c.k^{1/3} \tag{5}$$

From (1) $\dfrac{x^2}{\left(x^2/k\right)^{2/3}} + \dfrac{y^2}{\left(y^2/k\right)^{2/3}} = 1$

i.e., $\left[x^{2/3} + y^{2/3}\right]k^{2/3} = 1$

i.e., $\left[x^{2/3} + y^{2/3}\right]\left[\dfrac{x^{2/3} + y^{2/3}}{c}\right]^2 = 1$

i.e., $\left[x^{2/3} + y^{2/3}\right]^3 = c^2$

i.e., $x^{2/3} + y^{2/3} = c^{2/3}$ is the envelope of the given family of ellipses.

Method 2.

We have $\dfrac{x^2}{a^2} + \dfrac{y^2}{b^2} = 1$ and $b = c - a$

$$\therefore \frac{x^2}{a^2} + \frac{y^2}{(c-a)^2} = 1 \quad \text{is a family of curves with parameter a.} \tag{1}$$

Differentiating w.r.t a, we get,

$$-\frac{2x^2}{a^3} - \frac{2y^2}{(c-a)^3}.(-1) = 0 \tag{2}$$

Eliminate a between (1) and (2)

From (2), $\dfrac{(c-a)^3}{a^3} = \dfrac{y^2}{x^2}$

$$\frac{c-a}{a} = \frac{y^{2/3}}{x^{2/3}}$$

$$\frac{x^{2/3}}{a} = \frac{y^{2/3}}{c-a} = \frac{x^{2/3} + y^{2/3}}{c}$$

$$\therefore \frac{1}{a} = \frac{x^{2/3} + y^{2/3}}{c.x^{2/3}} \quad \text{and} \quad \frac{1}{c-a} = \frac{x^{2/3} + y^{2/3}}{c.y^{2/3}}$$

Substituting for $\frac{1}{a^2}$ and $\frac{1}{(c-a)^2}$ in (1) we get

$$x^2 . \frac{\left(x^{2/3} + y^{2/3}\right)^2}{c^2.x^{4/3}} + y^2 . \frac{\left(x^{2/3} + y^{2/3}\right)^2}{c^2.y^{4/3}} = 1$$

i.e., $\left(x^{2/3} + y^{2/3}\right)^2 . \left(x^{2/3} + y^{2/3}\right) = c^2$

i.e., $\left(x^{2/3} + y^{2/3}\right)^3 = c^2$

i.e., $x^{2/3} + y^{2/3} = c^{2/3}$

is the envelope of the given family of ellipses.

Example 3.3.12   Prove that the envelope of the family of ellipses $\dfrac{x^2}{a^2} + \dfrac{y^2}{b^2} = 1$, where a and b are connected by the relation $a^2 + b^2 = c^2$ and c is a constant is $x \pm y = \pm c$.

Solution:

Given $\dfrac{x^2}{a^2} + \dfrac{y^2}{b^2} = 1$ and $b^2 = c^2 - a^2$.

$$\therefore \frac{x^2}{a^2} + \frac{y^2}{c^2 - a^2} = 1$$

i.e., $(c^2 - a^2) x^2 + a^2 y^2 = a^2 (c^2 - a^2)$ (1)

Equation (1) is a family of curves with parameter a^2. Rewriting (1) as a quadratic in a^2, we get, $a^4 + (y^2 - x^2 - c^2)a^2 + c^2 x^2 = 0$

The equation of the envelope is given by

$\qquad B^2 = 4AC$ where $A = 1$, $B = y^2 - x^2 - c^2$ and $C = c^2 x^2$.

Substituting for A, B, C we get the equation of the envelope as,

$\qquad (y^2 - x^2 - c^2)^2 = 4.1\ c^2 x^2$

i.e., $y^2 - x^2 - c^2 = \pm 2cx$

i.e., $y^2 - x^2 - c^2 = 2cx; \qquad y^2 - x^2 - c^2 = -2cx$

i.e., $x^2 + 2cx + c^2 - y^2 = 0; \ x^2 - 2cx + c^2 - y^2 = 0$

i.e., $(x + c)^2 - y^2 = 0; \qquad (x - c)^2 - y^2 = 0$

i.e., $x + c = \pm y; \qquad\qquad x - c = \pm y$

i.e., $x \pm c = \pm y$

i.e., $x \pm y = \pm c$ is the envelope of the given family of ellipses.

Example 3.3.13 From any point of the ellipse $\dfrac{x^2}{a^2}+\dfrac{y^2}{b^2}=1$ perpendiculars are drawn to the coordinate axes. Prove that the envelope of the straight lines joining the feet of these perpendiculars is the curve $\left(\dfrac{x}{a}\right)^{2/3}+\left(\dfrac{y}{b}\right)^{2/3}=1$

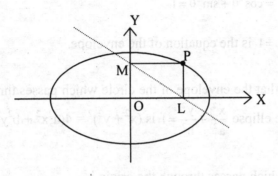

Fig.3.13

Solution:

Let P (a cos θ, b sin θ) be any point on the ellipse. Draw PL, PM perpendiculars to the axes. Then L and M are (a cos θ, 0) and (0, b sin θ). The equation of the line ML is

$$\frac{x}{a\cos\theta}+\frac{y}{b\sin\theta}=1 \qquad\qquad (1)$$

Equation (1) is a family of straight lines with θ as parameter. It is required to find the envelope of this family of straight lines.

Differentiating (1) w.r.t θ, we get,

$$\frac{x}{a}\sec\theta\tan\theta-\frac{y}{b}\csc\theta\cot\theta=0$$

$$\frac{x}{a}\frac{\sin\theta}{\cos^2\theta}-\frac{y}{b}\frac{\cos\theta}{\sin^2\theta}=0$$

$$\frac{x}{a}\sin^3\theta=\frac{y}{b}\cos^3\theta \qquad\qquad (2)$$

Eliminate θ between (1) and (2)

From (2), $\dfrac{x}{a}=\dfrac{y}{b}\dfrac{\cos^3\theta}{\sin^3\theta}$

Substituting in (1), we get, $\dfrac{y}{b}\dfrac{\cos^3\theta}{\sin^3\theta.\cos\theta}+\dfrac{y}{b\sin\theta}=1$

i.e., $\dfrac{y}{b}\left[\dfrac{\cos^2\theta+\sin^2\theta}{\sin^3\theta}\right]=1$

$$\therefore \frac{y}{b} = \sin^3\theta$$

From (2) we get, $\frac{x}{a}\sin^3\theta = \sin^3\theta.\cos^3\theta$

$$\therefore \frac{x}{a} = \cos^3\theta$$

$$\therefore \left(\frac{x}{a}\right)^{2/3} + \left(\frac{y}{b}\right)^{2/3} = \cos^2\theta + \sin^2\theta = 1$$

$$\therefore \left(\frac{x}{a}\right)^{2/3} + \left(\frac{y}{b}\right)^{2/3} = 1 \text{ is the equation of the envelope.}$$

Example 3.3.14 Show that the envelope of the circle which passes through the origin and whose centre lies on the ellipse $\frac{x^2}{a^2} + \frac{y^2}{b^2} = 1$ is $(x^2 + y^2)^2 = 4(a^2x^2 + b^2y^2)$.

Solution:
Equation of the circle which passes through the origin is
$$x^2 + y^2 + 2gx + 2fy = 0 \tag{1}$$

The centre (-g, -f) lies on the ellipse $\frac{x^2}{a^2} + \frac{y^2}{b^2} = 1$

$$\therefore \frac{g^2}{a^2} + \frac{f^2}{b^2} = 1 \tag{2}$$

Equation (1) is a family of circles with parameter g and f. Differentiating (1) w.r.t g
$$2x + 2y.\frac{df}{dg} = 0 \tag{3}$$

Differentiating (2) w.r.t g
$$\frac{2g}{a^2} + \frac{2f}{b^2}.\frac{df}{dg} = 0 \tag{4}$$

Comparing (3) and (4)
$$\frac{x}{g/a^2} = \frac{y}{f/b^2} = \frac{1}{k} \text{ (say)}$$

i.e., $\frac{xa^2}{g} = \frac{yb^2}{f} = \frac{1}{k}$

$$\therefore \quad g = ka^2x,$$
$$f = kb^2y \tag{5}$$

Eliminate g and f between (1), (2) and (5)

From (2) and (5) $\frac{k^2a^4x^2}{a^2} + \frac{k^2b^4x^2}{b^2} = 1$

$$\text{i.e., } k^2 (a^2x^2 + b^2y^2) = 1$$

$$\text{i.e., } k = \frac{\pm 1}{\sqrt{a^2x^2 + b^2y^2}}$$

From (1) and (5)

$$x^2 + y^2 + 2.ka^2x^2 + 2kb^2y^2 = 0$$

i.e., $\quad x^2 + y^2 + 2(a^2x^2 + b^2y^2).k = 0$

i.e., $\quad x^2 + y^2 + 2.(a^2x^2 + b^2y^2).\dfrac{(\pm 1)}{\sqrt{a^2x^2 + b^2y^2}} = 0$

i.e., $\quad x^2 + y^2 = \pm 2.\sqrt{a^2x^2 + b^2y^2}$

i.e.,$(x^2 + y^2)^2 = 4 (a^2x^2 + b^2y^2)$

Example 3.3.15 Show that all circles having for their diameters the radii vectors of a parabola touch a straight line or the curve r cos θ + a sin² θ = 0 according as the radii vectors are drawn from the focus or the vertex.

Solution:
The parabola $y^2 = 4ax$ has its vertex at A (0, 0) and focus at S (a, 0). P $(at^2, 2at)$ is any point on the parabola. The equation of the circle on SP as diameter is

$$(x - a) (x - at^2) + (y - 0) (y - 2at) = 0$$

i.e., $\quad x^2 - xat^2 - ax + a^2t^2 + y^2 - 2aty = 0$

i.e., $\quad t^2 (a^2 - ax) - 2ay.t + x^2 + y^2 - ax = 0 \qquad\qquad (1)$

The family of circles (1) is a quadratic in the parameter t.

Hence, the envelope is $B^2 = 4AC$ where $A = a^2 - x$, $B = -2ay$, $C = x^2 + y^2 - ax$

$$4a^2y^2 = 4(a^2 - ax)(x^2 + y^2 - ax)$$

$$a^2y^2 = a^2x^2 + a^2y^2 - a^3x - ax^3 - axy^2 + a^2x^2$$

i.e., $\quad ax(x^2 - 2ax + y^2 + a^2) = 0$

i.e., $\quad ax [(x - a)^2 + y^2] = 0 \quad$ i.e., x = 0 or $(x - a)^2 + y^2 = 0$

$(x - a)^2 + y^2 = 0$ can not be the envelope as (a, 0) is the only point on it.

∴ The envelope is the straight line x = 0, which is the y-axis or the tangent at the vertex of the parabola.

The equation of the circle on AP as diameter is $(x - 0)(x - at^2) + (y - 0)(y - 2at) = 0$

$$x^2 - axt^2 + y^2 - 2ayt = 0$$

$$axt^2 + 2ayt - (x^2 + y^2) = 0 \qquad\qquad (2)$$

Equation (1) is a family of circles with parameter t.

The envelope is $B^2 = 4AC$ where $A = ax$, $B = 2ay$ and $C = -(x^2 + y^2)$

i.e., $\quad 4a^2y^2 = -4ax (x^2 + y^2)$

$$a^2y^2 + x(x^2 + y^2) = 0$$

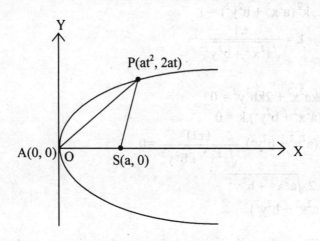

Fig.3.14

$\therefore$ The equation of the envelope is $ay^2 + x(x^2 + y^2) = 0$.

Transforming to polar coordinates by putting $x = r\cos\theta$, $y = r\sin\theta$ we get,

$$a.r^2. \sin^2\theta + r\cos\theta (r^2\cos^2\theta + r^2\sin^2\theta) = 0.$$

i.e., $ar^2\sin^2\theta + r^3\cos\theta = 0.$ i.e., $r\cos\theta + a\sin^2\theta = 0.$

Example 3.3.16 Given that $x^{2/3} + y^{2/3} = c^{2/3}$ is the envelope of $\dfrac{x}{a} + \dfrac{y}{b} = 1$;

Find the relation between a and b.

Solution:

Given $x^{2/3} + y^{2/3} = c^{2/3}$ (1)

Differentiating w.r.t. x, we get,

$$\frac{2}{3}x^{2/3-1} + \frac{2}{3}y^{2/3-1}.y_1 = 0$$

$$\frac{1}{x^{1/3}} + \frac{1}{y^{1/3}}.y_1 = 0$$ (2)

Also $\dfrac{x}{a} + \dfrac{y}{b} = 1$ (3)

Differentiating w.r.t. x, we get,

$$\frac{1}{a} + \frac{y_1}{b} = 0$$ (4)

The envelope touches each member of the family.

$\therefore$ From (2) and (4), equating the slopes, we get,

$$y_1 = -\frac{y^{1/3}}{x^{1/3}} = \frac{-b}{a}$$

i.e., $\frac{x^{1/3}}{a} = \frac{y^{1/3}}{b} = \lambda$ (say)

$\therefore \quad x^{1/3} = a\lambda$ and $y^{1/3} = b\lambda$

From (1) $(a\lambda)^2 + (b\lambda)^2 = c^{2/3}$

i.e., $(a^2 + b^2)\lambda^2 = c^{2/3}$ (5)

Also From (3) $\quad \frac{a^3\lambda^3}{a} + \frac{b^3\lambda^3}{b} = 1$

$\qquad\qquad (a^2 + b^2)\lambda^3 = 1 \qquad \therefore \lambda = (a^2 + b^2)^{-1/3}$

From (5) $(a^2 + b^2)(a^2 + b^2)^{-2/3} = c^{2/3}$

$\qquad\qquad (a^2 + b^2)^{1/3} = c^{2/3} \qquad$ i.e., $a^2 + b^2 = c^2$

EXERCISE 3.3

PART –A

1. Define envelope of a family of curves.
2. Define evolute of a curve as an envelope.
3. How will you find the envelope of a family of curves $f(x, y, \alpha) = 0$, where α is a parameter ?
4. How will you find the envelope of a family of curves $A\lambda^2 + B\lambda + C = 0$, where λ is a parameter and A, B, C are functions of x and y only?
5. Find the envelope of $x \cos \theta + y \sin \theta = p$ where θ is a parameter.
6. Find the envelope of $ty - x = at^2$ where t is a parameter.
7. Find the envelope of $y = mx + a\sqrt{1 + m^2}$.
8. What is the envelope of $y - \alpha x = 2/\alpha$?
9. Find the envelope of $x + y \tan \alpha = a \sec^2\alpha$.
10. Find the envelope of $y = mx + a/m$, m being a parameter.
11. Find the envelope of the family of curves $y = mx + am^2$, m being the parameter.
12. Find the envelope of $\frac{x}{t} + yt = 2c$, t being a parameter.
13. Find the envelope of the family of lines $\frac{x}{a} \cos \theta + \frac{y}{b} \sin \theta = 1$.
14. Find the envelope of the family of lines $x \sec \theta - y \tan \theta = a$, where θ is a parameter.

PART – B

15. Find the envelope of the family of circles $x^2 + y^2 - 2ax \cos \theta - 2ay \sin \theta = c^2$, where θ is the parameter.

16. Find the envelope of the family of straight lines $\frac{x}{a} + \frac{y}{b} = 1$ where $a^2 + b^2 = k^2$ and k is a constant.

17. Find the envelope of the family of straight lines $\frac{x}{a} + \frac{y}{b} = 1$ where a and b are connected by the relation (i) $a + b = c$ (ii) $ab = c^2$ (iii) $a^m + b^m = k^m$.

18. Circles of constant radius c are drawn with centres lying on the circle $x^2 + y^2 = a^2$. Prove that the envelope is $x^2 + y^2 = (a \pm c)^2$.

19. From any point of the ellipse $\frac{x^2}{a^2} + \frac{y^2}{b^2} = 1$ perpendiculars are drawn to the coordinate axes. Prove that the envelope of the straight lines joining the feet of these perpendiculars is the curve $\left(\frac{x}{a}\right)^{\frac{2}{3}} + \left(\frac{y}{b}\right)^{\frac{2}{3}} = 1$

20. Find the envelope of the family of straight lines $y \cos \alpha - x \sin \alpha = a \cos 2\alpha$ where α is a parameter.

21. Find the envelope of $\frac{x^2}{a^2} + \frac{y^2}{b^2} = 1$ given $a^n + b^n = c^n$ where c is a known constant.

22. Find the envelope of the system of lines $\frac{x}{l} + \frac{y}{m} = 1$ connected by $\frac{l}{a} + \frac{m}{b} = 1$

23. Find the envelope of the circles described on the radii vector of the ellipse $\frac{x^2}{a^2} + \frac{y^2}{b^2} = 1$ as diameters.

24. Show that the envelope of a family of concentric ellipses with their major and minor axes lying along the coordinates axes and of constant area k is $2xy = \frac{k}{\pi}$

25. Show that the envelope of the family of parabola $\sqrt{\frac{x}{a}} + \sqrt{\frac{y}{b}} = 1$ under the condition (i) $ab = c^2$ is hyperbola (ii) $a + b = c$ is an astroid.

26. Circles are described on the double ordinates of $y^2 = 4ax$ as diameters. Show that the envelope is $y^2 = 4a(x + a)$.

27. Find the envelope of a family of curves $A.\cos^n\theta + B.\sin^n\theta = C$ where θ is a parameter and A, B, C are functions of x and y.

28. Show that the envelope of a family of circles whose centre lies on the parabola $y^2 = 4ax$ and which passes through its vertex is the cissoid $y^2(x + 2a) + x^3 = 0$.

29. Show that the envelope of straight lines joining the ends of a pair of semi-conjugate diameters of the ellipse $\frac{x^2}{a^2} + \frac{y^2}{b^2} = 1$ is again an ellipse.

30. Show that the envelope of a family of circles whose centre lies on $xy = c^2$ and which passes through the origin is the lemniscate $r^2 = 8c^2 \sin 2\theta$

31. Given that the envelope of $\dfrac{x}{a} + \dfrac{y}{b} = 1$ is $x^p\, y^q = k^{p+q}$; find the relation between a and b.

Find the evolute of the following curves considering it as the envelope of the normals.

32. $xy = c^2$.

33. $x^2 = 4ay$.

34. $x = a\,(\cos\theta + \log\tan\theta/2)$, $y = a\sin\theta$

35. $x = a\,(3\cos t - 2\cos^3 t)$, $y = a\,(3\sin t - 2\sin^3 t)$

36. $x = a\,(\cos\theta + \theta\sin\theta)$, $y = a\,(\sin\theta - \theta\cos\theta)$.

37. Show that the equation of the normal to the curve $x^{2/3} + y^{2/3} = a^{2/3}$ can be written as

$y\cos\theta - x\sin\theta = a\cos 2\theta$. Hence show that the evolute of the curve is

$(x + y)^{2/3} + (x - y)^{2/3} = 2a^{2/3}$.

ANSWERS

EXERCISE 3.1

7) $\dfrac{\sqrt{2}}{3}$

8) $2\sqrt{2}$

9) $\dfrac{(109)^{3/2}}{60}$

10) $\dfrac{a}{\sqrt{2}}$

11) $\dfrac{(10)^{3/2}\,a}{12}$

12) $\sec(x/c)$

13) $\dfrac{5.\sqrt{5}}{4}$

14) $\sin x$

15) $3/8$

16) $a/2$

17) $\dfrac{\theta^4}{a.(1+\theta^2)^{3/2}}$

18) $\dfrac{1}{\sqrt{2}.r}$

21) c/y^2

22) $\dfrac{1}{c.\sqrt{2}}$

23) 6

24) $\dfrac{1}{\sqrt{2}}$

25) $\dfrac{a^2}{3.r}$

26) $\dfrac{2.\sqrt{2}}{3}$

27) $4/5$

28) $\dfrac{3.a.\sqrt{2}}{16}$

29) $3a/2$

30) $\dfrac{2.(ax+by)^{3/2}}{ab}$

31) $4a\cos\theta/2$

32) $2a\,e^\theta$

33) $a\cot\theta$

34) $3a\sin\theta$

35) $a\tan\theta.\sec\theta$

36) $2a.(1+t^2)^{3/2}$

37) $6t.(1+t^2)^2$

38) $a\sec^2\theta$

39) at

40) $na/2$

41) $\dfrac{\sin\alpha}{r}$

42) r^3/a^2

52) $(0, \tfrac{1}{2})$

53) $(1/e, 0)$

54) $(2c, 2c)$

55) $(-36, -43/6)$

56) $x^2 + y^2 - 30x + 12y - 27 = 0$

57) $\left(x - \dfrac{21a}{16}\right)^2 + \left(y - \dfrac{21a}{16}\right)^2 = \dfrac{9.a^2}{128}$

58) $\left(x - \dfrac{3a}{4}\right)^2 + \left(y - \dfrac{3a}{4}\right)^2 = \dfrac{a^2}{2}$

EXERCISE 3.2

(5) $x^2 + y^2 = a^2$ (6) $\left(\dfrac{x}{a}\right)^{2/3} + \left(\dfrac{y}{b}\right)^{2/3} = 1$

(7) $9a^2y^2 = 4a^2(x - 2a)^2$ (8) $a^{2/3} x^{2/3} + b^{2/3} y^{2/3} = \left(a^2 - b^2\right)^{2/3}$

9) $x = \dfrac{2}{3}\cos t - \dfrac{1}{3}\cos 2t, \; y = \dfrac{2}{3}\sin t - \dfrac{1}{3}\sin 2t$

10) $(x + y)^{2/3} + (x - y)^{2/3} = 2a^{2/3}$

11) $(x + y)^{2/3} - (x - y)^{2/3} = (4c)^{2/3}$

12) $x = a(\theta - \sin \theta), \; y = a(3 + \cos \theta)$

13) $(ax)^{2/3} - (by)^{2/3} = (a^2 + b^2)^{2/3}$

14) $4(y - 2a)^3 = 27.ax^2$ 15) $(x + y - a)^2 = 4xy$

EXERCISE 3.3

5) $x^2 + y^2 = a^2$ 6) $y^2 = 4ax$ 7) $x^2 + y^2 = a^2$ 8) $y^2 = 8x$

9) $y^2 = 4a(a - x)$ 10) $y^2 = 4ax$ 11) $x^2 + 4ay = 0$ 12) $xy = c^2$

13) $\dfrac{x^2}{a^2} + \dfrac{y^2}{b^2} = 1$ 14) $x^2 - y^2 = a^2$

15) $4a^2(x^2 + y^2) = (x^2 + y^2 - c^2)^2$ 16) $x^{2/3} + y^{2/3} = k^{2/3}$

17) (i) $\sqrt{x} + \sqrt{y} = c$ (ii) $4xy = c^2$ (iii) $x^{\frac{m}{m+1}} + y^{\frac{m}{m+1}} = k^{\frac{m}{m+1}}$

20) $(x + y)^{2/3} + (x - y)^{2/3} = 2a^{2/3}$ 21) $x^{\frac{2n}{n+2}} + y^{\frac{2n}{n+2}} = c^{\frac{2n}{n+2}}$

22) $\sqrt{\dfrac{x}{a}} + \sqrt{\dfrac{x}{b}} = 1$ 23) $(x^2 + y^2)^2 = (a^2 x^2 + b^2 y^2)$

27) $A^{\frac{2}{2-n}} + B^{\frac{2}{2-n}} = C^{\frac{2}{2-n}}$ 29) $\dfrac{x^2}{a^2} + \dfrac{y^2}{b^2} = \dfrac{1}{2}$

32) $(x + y)^{2/3} + (x - y)^{2/3} = (4c)^{2/3}$ 33) $4.(y - 2a)^3 = 27.ax^2$

34) $y = a \cosh x/a$ 35) $x^{2/3} + y^{2/3} = (4a)^{2/3}$

36) $x^2 + y^2 = a^2$

CHAPTER 4

FUNCTIONS OF SEVERAL VARIABLES

4.0 INTRODUCTION

We frequently come across quantities which depend on two or more variables. If x and y are the length and breadth of a room, then the area A of the floor is given by A = xy. For a given pair of values of x and y, A has a unique value. We say that A is a function of x and y. Any change in the value of x or y changes the value of A. x and y are called the independent variables and A is the dependent variable. Similarly if x, y and z are the length, breadth and height of a room, then the volume V of the room is given by V= xyz. Here V is a function of three variables x, y and z.

Definition 4.0.1 If a quantity z has a unique value for every pair of values of x and y, then z is called a **function** of two independent variables x and y and we write z = f(x, y).

The pair (x, y) represents a point in the xy-plane and the equation z = f(x, y) represents a surface in space. z is the height of the surface at the point (x, y). If R is a region in the xy-plane and f(x, y) is defined for all points (x, y) in R, then f(x, y) is said to be a function in the reigion R.

Definition 4.0.2 A δ-**neighbourhood or a neighbourhood** of a point (a, b) in the xy-plane is the set of points in the plane lying with in a circle having centre at (a, b) and radius δ>0. i.e., $(x - a)^2 + (y - b)^2 < \delta^2$.

Definition 4.0.3 Let f(x, y) be a function defined in a region R and (a, b) be a point in R. f(x, y) is said to tend to a **limit** l as x → a and y → b if given ε > 0 there exists a δ >0 such that | f(x, y) - l | < ε whenever (x, y) is in the δ-neighbourhood of (a, b). We write, $\underset{\substack{x \to a \\ y \to b}}{\text{Lt}} f(x,y) = l$

Definition 4.0.4 A function f(x, y) defined in a region R is said to be **continuous** at a point (a, b) in R if $\underset{\substack{x \to a \\ y \to b}}{\text{Lt}} f(x,y)$ exists and is equal to f(a, b).

A function f(x, y) that is defined and continuous at all points of a region R is said to be a **continuous function** in R.

4.1   PARTIAL DERIVATIVES

Let z = f(x, y) be a function of two variables x and y. If we keep y a constant and x a variable quantity, then z can be treated as a function of one variable x only. The derivative of z with respect to x, keeping y as a constant is called the partial derivative of z with respect to x and is denoted by $\dfrac{\partial z}{\partial x}$.

Definiton 4.1.1 Let z = f(x, y) be a function of two variables x and y. The **partial derivative of z with respect to x** is given by $\dfrac{\partial z}{\partial x} = \underset{\Delta x \to 0}{Lt} \dfrac{f(x + \Delta x, y) - f(x, y)}{\Delta x}$

Similarly the partial derivative of z with respect to y is obtained by differentiating z with respect to y, keeping x as constant.

i.e., $\dfrac{\partial z}{\partial y} = \underset{\Delta y \to 0}{Lt} \dfrac{f(x, y + \Delta y) - f(x, y)}{\Delta y}$

Note:

(i) $\dfrac{\partial z}{\partial x}$ is also denoted by $\dfrac{\partial f}{\partial x}$, $f_x(x, y)$, f_x or $D_x f$

$\dfrac{\partial z}{\partial y}$ is also denoted by $\dfrac{\partial f}{\partial y}$, $f_y(x, y)$, f_y or $D_y f$

(ii)   If $z = f(x_1, x_2 \ldots x_n)$, then z is a function of n variables $x_1, x_2 \ldots x_n$. The partial derivative of z with respect to x_i is obtained by differentiating z with respect to x_i, keeping all other variables constant and is written as $\dfrac{\partial z}{\partial x_i}$, i = 1, 2 ... n.

4.1.1   HIGHER ORDER PARTIAL DERIVATIVES

In general $\dfrac{\partial f}{\partial x}$ and $\dfrac{\partial f}{\partial x}$ are again functions of x and y and hence they can be differentiated further with respect to x or y. Thus we have the following second order partial derivatives,

$$\frac{\partial}{\partial x}\left(\frac{\partial z}{\partial x}\right) = \frac{\partial^2 z}{\partial x^2} \text{ or } \frac{\partial^2 f}{\partial x^2} \text{ or } f_{xx}$$

$$\frac{\partial}{\partial y}\left(\frac{\partial z}{\partial x}\right) = \frac{\partial^2 z}{\partial y \partial x} \text{ or } \frac{\partial^2 f}{\partial y \partial x} \text{ or } f_{yx}$$

$$\frac{\partial}{\partial x}\left(\frac{\partial z}{\partial y}\right) = \frac{\partial^2 z}{\partial x \partial y} \text{ or } \frac{\partial^2 f}{\partial x \partial y} \text{ or } f_{xy}$$

$$\frac{\partial}{\partial y}\left(\frac{\partial z}{\partial y}\right) = \frac{\partial^2 z}{\partial y^2} \text{ or } \frac{\partial^2 f}{\partial y^2} \text{ or } f_{yy}$$

Note:

In general, $\dfrac{\partial^2 f}{\partial x \partial y} \neq \dfrac{\partial^2 f}{\partial y \partial x}$. If $\dfrac{\partial f}{\partial x}, \dfrac{\partial f}{\partial y}, \dfrac{\partial^2 f}{\partial x \partial y}, \dfrac{\partial^2 f}{\partial y \partial x}$ exists and are continuous

functions in the neighbourhood of a point (a, b) then $\dfrac{\partial^2 f}{\partial x \partial y} = \dfrac{\partial^2 f}{\partial y \partial x}$ at (a, b)

Example 4.1.1 If $z = \log \dfrac{x^2 + y^2}{xy}$, prove that $\dfrac{\partial^2 f}{\partial x \partial y} = \dfrac{\partial^2 f}{\partial y \partial x}$

Solution:
We have $z = \log(x^2 + y^2) - \log x - \log y$

$$\frac{\partial z}{\partial x} = \frac{2x}{x^2 + y^2} - \frac{1}{x} \qquad \frac{\partial z}{\partial y} = \frac{2y}{x^2 + y^2} - \frac{1}{y}$$

$$\frac{\partial^2 z}{\partial y \partial x} = \frac{\partial}{\partial y}\left(\frac{2x}{x^2 + y^2} - \frac{1}{x}\right) = \frac{-4xy}{\left(x^2 + y^2\right)^2}$$

$$\frac{\partial^2 z}{\partial x \partial y} = \frac{\partial}{\partial x}\left(\frac{2y}{x^2 + y^2} - \frac{1}{y}\right) = \frac{-4xy}{\left(x^2 + y^2\right)^2}$$

Hence $\dfrac{\partial^2 f}{\partial x \partial y} = \dfrac{\partial^2 f}{\partial y \partial x}$.

Example 4.1.2 If $u = (x^2 + y^2 + z^2)^{-1/2}$, prove that $\dfrac{\partial^2 u}{\partial x^2} + \dfrac{\partial^2 u}{\partial y^2} + \dfrac{\partial^2 u}{\partial z^2} = 0$

Solution:

$$\frac{\partial u}{\partial x} = -\frac{1}{2}(x^2 + y^2 + z^2)^{-3/2}.2x = -x(x^2 + y^2 + z^2)^{-3/2}$$

$$\frac{\partial^2 u}{\partial x^2} = -1.(x^2 + y^2 + z^2)^{-3/2} - x.\left(\frac{-3}{2}\right)(x^2 + y^2 + z^2)^{-5/2}.2x$$

$$= (x^2 + y^2 + z^2)^{-5/2}.(-x^2 - y^2 - z^2 + 3x^2)$$

$$= (x^2 + y^2 + z^2)^{-5/2}.(2x^2 - y^2 - z^2)$$

Similarly, $\dfrac{\partial^2 u}{\partial y^2} = (x^2 + y^2 + z^2)^{-5/2}.(2y^2 - y^2 - z^2)$

and $\dfrac{\partial^2 u}{\partial z^2} = (x^2 + y^2 + z^2)^{-5/2}.(2z^2 - y^2 - z^2)$

$\therefore \quad \dfrac{\partial^2 u}{\partial x^2} + \dfrac{\partial^2 u}{\partial y^2} + \dfrac{\partial^2 u}{\partial z^2} = (x^2 + y^2 + z^2)^{-5/2}.0 = 0$

Note:

The equation $\dfrac{\partial^2 u}{\partial x^2} + \dfrac{\partial^2 u}{\partial y^2} + \dfrac{\partial^2 u}{\partial z^2} = 0$ is called the **Laplace equation** for the function

$u(x, y, z)$. A function $u(x, y, z)$ satisfying the Laplace equation is called a **harmonic function**.

Example 4.1.3 If $x = r\cos\theta$, $y = r\sin\theta$, prove that

(i) $\dfrac{\partial^2 r}{\partial x^2} + \dfrac{\partial^2 r}{\partial y^2} = \dfrac{1}{r}\left[\left(\dfrac{\partial r}{\partial x}\right)^2 + \left(\dfrac{\partial r}{\partial y}\right)^2\right]$

(ii) $\dfrac{\partial^2 \theta}{\partial x^2} + \dfrac{\partial^2 \theta}{\partial y^2} = 0 \quad (x \neq 0, y \neq 0)$

Solution:

$x = r\cos\theta$, $y = r\sin\theta$.

$\therefore \quad x^2 + y^2 = r^2$ and $\tan\theta = y/x$

Differentiating $r^2 = x^2 + y^2$ partially w.r.t x, we get

$2r.\dfrac{\partial r}{\partial x} = 2x \qquad$ i.e., $\dfrac{\partial r}{\partial x} = \dfrac{x}{r}$ \hfill (1)

Differentiating $r^2 = x^2 + y^2$ partially w.r.t y, we get

$2r.\dfrac{\partial r}{\partial y} = 2y \qquad$ i.e., $\dfrac{\partial r}{\partial y} = \dfrac{y}{r}$ \hfill (2)

$\therefore \quad \dfrac{1}{r}\left[\left(\dfrac{\partial r}{\partial x}\right)^2 + \left(\dfrac{\partial r}{\partial y}\right)^2\right] = \dfrac{1}{r}\left[\dfrac{x^2}{r^2} + \dfrac{y^2}{r^2}\right]$

$\qquad\qquad\qquad\qquad\qquad = \dfrac{1}{r}.\dfrac{1}{r^2}(x^2 + y^2)$

$\qquad\qquad\qquad\qquad\qquad = \dfrac{1}{r}$ \hfill (3)

Differentiating (1) partially w.r.t x, we get

$\dfrac{\partial^2 r}{\partial x^2} = x\left(\dfrac{-1}{r^2}\right)\dfrac{\partial r}{\partial x} + 1.\dfrac{1}{r}$

$$= \left(\frac{-x}{r^2}\right) \cdot \frac{x}{r} + \frac{1}{r}$$

Simarly from (2), we get,

$$\frac{\partial^2 r}{\partial y^2} = y\left(\frac{-1}{r^2}\right)\frac{\partial r}{\partial y} + 1 \cdot \frac{1}{r}$$

$$= \left(\frac{-y}{r^2}\right) \cdot \frac{y}{r} + \frac{1}{r}$$

$$\therefore \quad \frac{\partial^2 r}{\partial x^2} + \frac{\partial^2 r}{\partial y^2} = -\frac{1}{r^3}(x^2 + y^2) + \frac{2}{r}$$

$$= -\frac{1}{r} + \frac{2}{r}$$

$$= \frac{1}{r} \tag{4}$$

From (3) and (4), we get,

$$\frac{\partial^2 r}{\partial x^2} + \frac{\partial^2 r}{\partial y^2} = \frac{1}{r}\left[\left(\frac{\partial r}{\partial x}\right)^2 + \left(\frac{\partial r}{\partial y}\right)^2\right]$$

Example 4.1.4 If $z = f(x + ay) + g(x - ay) - \dfrac{x}{2a^2}\cos(x + ay)$, show that

$$a^2 \frac{\partial^2 z}{\partial x^2} - \frac{\partial^2 z}{\partial y^2} = \sin(x + ay)$$

Solution:
Differentiating the given equation partially w.r.t x, we get,

$$\frac{\partial z}{\partial x} = f'(x + ay).1 + g'(x - ay).1 - \frac{1}{2a^2}\cos(x + ay) + \frac{x}{2a^2}\sin(x + ay).1$$

Again partially differentiating w.r.t x, we get,

$$\frac{\partial^2 z}{\partial x^2} = f''(x + ay) + g''(x - ay) + \frac{1}{2a^2}\sin(x + ay) + \frac{1}{2a^2}\sin(x + ay)$$

$$+ \frac{x}{2a^2}\cos(x + ay)$$

$$a^2 \frac{\partial^2 z}{\partial x^2} = a^2 f''(x + ay) + a^2 g''(x - ay) + \frac{x}{2}\cos(x + ay) + \sin(x + ay) \tag{1}$$

Similarly differentiating the given equation twice partially w.r.t y, we get

$$\frac{\partial z}{\partial y} = f'(x + ay).a + g'(x - ay).(-a) + \frac{x}{2a^2}\sin(x + ay).a$$

$$\frac{\partial^2 z}{\partial y^2} = f''(x+ay).a^2 + g''(x-ay).(-a)^2 + \frac{x}{2a^2}\cos(x+ay).a^2$$

$$= a^2 f''(x+ay) + a^2 g''(x-ay) + \frac{x}{2}\cos(x+ay) \tag{2}$$

From (1) and (2), we get,

$$a^2 \frac{\partial^2 z}{\partial x^2} - \frac{\partial^2 z}{\partial y^2} = \sin(x+ay)$$

4.1.2   HOMOGENEOUS FUNCTIONS

Definition 4.1.2   A function $f(x, y)$ is said to be homogeneous of degree n in x and y if $f(tx, ty) = t^n.f(x, y)$.

A polynomial $a_0 x^n + a_1 x^{n-1}.y + a_2 x^{n-2}y^2 + \ldots + a_n y^n$ in which every term is of degree n in x and y is homogeneous and is called a **homogeneous polynomial of degree n in x and y**.

$$f(x, y) = a_0 x^n + a_1 x^{n-1}.y + a_2 x^{n-2}y^2 + \ldots + a_n y^n$$
$$= x^n [a_0 + a_1 (y/x) + a_2 (y/x)^2 + \ldots + a_n (y/x)^n]$$
$$= x^n \varphi(y/x).$$

A function $f(x, y)$ is homogeneous of degree n if and only if it can be expressed in the form $x^n. \varphi(y/x)$.

Note:

A function $f(x_1, x_2 \ldots x_n)$ is homogeneous of degree n in $x_1, x_2 \ldots x_n$ if and only if it can be expressed in the form $x_r^n \varphi\left(\dfrac{x_1}{x_r}, \dfrac{x_2}{x_r}, \ldots, \dfrac{x_n}{x_r}\right)$

4.1.3   EULER'S THEOREM

If $f(x, y)$ is a homogeneous function of degree n in x and y, then $x.\dfrac{\partial f}{\partial x} + y.\dfrac{\partial f}{\partial y} = n.f$

Proof: Since $f(x, y)$ is homogeneous of degree n, $f(x, y) = x^n.\varphi(y/x)$
Differentiating partially w.r.t x

$$\frac{\partial f}{\partial x} = n.x^{n-1}.\varphi(y/x) + x^n \varphi'(y/x).(-y/x^2)$$

$$\therefore \quad x.\frac{\partial f}{\partial x} = n.x^n.\varphi(y/x) + x^n \varphi'(y/x).(-y/x) \tag{1}$$

Similarly differentiating partially w.r.t y

$$\frac{\partial f}{\partial y} = x^n \varphi'(y/x)(1/x)$$

$$\therefore \quad y.\frac{\partial f}{\partial y} = x^n \varphi'(y/x)(y/x) \tag{2}$$

Adding (1) and (2)

$$x.\frac{\partial f}{\partial x} + y.\frac{\partial f}{\partial y} = n.x^n.\varphi\left(\frac{y}{x}\right) = n.f(x, y)$$

i.e., $\quad x.\dfrac{\partial f}{\partial x} + y.\dfrac{\partial f}{\partial y} = n.f$

Note:
(i) If $f(x_1, x_2 \ldots x_n)$ is a homogeneous function of degree n in $x_1, x_2 \ldots x_n$ then

$$x_1.\frac{\partial f}{\partial x_1} + x_2.\frac{\partial f}{\partial x_2} + \ldots + x_n.\frac{\partial f}{\partial x_n} = n.f$$

Example 4.1.5 If $f(x, y)$ is a homogeneous function of degree n in x and y, show that

$$x^2.\frac{\partial^2 f}{\partial x^2} + 2xy\frac{\partial^2 f}{\partial x \partial y} + y^2\frac{\partial^2 f}{\partial y^2} = n(n-1).f$$

Solution: As $f(x, y)$ is a homogeneous function of degree n, by Euler's theorem,

$$x.\frac{\partial f}{\partial x} + y.\frac{\partial f}{\partial y} = n.f \tag{1}$$

Differentiating (1) partially w.r.t x, we get,

$$x.\frac{\partial^2 f}{\partial x^2} + 1.\frac{\partial f}{\partial x} + y.\frac{\partial^2 f}{\partial x \partial y} = n.\frac{\partial f}{\partial x}$$

$$x.\frac{\partial^2 f}{\partial x^2} + y.\frac{\partial^2 f}{\partial x \partial y} = (n-1).\frac{\partial f}{\partial x} \tag{2}$$

Again differentiating (1) partially w.r.t y, we get,

$$x.\frac{\partial^2 f}{\partial y \partial x} + 1.\frac{\partial f}{\partial y} + y.\frac{\partial^2 f}{\partial y^2} = n.\frac{\partial f}{\partial y}$$

$$x.\frac{\partial^2 f}{\partial y \partial x} + y.\frac{\partial^2 f}{\partial y^2} = (n-1).\frac{\partial f}{\partial y} \tag{3}$$

Multiplying (2) by x and (3) by y and adding, we get,

$$x^2.\frac{\partial^2 f}{\partial x^2} + 2xy\frac{\partial^2 f}{\partial x \partial y} + y^2\frac{\partial^2 f}{\partial y^2} = (n-1)(x.\frac{\partial f}{\partial x} + y.\frac{\partial f}{\partial y})$$

$$= (n-1). n.f \qquad \text{(Using (1))}$$

$$x^2.\frac{\partial^2 f}{\partial x^2} + 2xy\frac{\partial^2 f}{\partial x \partial y} + y^2\frac{\partial^2 f}{\partial y^2} = n(n-1).f$$

Example 4.1.6 Show that

$$x.\frac{\partial u}{\partial x} + y.\frac{\partial u}{\partial y} + z.\frac{\partial u}{\partial z} = 2 \tan u \text{ where } u = \sin^{-1}\left[\frac{x^3 + y^3 + z^3}{ax + by + cz}\right]$$

Solution: We have, $\sin u = \left[\dfrac{x^3 + y^3 + z^3}{ax + by + cz}\right]$

Let $f(x, y, z) = \dfrac{x^3 + y^3 + z^3}{ax + by + cz}$ \hfill (1)

$$f(tx, ty, tz) = \frac{t^3 x^3 + t^3 y^3 + t^3 z^3}{atx + bty + ctz} = t^2 f(x, y, z)$$

$\therefore$ $f(x, y, z)$ is a homogeneous function of degree 2.

$\therefore$ By Euler's theorem,

$$x.\frac{\partial f}{\partial x} + y.\frac{\partial f}{\partial y} + z.\frac{\partial f}{\partial z} = 2.f \hspace{2cm} (2)$$

From (1), we have, $f = \sin u$

$$\therefore \quad \frac{\partial f}{\partial x} = \cos u.\frac{\partial u}{\partial x} \qquad \frac{\partial f}{\partial y} = \cos u.\frac{\partial u}{\partial y} \quad \text{and} \quad \frac{\partial f}{\partial z} = \cos u.\frac{\partial u}{\partial z}$$

Substituting these in (2), we get,

$$x.\cos u.\frac{\partial u}{\partial x} + y.\cos u.\frac{\partial u}{\partial y} + z.\cos u.\frac{\partial u}{\partial z} = 2.\sin u$$

$$x.\frac{\partial u}{\partial x} + y.\frac{\partial u}{\partial y} + z..\frac{\partial u}{\partial z} = 2.\tan u$$

Example 4.1.7 Given $z = x^n.f_1(y/x) + y^{-n}.f_2(x/y)$, prove that

$$x^2.\frac{\partial^2 z}{\partial x^2} + 2xy\frac{\partial^2 z}{\partial x \partial y} + y^2\frac{\partial^2 z}{\partial y^2} + x.\frac{\partial z}{\partial x} + y.\frac{\partial z}{\partial y} = n^2 z$$

Solution:

Let $u(x, y) = x^n.f_1(y/x)$ \hfill (1)

$ v(x, y) = y^{-n}.f_2(x/y)$ \hfill (2)

then $z = u + v$ \hfill (3)

$ u(tx, ty) = t^n x^n .f_1(y/x) = t^n.u(x, y)$

$\therefore$ (1) is homogeneous of degree n

$$x^2.\frac{\partial^2 u}{\partial x^2} + 2xy\frac{\partial^2 u}{\partial x \partial y} + y^2\frac{\partial^2 u}{\partial y^2} = n(n-1).u \hspace{1.5cm} (4)$$

and $\quad x.\dfrac{\partial u}{\partial x} + y.\dfrac{\partial u}{\partial y} = n.u$ \hfill (5)

$$v(tx, ty) = t^{-n} y^{-n} . f_2(x/y) = t^{-n} . v(x, y)$$

$\therefore$ (2) is homogeneous of degree -n

Hence $x^2 . \dfrac{\partial^2 v}{\partial x^2} + 2xy \dfrac{\partial^2 v}{\partial x \partial y} + y^2 \dfrac{\partial^2 v}{\partial y^2} = -n(-n-1).v$ (6)

and $\quad x . \dfrac{\partial v}{\partial x} + y . \dfrac{\partial v}{\partial y} = (-n).v$ (7)

adding (4) and (6) we get

$$x^2 . \dfrac{\partial^2 (u+v)}{\partial x^2} + 2xy \dfrac{\partial^2 (u+v)}{\partial x \partial y} + y^2 \dfrac{\partial^2 (u+v)}{\partial y^2} = n.(n-1).u - n(-n-1).v$$

i.e., $x^2 . \dfrac{\partial^2 z}{\partial x^2} + 2xy \dfrac{\partial^2 z}{\partial x \partial y} + y^2 \dfrac{\partial^2 z}{\partial y^2} = n^2.(u+v) + (-n)u + n.v$

i.e., $x^2 . \dfrac{\partial^2 z}{\partial x^2} + 2xy \dfrac{\partial^2 z}{\partial x \partial y} + y^2 \dfrac{\partial^2 z}{\partial y^2} + x.\dfrac{\partial u}{\partial x} + y.\dfrac{\partial u}{\partial y} + x.\dfrac{\partial v}{\partial x} + y.\dfrac{\partial v}{\partial y} = n^2 z$ (Using (5), (7))

i.e., $x^2 . \dfrac{\partial^2 z}{\partial x^2} + 2xy \dfrac{\partial^2 z}{\partial x \partial y} + y^2 \dfrac{\partial^2 z}{\partial y^2} + x.\dfrac{\partial z}{\partial x} + y.\dfrac{\partial z}{\partial y} = n^2 z$

EXERCISE 4.1

PART –A

1. Define partial derivatives $z = f(x, y)$.

2. State the conditions for the equality of $\dfrac{\partial^2 f}{\partial x \partial y}$ and $\dfrac{\partial^2 f}{\partial y \partial x}$ at (a, b).

3. Write the Laplace equations for the function $u(x, y, z)$.

4. Define 'harmonic function'.

5. State Euler's theorem for homogenous functions.

6. If $u = x^3 + y^3 + z^3 + 3xyz$, what is $x.\dfrac{\partial u}{\partial x} + y.\dfrac{\partial u}{\partial y} + z.\dfrac{\partial u}{\partial z}$

7. If $u = \dfrac{xy}{x+y}$, what is $x.\dfrac{\partial u}{\partial x} + y.\dfrac{\partial u}{\partial y}$?

8. If $\sin u = \dfrac{x^2 + y^2}{x+y}$, what is $x.\dfrac{\partial u}{\partial x} + y.\dfrac{\partial u}{\partial y}$?

9. Find the value of $x.u_x + y.u_y + z.u_z$ where $u = \sin^{-1}\left[\dfrac{x^3 + y^3 + z^3}{ax + by + cz}\right]$.

10. If $x = r \cos \theta$, $y = r \sin \theta$, find the value of $\dfrac{\partial^2 \theta}{\partial x^2} + \dfrac{\partial^2 \theta}{\partial y^2}$.

11. If $z = f(x^2 + y^2)$, what is the value of $x.\dfrac{\partial z}{\partial y} - y.\dfrac{\partial z}{\partial x}$?

12. If $u = (x^2 + y^2)^n$, what is the value of $x.\dfrac{\partial u}{\partial x} + y.\dfrac{\partial u}{\partial y}$?

13. If $u = e^{x^3 + y^3}$, find the value of $x.\dfrac{\partial u}{\partial x} + y.\dfrac{\partial u}{\partial y}$

14. If $u = f(y/x)$, what is the value of $x.\dfrac{\partial u}{\partial x} + y.\dfrac{\partial u}{\partial y}$?

15. If $z = f(x^2/y)$, show that $x.\dfrac{\partial z}{\partial x} + 2y.\dfrac{\partial z}{\partial y} = 0$

PART – B

16. If $u = f(r)$ and $x = r\cos\theta$, $y = r\sin\theta$, prove that $\dfrac{\partial^2 u}{\partial x^2} + \dfrac{\partial^2 u}{\partial y^2} = f''(r) + \dfrac{1}{r}f'(r)$.

17. If $u = f(r)$ and $r = \sqrt{x^2 + y^2 + z^2}$, show that

$$\frac{\partial^2 u}{\partial x^2} + \frac{\partial^2 u}{\partial y^2} + \frac{\partial^2 u}{\partial z^2} = f''(r) + \frac{2}{r}f'(r).$$

18. If $u = \cos^{-1}\dfrac{x+y}{\sqrt{x}+\sqrt{y}}$, prove that $x.\dfrac{\partial u}{\partial x} + y.\dfrac{\partial u}{\partial y} = \dfrac{-1}{2}.\cot u$.

19. If $z = x.\varphi(y/x) + \chi(y/x)$, prove that $x^2.\dfrac{\partial^2 z}{\partial x^2} + 2xy.\dfrac{\partial^2 z}{\partial x\partial y} + y^2\dfrac{\partial^2 z}{\partial y^2} = 0$

20. If $u = x^2.\tan^{-1}(y/x) - y^2\tan^{-1}(x/y)$, evaluate $x^2.\dfrac{\partial^2 u}{\partial x^2} + 2xy.\dfrac{\partial^2 u}{\partial x\partial y} + y^2\dfrac{\partial^2 u}{\partial y^2}$

21. If $V = \dfrac{xz}{x^2 + y^2}$, prove that V satisfies the Laplace's equation.

22. If $u = \dfrac{1}{r}$ and $r^2 = (x - a)^2 + (y - b)^2 + (z - c)^2$, prove that u is a harmonic function.

23. If $z = x.\tan^{-1}(y/x) + x.e^{x/y}$, show that $x^2.\dfrac{\partial^2 z}{\partial x^2} + 2xy.\dfrac{\partial^2 z}{\partial x\partial y} + y^2\dfrac{\partial^2 z}{\partial y^2} = 0$.

24. If $v = x^2 + y^2 + \varphi(xy) + f(y/x)$, then prove that

$$x^2.\frac{\partial^2 v}{\partial x^2} - y^2\frac{\partial^2 v}{\partial y^2} + x.\frac{\partial v}{\partial x} - y.\frac{\partial v}{\partial y} = 4(x^2 - y^2).$$

25. If $u = x^y$, prove that $\dfrac{\partial^3 u}{\partial x^2 \partial y} = \dfrac{\partial^3 u}{\partial x\partial y\partial x}$.

26. If $V = r^n$, where $r^2 = x^2 + y^2 + z^2$, show that $\dfrac{\partial^2 V}{\partial x^2} + \dfrac{\partial^2 V}{\partial y^2} + \dfrac{\partial^2 V}{\partial z^2} = n(n+1).r^{n-2}$.

27. If $\varphi = e^{-2x}.\sin x. \sin t$, show that $\dfrac{\partial^2 \varphi}{\partial x^2} + 4.\dfrac{\partial \varphi}{\partial x} = 5.\dfrac{\partial^2 \varphi}{\partial t^2}$.

28. If $V = \left(Ar^n + \dfrac{B}{r^n} \right) \cos(n\theta - \alpha)$, where A, B, n, r are arbitrary constants, show that

$$\dfrac{\partial^2 V}{\partial r^2} + \dfrac{1}{r}.\dfrac{\partial V}{\partial r} + \dfrac{1}{r^2}.\dfrac{\partial^2 V}{\partial \theta^2} = 0.$$

29. If $p^2 = c^2 q - \dfrac{1}{4}k^2$ and $\varphi = Ae^{\frac{1}{2kt}} \sin pt. \cos qt$, show that φ satisfies the equation

$$c^2 \dfrac{\partial^2 \varphi}{\partial x^2} = \dfrac{\partial^2 \varphi}{\partial t^2} + k.\dfrac{\partial \varphi}{\partial t}.$$

30. If $u = r^n (\cos 2\theta + \cos^2 \theta)$ is a solution of $\dfrac{\partial}{\partial r}\left(r^2.\dfrac{\partial u}{\partial r} \right) + \dfrac{1}{\sin \theta}.\dfrac{\partial}{\partial \theta}\left(\sin \theta.\dfrac{\partial u}{\partial \theta} \right) = 0$ prove

that u = 2 or -3.

4.2 TOTAL DERIVATIVES

Definition 4.2.1 Let z = f(x, y) where x and y are continuous functions of another variable t. Substituting for x and y in terms of t, we get z as a function of t alone. Then we can find $\dfrac{dz}{dt}$, which is called the **total differential or the total derivative of z w.r.t t.**

Let Δt be a small change in t. The corresponding changes in x, y and z be Δx, Δy and Δz respectively.

Then $\Delta z = f(x + \Delta x, y + \Delta y) - f(x, y)$

$\qquad = f(x + \Delta x, y + \Delta y) - f(x, y + \Delta y) + f(x, y + \Delta y) - f(x, y)$

$\dfrac{\Delta z}{\Delta t} = \dfrac{f(x + \Delta x, y + \Delta y) - f(x, y + \Delta y)}{\Delta x}.\dfrac{\Delta x}{\Delta t} + \dfrac{f(x, y + \Delta y) - f(x, y)}{\Delta y}.\dfrac{\Delta y}{\Delta t}$

When $\Delta t \to 0$, Δx and Δy also $\to 0$

$\therefore$ Taking limit as $\Delta t \to 0$, since f is a function of x and y, whereas x and y are functions of t only, we get,

$$\dfrac{dz}{dt} = \dfrac{\partial f}{\partial x}.\dfrac{dx}{dt} + \dfrac{\partial f}{\partial y}.\dfrac{dy}{dt} \qquad\qquad (1)$$

Note:

(i) $dz = \dfrac{\partial z}{\partial x}.dx + \dfrac{\partial z}{\partial y}.dy$ is called **total derivative or total differential of z.**

(ii)   When x = t, (1) becomes $\dfrac{dz}{dx} = \dfrac{\partial z}{\partial x} + \dfrac{\partial z}{\partial y}.\dfrac{dy}{dx}$ (2)

(iii)   When f(x, y) = c (Implicit relation between x and y) (2) becomes

$$0 = \frac{df}{dx} = \frac{\partial f}{\partial x} + \frac{\partial f}{\partial y}.\frac{dy}{dx}$$

i.e., $\dfrac{dy}{dx} = - \dfrac{\partial f}{\partial x} \Big/ \dfrac{\partial f}{\partial y}$

Example 4.2.1 If $u = x^3 y^4 z^2$, $x = t^2$, $y = t^3$, $z = t^4$, find $\dfrac{du}{dt}$

Solution:

We have $\dfrac{du}{dt} = \dfrac{\partial u}{\partial x}.\dfrac{dx}{dt} + \dfrac{\partial u}{\partial y}.\dfrac{dy}{dt} + \dfrac{\partial u}{\partial z}.\dfrac{dz}{dt}$ (1)

$$\frac{\partial u}{\partial x} = 3x^2 y^4 z^2, \quad \frac{\partial u}{\partial y} = 4y^3 x^3 z^2, \quad \frac{\partial u}{\partial z} = 2zx^3 y^4$$

$$\frac{dx}{dt} = 2t, \quad \frac{dy}{dt} = 3t^2, \quad \frac{dz}{dt} = 4t^3$$

∴   From (1), we get $\dfrac{du}{dt} = 3x^2 y^4 z^2.2t + 4x^3 y^3 z^2.3t^2 + 2x^3 y^4 z.4t^3$

$$= 3.t^4 t^{12} t^8.2t + 4.t^6 t^9 t^8.3t^2 + 2.t^6 t^{12} t^4.4t^3$$
$$= 26t^5.$$

Example 4.2.2 Find $\dfrac{du}{dx}$ if $u = \tan^{-1}\left(\dfrac{y}{x}\right)$ and $y = \tan^2 x$

Solution:

We have $\dfrac{du}{dx} = \dfrac{\partial u}{\partial x} + \dfrac{\partial u}{\partial y}.\dfrac{dy}{dx}$

$$\frac{\partial u}{\partial x} = \frac{1}{1 + \left(\dfrac{y}{x}\right)^2}.y\left(\frac{-1}{x^2}\right) = \frac{-y}{x^2 + y^2}$$

$$\frac{\partial u}{\partial y} = \frac{1}{1 + \left(\dfrac{y}{x}\right)^2}.\left(\frac{1}{x}\right) = \frac{x}{x^2 + y^2}$$

$$\frac{dy}{dx} = 2\tan x.\sec^2 x$$

∴ $\dfrac{du}{dx} = \dfrac{-y}{x^2 + y^2} + \dfrac{x}{x^2 + y^2}.2\tan x.\sec^2 x$

$$= \frac{2x.\tan x.\sec^2 x - y}{x^2 + y^2}$$

Example 4.2.3 If $f(x, y) = 0$, show that $\dfrac{d^2y}{dx^2} = -\dfrac{q^2r - 2pqs + p^2t}{q^3}$

where $p = \dfrac{\partial f}{\partial x}, q = \dfrac{\partial f}{\partial y}, r = \dfrac{\partial^2 f}{\partial x^2}, s = \dfrac{\partial^2 f}{\partial x \partial y}, t = \dfrac{\partial^2 f}{\partial y^2}$

Solution: We have, $\dfrac{dy}{dx} = -\dfrac{\partial f}{\partial x} \bigg/ \dfrac{\partial f}{\partial y} = \dfrac{-p}{q}$

$$\therefore \quad \frac{d^2y}{dx^2} = -\frac{d}{dx}\left(\frac{p}{q}\right)$$

$$= -\frac{q.\dfrac{dp}{dx} - p.\dfrac{dq}{dx}}{q^2}$$

$$= -\frac{q\left[\dfrac{\partial p}{\partial x} + \dfrac{\partial p}{\partial y}.\dfrac{dy}{dx}\right] - p\left[\dfrac{\partial q}{\partial x} + \dfrac{\partial q}{\partial y}.\dfrac{dy}{dx}\right]}{q^2}$$

$$= -\frac{q\left[r + s.\left(-\frac{p}{q}\right)\right] - p\left[s + t.\left(-\frac{p}{q}\right)\right]}{q^2}$$

$$= -\frac{q^2r - 2pqs + p^2t}{q^3} = -\frac{p^2t - 2pqs - q^2r}{q^3}$$

Example 4.2.4 Use partial differentiation to find $\dfrac{d^2y}{dx^2}$ when $x^3 + y^3 - 3axy = 0$

Solution:

$$f(x, y) = x^3 + y^3 - 3axy$$

$$\frac{\partial f}{\partial x} = 3x^2 - 3ay \qquad \frac{\partial f}{\partial y} = 3y^2 - 3ax$$

$$\frac{dy}{dx} = -\frac{\partial f}{\partial x} \bigg/ \frac{\partial f}{\partial y} = -\frac{3x^2 - 3ay}{3y^2 - 3ax} = -\frac{x^2 - ay}{y^2 - ax}$$

$$\frac{d^2y}{dx^2} = -\frac{d}{dx}\left(\frac{x^2 - ay}{y^2 - ax}\right)$$

$$= \frac{(y^2 - ax)\dfrac{d}{dx}(x^2 - ay) - (x^2 - ay)\dfrac{d}{dx}(y^2 - ax)}{\left(y^2 - ax\right)^2}$$

i.e., $\dfrac{d^2y}{dx^2} = \dfrac{(y^2 - ax)\left[2x + (-a)\frac{dy}{dx}\right] - (x^2 - ay)\left[-a + 2y.\frac{dy}{dx}\right]}{\left(y^2 - ax\right)^2}$

$$= -\frac{(y^2 - ax)\left[2x + a\frac{x^2-ay}{y^2-ax}\right] - (x^2 - ay)\left[-a - 2y.\frac{x^2-ay}{y^2-ax}\right]}{\left(y^2 - ax\right)^2}$$

$$= \frac{(y^2 - ax)\left[2x(y^2 - ax) + a(x^2 - ay)\right] + (x^2 - ay)\left[a(y^2 - ax) + 2y.(x^2 - ay)\right]}{-\left(y^2 - ax\right)^3}$$

$$= \frac{2x(y^2 - ax)^2 + 2a(x^2 - ay)(y^2 - ax) + 2y(x^2 - ay)^2}{(ax - y^2)^3}$$

$$= \frac{2xy(x^3 + y^3 - 3axy) + 2a^3xy}{(ax - y^2)^3}$$

$$= \frac{2a^3xy}{(ax - y^2)^3}$$

Example 4.2.5 Find $\dfrac{du}{dx}$ if $u = x^2y$ and $x^2 + xy + y^2 = 1$

Solution:

We have $\dfrac{du}{dx} = \dfrac{\partial u}{\partial x} + \dfrac{\partial u}{\partial y}.\dfrac{dy}{dx}$

$$= 2xy + x^2.\dfrac{dy}{dx} \hspace{4cm} (1)$$

Let $f(x, y) = x^2 + xy + y^2 - 1$

Then $\dfrac{dy}{dx} = -\dfrac{\partial f / \partial x}{\partial f / \partial y} = \dfrac{-(2x + y)}{2y + x}$

$\therefore$ From (1) $\dfrac{du}{dx} = 2xy + x^2.\dfrac{-(2x + y)}{2y + x}$

$$= \frac{4xy^2 + 2x^2y - 2x^3 - x^2y}{x + 2y}$$

$$= \frac{x(4y^2 + xy - 2x^2)}{x + 2y}$$

4.2.1 PARTIAL DERIVATIVES OF A FUNCTION OF TWO FUNCTIONS

Let z = f(x, y) where x = g(s, t) and y = h(s, t). Substituting for x and y in f(x, y), z can be obtained as a function of s and t. Hence we can find $\frac{\partial z}{\partial s}$ and $\frac{\partial z}{\partial t}$. When t is kept constant, x, y and z are functions of s only. Therefore we have,

$$\frac{\partial z}{\partial s} = \frac{\partial z}{\partial x} \cdot \frac{\partial x}{\partial s} + \frac{\partial z}{\partial y} \cdot \frac{\partial y}{\partial s}$$

Similarly keeping s constant, we obtain

$$\frac{\partial z}{\partial t} = \frac{\partial z}{\partial x} \cdot \frac{\partial x}{\partial t} + \frac{\partial z}{\partial y} \cdot \frac{\partial y}{\partial t}$$

Example 4.2.6 If u = f(x, y) and x = rcos θ, y = rsin θ, prove that

$$\left(\frac{\partial u}{\partial x}\right)^2 + \left(\frac{\partial u}{\partial y}\right)^2 = \left(\frac{\partial u}{\partial r}\right)^2 + \frac{1}{r^2}\left(\frac{\partial u}{\partial \theta}\right)^2$$

Solution:

x = rcos θ

$\therefore \quad \frac{\partial x}{\partial r} = \cos\theta \qquad \frac{\partial x}{\partial \theta} = -r\sin\theta$

y = rsin θ

$\therefore \quad \frac{\partial y}{\partial r} = \sin\theta \qquad \frac{\partial y}{\partial \theta} = r\cos\theta$

We have $\dfrac{\partial u}{\partial r} = \dfrac{\partial u}{\partial x} \cdot \dfrac{\partial x}{\partial r} + \dfrac{\partial u}{\partial y} \cdot \dfrac{\partial y}{\partial r}$

i.e., $\quad \dfrac{\partial u}{\partial r} = \dfrac{\partial u}{\partial x} \cdot \cos\theta + \dfrac{\partial u}{\partial y} \cdot \sin\theta$ \hfill (1)

Also we have $\dfrac{\partial u}{\partial \theta} = \dfrac{\partial u}{\partial x} \cdot \dfrac{\partial x}{\partial \theta} + \dfrac{\partial u}{\partial y} \cdot \dfrac{\partial y}{\partial \theta}$

i.e., $\quad \dfrac{\partial u}{\partial \theta} = \dfrac{\partial u}{\partial x} \cdot (-r\sin\theta) + \dfrac{\partial u}{\partial y} \cdot r\cos\theta$

$\therefore \quad \dfrac{1}{r}\dfrac{\partial u}{\partial \theta} = -\dfrac{\partial u}{\partial x} \cdot \sin\theta + \dfrac{\partial u}{\partial y} \cdot \cos\theta$ \hfill (2)

Squaring and adding (1) and (2), we get,

$$\left(\frac{\partial u}{\partial r}\right)^2 + \frac{1}{r^2}\left(\frac{\partial u}{\partial \theta}\right)^2 = \left(\frac{\partial u}{\partial x}\right)^2\left(\cos^2\theta + \sin^2\theta\right) + \left(\frac{\partial u}{\partial y}\right)^2\left(\sin^2\theta + \cos^2\theta\right)$$

$$= \left(\frac{\partial u}{\partial x}\right)^2 + \left(\frac{\partial u}{\partial y}\right)^2$$

$$\therefore \quad \left(\frac{\partial u}{\partial x}\right)^2 + \left(\frac{\partial u}{\partial y}\right)^2 = \left(\frac{\partial u}{\partial r}\right)^2 + \frac{1}{r^2}\left(\frac{\partial u}{\partial \theta}\right)^2$$

Example 4.2.7 If $u = f(x, y)$ where $x = r\cos\theta$ and $y = r\sin\theta$, prove that

$$\frac{\partial^2 u}{\partial x^2} + \frac{\partial^2 u}{\partial y^2} = \frac{\partial^2 u}{\partial r^2} + \frac{1}{r^2}\frac{\partial^2 u}{\partial \theta^2} + \frac{1}{r}\frac{\partial u}{\partial r}$$

Solution:
We have $x = r\cos\theta$ and $y = r\sin\theta$.

$$\therefore \quad r = \sqrt{x^2 + y^2} \quad \text{and } \theta = \tan^{-1}(y/x)$$

Hence $\dfrac{\partial r}{\partial x} = \dfrac{x}{\sqrt{x^2 + y^2}} = \cos\theta$, $\dfrac{\partial r}{\partial y} = \dfrac{y}{\sqrt{x^2 + y^2}} = \sin\theta$

$$\frac{\partial \theta}{\partial x} = \frac{-y}{x^2 + y^2} = -\frac{\sin\theta}{r}, \qquad \frac{\partial \theta}{\partial y} = \frac{x}{x^2 + y^2} = \frac{\cos\theta}{r}$$

Now substituting for x and y, u is obtained as a function of r and θ, where r and θ are functions of x and y.

$$\frac{\partial u}{\partial x} = \frac{\partial u}{\partial r}\cdot\frac{\partial r}{\partial x} + \frac{\partial u}{\partial \theta}\cdot\frac{\partial \theta}{\partial x} = \cos\theta\cdot\frac{\partial u}{\partial r} - \frac{\sin\theta}{r}\frac{\partial u}{\partial \theta}$$

$$\therefore \quad \frac{\partial}{\partial x} = \cos\theta\frac{\partial}{\partial r} - \frac{\sin\theta}{r}\cdot\frac{\partial}{\partial \theta}$$

Similarly $\dfrac{\partial u}{\partial y} = \dfrac{\partial u}{\partial r}\cdot\dfrac{\partial r}{\partial y} + \dfrac{\partial u}{\partial \theta}\cdot\dfrac{\partial \theta}{\partial y} = \sin\theta\cdot\dfrac{\partial u}{\partial r} + \dfrac{\cos\theta}{r}\dfrac{\partial u}{\partial \theta}$

and $\dfrac{\partial}{\partial y} = \sin\theta\dfrac{\partial}{\partial r} + \dfrac{\cos\theta}{r}\cdot\dfrac{\partial}{\partial \theta}$

$$\therefore \quad \frac{\partial^2 u}{\partial x^2} = \frac{\partial}{\partial x}\left(\frac{\partial u}{\partial x}\right) = \left(\cos\theta\frac{\partial}{\partial r} - \frac{\sin\theta}{r}\frac{\partial}{\partial \theta}\right)\left(\cos\theta\frac{\partial u}{\partial r} - \frac{\sin\theta}{r}\frac{\partial u}{\partial \theta}\right)$$

$$= \cos^2\theta\cdot\frac{\partial^2 u}{\partial r^2} + \frac{\sin^2\theta}{r^2}\cdot\frac{\partial^2 u}{\partial \theta^2} - \frac{\sin\theta}{r}\frac{\partial}{\partial \theta}\left(\cos\theta\frac{\partial u}{\partial r}\right) - \cos\theta\frac{\partial}{\partial r}\left(\frac{\sin\theta}{r}\cdot\frac{\partial u}{\partial \theta}\right)$$

$$= \cos^2\theta\cdot\frac{\partial^2 u}{\partial r^2} + \frac{\sin^2\theta}{r^2}\cdot\frac{\partial^2 u}{\partial \theta^2} - \frac{\sin\theta}{r}\left[-\sin\theta\frac{\partial u}{\partial r} + \cos\theta\frac{\partial^2 u}{\partial \theta\partial r}\right]$$

$$-\cos\theta\left[\frac{-\sin}{r^2}\cdot\frac{\partial u}{\partial\theta}+\frac{\sin\theta}{r}\cdot\frac{\partial^2 u}{\partial r\partial\theta}\right]$$

$$=\cos^2\theta.\frac{\partial^2 u}{\partial r^2}+\frac{\sin^2\theta}{r^2}\cdot\frac{\partial^2 u}{\partial\theta^2}-\frac{2\sin\theta\cos\theta}{r}\cdot\frac{\partial^2 u}{\partial r\partial\theta}+\frac{\sin^2\theta}{r}\cdot\frac{\partial u}{\partial r}+\frac{\sin\theta\cos\theta}{r^2}\cdot\frac{\partial u}{\partial\theta}\qquad(1)$$

$$\therefore\frac{\partial^2 u}{\partial y^2}=\frac{\partial}{\partial y}\left(\frac{\partial u}{\partial y}\right)=\left(\sin\theta\frac{\partial}{\partial r}+\frac{\cos\theta}{r}\frac{\partial}{\partial\theta}\right)\left(\sin\theta\frac{\partial u}{\partial r}+\frac{\cos\theta}{r}\frac{\partial u}{\partial\theta}\right)$$

$$=\sin^2\theta.\frac{\partial^2 u}{\partial r^2}+\frac{\cos^2\theta}{r^2}\cdot\frac{\partial^2 u}{\partial\theta^2}+\sin\theta\frac{\partial}{\partial r}\left(\frac{\cos\theta}{r}\frac{\partial u}{\partial\theta}\right)+\frac{\cos\theta}{r}\frac{\partial}{\partial\theta}\left(\sin\theta.\frac{\partial u}{\partial r}\right)$$

$$=\sin^2\theta.\frac{\partial^2 u}{\partial r^2}+\frac{\cos^2\theta}{r^2}\cdot\frac{\partial^2 u}{\partial\theta^2}+\frac{2\sin\theta.\cos\theta}{r}\cdot\frac{\partial^2 u}{\partial r\partial\theta}+\frac{\cos^2\theta}{r}\frac{\partial u}{\partial r}-\frac{\sin\theta.\cos\theta}{r^2}\cdot\frac{\partial u}{\partial\theta}\qquad(2)$$

Adding (1) and (2) we get

$$\frac{\partial^2 u}{\partial x^2}+\frac{\partial^2 u}{\partial y^2}=\frac{\partial^2 u}{\partial r^2}+\frac{1}{r^2}\frac{\partial^2 u}{\partial\theta^2}+\frac{1}{r}\frac{\partial u}{\partial r}$$

Example 4.2.8 Given that $u(x, y, z) = f(x^2 + y^2 + z^2)$ where $x = r\cos\theta\cos\varphi$, $y = r\cos\theta\sin\varphi$, $y = r\sin\theta$. Find $\dfrac{\partial u}{\partial\theta}$ and $\dfrac{\partial u}{\partial\varphi}$.

Solution:

$$\frac{\partial x}{\partial\theta}=-r\sin\theta\cos\varphi,\quad\frac{\partial x}{\partial\varphi}=-r\cos\theta\sin\varphi$$

$$\frac{\partial y}{\partial\theta}=-r\sin\theta\sin\varphi,\quad\frac{\partial y}{\partial\varphi}=r\cos\theta\cos\varphi$$

$$\frac{\partial z}{\partial\theta}=r\cos\theta,\quad\frac{\partial z}{\partial\varphi}=0$$

Also substituting $x^2 + y^2 + z^2 = w$, we have $u(x, y, z) = f(w)$

$$\therefore\quad\frac{\partial u}{\partial x}=f'(w).\frac{\partial w}{\partial x}=f'(w).2x$$

Similarly $\dfrac{\partial u}{\partial y}=f'(w).2y$ and $\dfrac{\partial u}{\partial z}=f'(w).2z$

Now, $\dfrac{\partial u}{\partial\theta}=\dfrac{\partial u}{\partial x}\cdot\dfrac{\partial x}{\partial\theta}+\dfrac{\partial u}{\partial y}\cdot\dfrac{\partial y}{\partial\theta}+\dfrac{\partial u}{\partial z}\dfrac{\partial z}{\partial\theta}$

$$=f'(w).2x(-r\sin\theta\cos\varphi)+f'(w).2y(-r\sin\theta\sin\varphi)+f'(w).2z.r\cos\theta$$

$$= 2f'(w)[-r^2 \cos\theta \sin\theta \cos^2\varphi - r^2 \cos\theta \sin\theta \sin^2\varphi + r^2 \sin\theta \cos\theta]$$

$$= 2f'(w)[-r^2 \cos\theta \sin\theta + r^2 \sin\theta \cos\theta]$$

$$= 2f'(w)[0] = 0$$

$$\frac{\partial u}{\partial \varphi} = \frac{\partial u}{\partial x}\cdot\frac{\partial x}{\partial \varphi} + \frac{\partial u}{\partial y}\cdot\frac{\partial y}{\partial \varphi} + \frac{\partial u}{\partial z}\frac{\partial z}{\partial \varphi}$$

$$= f'(w).2x(-r \cos\theta \sin\varphi) + f'(w).2y(-r\cos\theta \cos\varphi) + f'(w).0$$

$$= 2f'(w)[-r^2 \cos^2\theta \cos\varphi\sin\varphi + r^2 \cos^2\theta\sin\varphi \cos\varphi]$$

$$= 2f'(w)[0] = 0$$

Example 4.2.9 If $u = u(x, y)$ and $x = e^r\cos\theta$, $y = e^r\sin\theta$ show that

$$\left(\frac{\partial u}{\partial x}\right)^2 + \left(\frac{\partial u}{\partial y}\right)^2 = e^{-2r}\cdot\left[\left(\frac{\partial u}{\partial r}\right)^2 + \left(\frac{\partial u}{\partial \theta}\right)^2\right]$$

Solution:

We have $\dfrac{\partial u}{\partial r} = \dfrac{\partial u}{\partial x}\cdot\dfrac{\partial x}{\partial r} + \dfrac{\partial u}{\partial y}\cdot\dfrac{\partial y}{\partial r}$

$$= \frac{\partial u}{\partial x}\left(e^r\cos\theta\right) + \frac{\partial u}{\partial y}\left(e^r\sin\theta\right) \qquad (1)$$

$$\frac{\partial u}{\partial \theta} = \frac{\partial u}{\partial x}\cdot\frac{\partial x}{\partial \theta} + \frac{\partial u}{\partial y}\cdot\frac{\partial y}{\partial \theta}$$

$$= \frac{\partial u}{\partial x}\left(-e^r\sin\theta\right) + \frac{\partial u}{\partial y}\left(e^r\cos\theta\right) \qquad (2)$$

Squaring and adding (1) and (2), we get

$$\left(\frac{\partial u}{\partial r}\right)^2 + \left(\frac{\partial u}{\partial \theta}\right)^2 = e^{2r}\cdot\left(\frac{\partial u}{\partial x}\right)^2 + e^{2r}\cdot\left(\frac{\partial u}{\partial y}\right)^2$$

i.e., $\left(\dfrac{\partial u}{\partial x}\right)^2 + \left(\dfrac{\partial u}{\partial y}\right)^2 = e^{-2r}\left[\left(\dfrac{\partial u}{\partial r}\right)^2 + \left(\dfrac{\partial u}{\partial \theta}\right)^2\right]$

Example 4.2.10 If $z = u^2 + v^2$, $x = u^2 - v^2$, $y = uv$, find $\dfrac{\partial z}{\partial x}$ and $\dfrac{\partial z}{\partial y}$.

Solution:

We have $\dfrac{\partial z}{\partial x} = \dfrac{\partial z}{\partial u}\cdot\dfrac{\partial u}{\partial x} + \dfrac{\partial z}{\partial v}\cdot\dfrac{\partial v}{\partial x}$

i.e., $\dfrac{\partial z}{\partial x} = 2u\cdot\dfrac{\partial u}{\partial x} + 2v\cdot\dfrac{\partial v}{\partial x} \qquad (1)$

Differentiating $x = u^2 - v^2$ and $y = uv$ partially w.r.t x, we get

$$1 = 2u.\frac{\partial u}{\partial x} - 2v.\frac{\partial v}{\partial x} \qquad (2)$$

$$0 = v.\frac{\partial u}{\partial x} + u.\frac{\partial v}{\partial x} \qquad (3)$$

Solving (2) and (3) we get

$$\frac{\partial u}{\partial x} = \frac{u}{2(u^2 + v^2)}, \quad \frac{\partial v}{\partial x} = \frac{-v}{2(u^2 + v^2)}$$

$$\therefore \text{from (1)} \quad \frac{\partial z}{\partial x} = \frac{2u^2}{2(u^2 + v^2)} + \frac{(-2v^2)}{2(u^2 + v^2)}$$

$$= \frac{u^2 - v^2}{u^2 + v^2} = \frac{x}{z}$$

We have, $\quad \dfrac{\partial z}{\partial y} = \dfrac{\partial z}{\partial u}.\dfrac{\partial u}{\partial y} + \dfrac{\partial z}{\partial y}.\dfrac{\partial v}{\partial y}$

i.e., $\quad \dfrac{\partial z}{\partial y} = 2u.\dfrac{\partial u}{\partial y} + 2v.\dfrac{\partial v}{\partial y} \qquad (4)$

Differentiating $x = u^2 - v^2$ and $y = uv$ partially w.r.t y, we get

$$0 = 2u.\frac{\partial u}{\partial y} - 2v.\frac{\partial v}{\partial y} \qquad (5)$$

$$1 = v.\frac{\partial u}{\partial y} + u.\frac{\partial v}{\partial y} \qquad (6)$$

Solving (5) and (6) we get

$$\frac{\partial u}{\partial y} = \frac{v}{u^2 + v^2} \quad \text{and} \quad \frac{\partial v}{\partial y} = \frac{u}{u^2 + v^2}$$

$$\therefore \text{from (4)} \quad \frac{\partial z}{\partial y} = \frac{2uv}{u^2 + v^2} + \frac{2uv}{u^2 + v^2}$$

$$= \frac{4uv}{u^2 + v^2}$$

$$= \frac{4.y}{z}$$

Note: The above problem can be solved by first eliminating u and v from the given equations.

$$z^2 = (u^2 + v^2)^2 = (u^2 - v^2)^2 + 4u^2v^2 = x^2 + 4y^2$$

$$\therefore 2z\frac{\partial z}{\partial x} = 2x \text{ and } 2z\frac{\partial z}{\partial y} = 8y.$$

i.e., $\quad \dfrac{\partial z}{\partial x} = \dfrac{x}{z} \text{ and } \dfrac{\partial z}{\partial y} = \dfrac{4y}{z}$

EXERCISE 4.2

PART - A

1. Define 'total derivative'.

2. If $u = \sin(xy^2)$, $x = \log t$, $y = e^t$ find $\dfrac{du}{dt}$.

3. If $u = \sin(xy^2)$, express the total differential of u in terms of those of x and y.

4. If $u = xy \log xy$, express du in terms of dx and dy.

5. Find $\dfrac{du}{dt}$, if $u = x^3y^2 + x^2y^3$, where $x = at^2$, $y = 2at$.

6. Find $\dfrac{du}{dx}$, if $u = x^2 + y^2 + a^2$ and $x^3 + y^3 = a^3$.

7. Find $\dfrac{du}{dx}$, if $u = x \log(xy)$ where $x^3 + y^3 - 3axy = 0$.

8. If $xy = c^2$, use partial differentiation to find $\dfrac{du}{dx}$.

9. If $x^3 + 3x^2y + 6xy^2 + y^3 = 1$, find $\dfrac{dy}{dx}$ using partial differentiation.

10. Find $\dfrac{du}{dx}$ if $u = \sin(x^2 + y^2)$ and $\dfrac{x^2}{a^2} + \dfrac{y^2}{b^2} = 1$.

PART - B

11. If $V = f\left(\dfrac{x}{z}, \dfrac{y}{z}\right)$, prove that $x.\dfrac{\partial V}{\partial x} + y.\dfrac{\partial V}{\partial y} + z.\dfrac{\partial V}{\partial z} = 0$.

12. If $V = f(x - y, y - z, z - x)$, prove that $\dfrac{\partial V}{\partial x} + \dfrac{\partial V}{\partial y} + \dfrac{\partial V}{\partial z} = 0$.

13. If $V = f(x, y)$ and if $x = u^2 - v^2$, $y = 2uv$, prove that (i) $u.\dfrac{\partial V}{\partial u} - v.\dfrac{\partial V}{\partial v} = 2(u^2 + v^2).\dfrac{\partial V}{\partial x}$

    (ii) $\dfrac{\partial^2 V}{\partial x^2} + \dfrac{\partial^2 V}{\partial y^2} = \dfrac{1}{4(u^2 + v^2)}\left[\dfrac{\partial^2 V}{\partial u^2} + \dfrac{\partial^2 V}{\partial v^2}\right]$

14. If z is a function of x and y and if $x = e^u \sin v$, $y = e^u \cos v$, prove that

    (i) $\dfrac{\partial^2 z}{\partial u^2} + \dfrac{\partial^2 z}{\partial v^2} = (x^2 + y^2)\left[\dfrac{\partial^2 z}{\partial x^2} + \dfrac{\partial^2 z}{\partial y^2}\right]$

    (ii) $\dfrac{\partial z}{\partial x} = e^{-u}\left(\sin v.\dfrac{\partial z}{\partial u} + \cos v.\dfrac{\partial z}{\partial v}\right)$

15. If $f(x, y) = \theta(u, v)$ where $u = x^2 - y^2$, $v = 2xy$ show that

$$\frac{\partial^2 f}{\partial x^2} + \frac{\partial^2 f}{\partial y^2} = 4.(x^2 + y^2)\left[\frac{\partial^2 \theta}{\partial u^2} + \frac{\partial^2 \theta}{\partial v^2}\right]$$

16. If $z = z(x, y)$ where $x = e^u + e^{-v}$ and $y = e^{-u} - e^v$ prove $\dfrac{\partial z}{\partial u} - \dfrac{\partial z}{\partial v} = x.\dfrac{\partial z}{\partial x} - y.\dfrac{\partial z}{\partial y}$.

17. If $x = u + v$, $y = uv$ and z is a function of x and y, prove that

$$u.\frac{\partial z}{\partial u} + v.\frac{\partial z}{\partial v} = x.\frac{\partial z}{\partial x} + 2y.\frac{\partial z}{\partial y}.$$

18. If $x = u^2 + v^2$, $y = 2uv$ and $z = f(x, y)$, prove that $u.\dfrac{\partial z}{\partial u} - v.\dfrac{\partial z}{\partial v} = 2\sqrt{x^2 - y^2}.\dfrac{\partial z}{\partial x}$.

19. If $x = u + v + w$, $y = uv + vw + wu$, $z = uvw$ and $f = f(x, y, z)$, prove that

$$x.\frac{\partial f}{\partial x} + 2y.\frac{\partial f}{\partial y} + 3z.\frac{\partial f}{\partial z} = u.\frac{\partial f}{\partial u} + v.\frac{\partial f}{\partial v} + w.\frac{\partial f}{\partial w}.$$

20. If $V = f(x/y, y/z, z/x)$, show that $x.\dfrac{\partial V}{\partial x} + y.\dfrac{\partial V}{\partial y} + z.\dfrac{\partial V}{\partial z} = 0$

21. If $x = u \cos \alpha - v \sin \alpha$, $y = u \sin \alpha + v \cos \alpha$, show that $\dfrac{\partial^2 z}{\partial x^2} + \dfrac{\partial^2 z}{\partial y^2} = \dfrac{\partial^2 z}{\partial u^2} + \dfrac{\partial^2 z}{\partial v^2}$.

22. If $x = e^r \cos \theta$, $y = e^r \sin \theta$ and u is a function of x and y, prove that

$$\frac{\partial^2 u}{\partial x^2} + \frac{\partial^2 u}{\partial y^2} = e^{-2r}\left[\frac{\partial^2 u}{\partial r^2} + \frac{\partial^2 u}{\partial \theta^2}\right].$$

23. If $u = f(x^2 - y^2, y^2 - z^2, z^2 - x^2)$, prove that $\dfrac{1}{x}.\dfrac{\partial u}{\partial x} + \dfrac{1}{y}.\dfrac{\partial u}{\partial y} + \dfrac{1}{z}.\dfrac{\partial u}{\partial z} = 0$.

24. If z is a function of x and y and u and v are such that $u = lx + my$, $v = ly - mx$, show

25. that $\dfrac{\partial^2 z}{\partial x^2} + \dfrac{\partial^2 z}{\partial y^2} = (l^2 + m^2)\left[\dfrac{\partial^2 z}{\partial u^2} + \dfrac{\partial^2 z}{\partial v^2}\right]$.

26. If $z = uv$, $u^2 + v^2 - x - y = 0$, $u^2 - v^2 + 3x + y = 0$, find $\dfrac{\partial z}{\partial x}$.

27. If $f = f\left(\dfrac{y - x}{yx}, \dfrac{z - x}{zx}\right)$, show that $x^2.\dfrac{\partial f}{\partial x} + y^2.\dfrac{\partial f}{\partial y} + z^2.\dfrac{\partial f}{\partial z} = 0$.

28. If $z = f(u, v)$, where $u = x^2 - 2xy - y^2$ and $v = y$, show that the equation

$$(x + y).\frac{\partial z}{\partial x} + (x - y).\frac{\partial z}{\partial y} = 0 \text{ is equivalent to } \frac{\partial z}{\partial v} = 0.$$

29. Transform the equation $\dfrac{\partial^2 z}{\partial x^2} - \dfrac{\partial^2 z}{\partial y^2} = 0$ by changing the independent variables using

$y = x + y$ and $v = x - y$

30. Transform the equation $x^2 \cdot \dfrac{\partial^2 z}{\partial x^2} + 2xy \cdot \dfrac{\partial^2 z}{\partial x \partial y} + y^2 \dfrac{\partial^2 z}{\partial y^2} = 0,$ by changing the

   independent variables using $u = x$ and $v = \dfrac{y^2}{x}$.

31. Transform the equation $\dfrac{\partial^2 u}{\partial x^2} + \dfrac{\partial^2 u}{\partial y^2} = 0$ by changing the independent variables

   using $z = x + iy$, $\bar{z} = x - iy$.

32. If $u = \sin^{-1}(x - y)$, $x = 3t$, $y = 4t^3$, find $\dfrac{du}{dt}$.

33. Find $\dfrac{du}{dx}$ if $u = \sin(x^2 + y^2)$ and $a^2x^2 + b^2y^2 = c^2$.

34. Find $\dfrac{du}{dt}$ if $u = \log(x^2 + y^2)$, $x = \sqrt{1+t}$, $y = \sqrt{1-t}$

35. Use partial differentiation to find $\dfrac{d^2y}{dx^2}$, when $x^4 + y^4 = 4a^2xy$.

36. Use partial differentiation to find $\dfrac{dy}{dx}$ when $x^m \cdot y^m = (x + y)^{m+n}$

37. If $ay^2 + 2by + c = x^2$, use partial differentiation to find $\dfrac{d^2y}{dx^2}$.

4.3 TAYLORE'S EXPANSION FOR FUNCTIONS OF TWO VARIABLES

When we have a function $f(x)$ of a single variable x with continuous derivatives $f'(x)$, $f''(x)$, $f'''(x)$ ... then the Taylor's expansion of $f(x + h)$ about the point x is given by

$$f(x + h) = f(x) + \frac{h}{1!}f'(x) + \frac{h^2}{2!}f''(x) + \cdots .$$ A similar expansion is possible when we have a function $f(x, y)$ of two variables x and y.

4.3.1 TAYLOR'S THEOREM

Theorem 4.3.1 If $f(x, y)$ and all its partial derivatives are continuous in any neighbourhood of the point (x, y), then,

$$f(x + h, y + k) = f(x, y) + \frac{1}{1!}\left(h\frac{\partial}{\partial x} + k\frac{\partial}{\partial y} \right)f + \frac{1}{2!}\left(h.\frac{\partial}{\partial x} + k\frac{\partial}{\partial y} \right)^2 f$$

$$+ \frac{1}{3!}\left(h\frac{\partial}{\partial x} + k\frac{\partial}{\partial y} \right)^3 f + \cdots$$

Proof: Considering $f(x + h, y + k)$ as a function of a single variables x and applying Taylor's series expansion of a function of a single variable, we have,

$$f(x+h, y+k) = f(x, y+k) + \frac{h}{1!}\frac{\partial f(x, y+k)}{\partial x} + \frac{h^2}{2!}\frac{\partial^2 f(x, y+k)}{\partial x^2} + \cdots \qquad (1)$$

Now treating $f(x, y + k)$ as a function of y only, we have

$$f(x, y+k) = f(x, y) + \frac{k}{1!}\frac{\partial f(x, y)}{\partial y} + \frac{k^2}{2!}\frac{\partial^2 f(x, y)}{\partial y^2} + \cdots$$

∴ (1) becomes

$$f(x+h, y+k) = f(x, y) + \frac{k}{1!}\frac{\partial f(x, y)}{\partial y} + \frac{k^2}{2!}\frac{\partial^2 f(x, y)}{\partial y^2} + \cdots$$

$$\cdots + \frac{h}{1!}\frac{\partial}{\partial x}\left\{ f(x, y) + \frac{k}{1!}\frac{\partial f(x, y)}{\partial y} + \frac{k^2}{2!}\frac{\partial^2 f(x, y)}{\partial y^2} + \cdots \right\}$$

$$+ \frac{h^2}{2!}\frac{\partial^2}{\partial x^2}\left\{ f(x, y) + \frac{k}{1!}\frac{\partial f(x, y)}{\partial y} + \frac{k^2}{2!}\frac{\partial^2 f(x, y)}{\partial y^2} + \cdots \right\} + \cdots$$

$$= f(x, y) + \frac{1}{1!}\left(h.\frac{\partial f}{\partial x} + k.\frac{\partial f}{\partial y} \right) + \frac{1}{2!}\left(h^2\frac{\partial^2 f}{\partial x^2} + 2hk\frac{\partial^2 f}{\partial x\partial y} + k^2\frac{\partial^2 f}{\partial y^2} \right) + \cdots$$

$$= f(x, y) + \frac{1}{1!}\left(h.\frac{\partial}{\partial x} + k.\frac{\partial}{\partial y} \right)f + \frac{1}{2!}\left(h\frac{\partial}{\partial x} + k\frac{\partial}{\partial y} \right)^2 f + \cdots$$

Note:

(i)   Putting x = a, y = b, we get

$$f(a+h, b+k) = f(a,b) + \frac{1}{1!}\left(h\frac{\partial}{\partial x} + k\frac{\partial}{\partial y} \right)f(a, b) + \frac{1}{2!}\left(h.\frac{\partial}{\partial x} + k\frac{\partial}{\partial y} \right)^2 f(a, b) + \cdots \quad (2)$$

(ii) Putting a + h = x and b + k = y in (2), we get

$$f(x, y) = f(a, b) + \frac{1}{1!}\left((x-a)\frac{\partial}{\partial x} + (y-b)\frac{\partial}{\partial y} \right)f(a, b) +$$

$$\frac{1}{2!}\left((x-a).\frac{\partial}{\partial x} + (y-b)\frac{\partial}{\partial y} \right)^2 f(a, b) + \cdots \qquad (3)$$

This is the Taylor's expansion of $f(x, y)$ in the neighbourhood of (a, b).

(iii) Putting a = b = 0 in (3), we get

$$f(x, y) = f(0,0) + \frac{1}{1!}\left(x\frac{\partial}{\partial x} + y\frac{\partial}{\partial y} \right)f(0,0) + \frac{1}{2!}\left(x.\frac{\partial}{\partial x} + y\frac{\partial}{\partial y} \right)^2 f(0,0) + \cdots \qquad (4)$$

The series (4) is called the **Maclaurin's expansion** of $f(x, y)$ in powers of x and y. It is the expansion of $f(x, y)$ in the neighbourhood of the origin $(0, 0)$.

Thus the **Maclaurin's expansion** of f(x, y) is

$$f(x,y) = f(0,0) + \frac{1}{1!}\big(x.f_x(0,0) + y.f_y(0,0)\big) +$$

$$\frac{1}{2!}\big(x^2.f_{xx}(0,0) + 2xy.f_{xy}(0,0) + y^2.f_{yy}(0,0)\big) + \cdots$$

4.3.2   ERRORS AND APPROXIMATIONS

Let the quantity y be a function of the variable x. i.e., y = f(x). Let Δx be small error made while measuring x and Δy be the corresponding error produced in y. Then Δy can be computed in terms of Δx using Taylor's theorem.

$$\Delta y = f(x + \Delta x) - f(x) \tag{1}$$

Expanding f(x + Δx) by Taylor's theorem, we get

$$f(x + \Delta x) = f(x) + \frac{\Delta x}{1!} f'(x) \quad \text{(approximately, neglecting higher powers of } \Delta x)$$

$\therefore$ f(x + Δx) - f(x) = $\Delta x.f'(x)$

i.e., $\Delta y = \Delta x.f'(x)$ (Using (1))

i.e., $\Delta y = \dfrac{df}{dx}.\Delta x$ \hfill (2)

Similarly, when y is a function of two variables x_1 and x_2 and Δx_1 and Δx_2 are the small errors made while measuring x_1 and x_2, then the corresponding error Δy in y can be computed as follows. Let y = f(x_1, x_2).

Then Δy = f(x_1 + Δx_1, x_2 + Δx_2) - f(x_1, x_2) \hfill (3)

By Taylor's theorem, we have

$$f(x_1 + \Delta x_1, x_2 + \Delta x_2) = f(x_1, x_2) + \left(\frac{\partial f}{\partial x_1}.\Delta x_1 + \frac{\partial f}{\partial x_2}.\Delta x_2 \right) \text{(approximately neglecting the}$$

products, squares and higher powers of Δx_1 and Δx_2)

$\therefore$ Using (3)

$$\Delta y = \frac{\partial f}{\partial x_1}.\Delta x_1 + \frac{\partial f}{\partial x_2}.\Delta x_2 \tag{4}$$

From (4) Δy can be computed when f(x_1, x_2), Δx_1 and Δx_2 are known.

In general when y is a function of n variables x_1, x_2 ... x_n or y = f(x_1, x_2 ... x_n),

we have, $\Delta y = \dfrac{\partial f}{\partial x_1}.\Delta x_1 + \dfrac{\partial f}{\partial x_2}.\Delta x_2 + \cdots + \dfrac{\partial f}{\partial x_n}.\Delta x_n$ \hfill (5)

From (5) we can compute the error Δy produced in the variable y, as a result of the errors Δx_1, Δx_2 ... Δx_n in the variables x_1, x_2 ... x_n. Hence (5) is called the **error relation** of the function y = f(x_1, x_2 ... x_n).

Definition 4.3.1 If Δx is the small error in the measured value of x, then Δx is called the **absolute error** in x. Also $\dfrac{\Delta x}{x}$ is called the **relative error** in x and $\dfrac{\Delta x}{x} \times 100$ is called the **percentage error** in x.

Example 4.3.1 Expand $e^x \log (1 + y)$ in powers of x and y.

Solution: We have,

$$f(x, y) = e^x \log(1 + y) \quad \therefore \ f(0, 0) = 0$$

$$f_x(x,y) = e^x \log(1+y) \qquad\qquad \therefore f_x(0,0) = 0$$

$$f_y(x,y) = e^x \frac{1}{1+y} \qquad\qquad\quad f_y(0,0) = 1$$

$$f_{xx}(x,y) = e^x \log(1+y) \qquad\qquad f_{xx}(0,0) = 0$$

$$f_{xy}(x,y) = e^x \frac{1}{1+y} \qquad\qquad\quad f_{xy}(0,0) = 1$$

$$f_{yy}(x,y) = -e^x (1+y)^{-2} \qquad\qquad f_{yy}(0,0) = -1$$

$$f_{xxx}(x,y) = e^x \log(1+y) \qquad\qquad f_{xxx}(0,0) = 0$$

$$f_{xxy}(x,y) = e^x \frac{1}{1+y} \qquad\qquad f_{xxy}(0,0) = 1$$

$$f_{xyy}(x,y) = -e^x (1+y)^{-2} \qquad\qquad f_{xyy}(0,0) = -1$$

$$f_{yyy}(x,y) = 2e^x (1+y)^{-3} \qquad\qquad f_{yyy}(0,0) = 2$$

$$f(x,y) = f(0,0) + x.f_x(0,0) + y.f_y(0,0) + \frac{1}{2!}\left(x^2.f_{xx}(0,0) + 2xy.f_{xy}(0,0) + y^2.f_{yy}(0,0)\right)$$

$$+ \frac{1}{3!}\left(x^3.f_{xxx}(0,0) + 3x^2y.f_{xxy}(0,0) + 3xy^2.f_{xyy}(0,0) + y^3.f_{yyy}(0,0)\right) + \cdots \tag{6}$$

$$\therefore \ e^x \log(1+y) = 0 + x.0 + y.1 + \frac{1}{2!}\left(x^2.0 + 2xy.1 + y^2.(-1)\right)$$

$$+ \frac{1}{3!}\left(x^3.0 + 3x^2y.1 + 3xy^2.(-1) + y^3.2\right) + \cdots$$

$$= y + xy - \frac{1}{2}y^2 + \frac{1}{2}x^2y - \frac{1}{2}xy^2 + \frac{1}{3}y^3 + \cdots$$

Example 4.3.2 Expand $e^x \sin y$ in powers of x and y as far as terms of third degree

Solution:

$$f(x, y) = e^x \sin y \qquad\qquad \therefore f(0, 0) = 0$$

$$f_x(x,y) = e^x \sin y \qquad\qquad\quad f_x(0,0) = 0$$

$$f_y(x, y) = e^x \cos y \qquad\qquad f_y(0,0) = 1$$
$$f_{xx}(x, y) = e^x \sin y \qquad\qquad f_{xx}(0,0) = 0$$
$$f_{xy}(x, y) = e^x \cos y \qquad\qquad f_{xy}(0,0) = 1$$
$$f_{yy}(x, y) = -e^x \sin y \qquad\qquad f_{yy}(0,0) = 0$$
$$f_{xxx}(x, y) = e^x \sin y \qquad\qquad f_{xxx}(0,0) = 0$$
$$f_{xxy}(x, y) = e^x \cos y \qquad\qquad f_{xxy}(0,0) = 1$$
$$f_{xyy}(x, y) = -e^x \sin y \qquad\qquad f_{xyy}(0,0) = 0$$
$$f_{yyy}(x, y) = -e^x \cos y \qquad\qquad f_{yyy}(0,0) = -1$$

Substituting in (6) we get,

$$e^x \sin y = y + xy + \frac{1}{2}x^2 y - \frac{1}{6}y^3 + \cdots$$

Example 4.3.3 Expand sin (xy) in powers of x -1 and y-π/2 upto second degree terms.

Solution:
We have
$$f(x, y) = f(a, b) + (x - a).f_x(a, b) + (y - b).f_y(a, b) +$$

$$\frac{1}{2!}\left((x - a)^2.f_{xx}(a, b) + 2(x - a)(y - b).f_{xy}(a, b) + (y - b)^2.f_{yy}(a, b)\right) + \cdots \qquad (7)$$

Here a = 1, b = π/2.

$$f(x, y) = \sin(xy) \qquad\qquad \therefore f(1, \pi/2) = 1$$
$$f_x(x, y) = y \cos(xy) \qquad\qquad f_x(1, \pi/2) = 0$$
$$f_{xx}(x, y) = -y^2 \sin(xy) \qquad\qquad f_{xx}(1, \pi/2) = -\pi^2\big/4$$
$$f_y(x, y) = x \cos(xy) \qquad\qquad f_y(1, \pi/2) = 0$$
$$f_{yy}(x, y) = -x^2 \sin(xy) \qquad\qquad f_{yy}(1, \pi/2) = -1$$
$$f_{xy}(x, y) = \cos(xy) + x.(-\sin(xy).y)$$
$$\qquad\qquad = \cos(xy) - xy \sin(xy) \qquad\qquad f_{xy}(1, \pi/2) = -\pi\big/2$$

∴ Substituting in (7) we get,

$$\sin(xy) = 1 + (x - 1).0 + (y - \pi/2).0 +$$

$$\frac{1}{2!}\left((x-1)^2\left(-\pi^2\big/4\right) + 2(x-1)(y-\pi/2).(-\pi/2) + (y-\pi/2)^2.(-1)\right) + \cdots$$

$$= 1 - \frac{\pi^2}{8}(x-1)^2 - \frac{\pi}{2}(x-1)(y-\pi/2) - \frac{1}{2}(y-\pi/2)^2 , \quad \text{taking terms up to second}$$

degree only.

Example 4.3.4 Expand $x^2y + 3y - 2$ in powers of $(x -1)$ and $(y + 2)$ using Taylor's theorem

Solution:
Let $f(x, y) = x^2y + 3y - 2$

$$f(x,y) = f(a,b) + (x - a).f_x(a,b) + (y - b).f_y(a,b) +$$

$$\frac{1}{2!}\Big((x-a)^2..f_{xx}(a,b) + 2(x-a)(y-b)..f_{xy}(a,b) + (y-b)^2.f_{yy}(a,b)\Big) +$$

$$\frac{1}{3!}\Big\{(x-a)^3.f_{xxx}(a,b) + 3(x-a)^2(y-b).f_{xxy}(a,b) +$$

$$+ 3(x-a)(y-b)^2 f_{xyy} + (y-b)^3.f_{yyy}(a,b)\Big\} + \cdots$$

$$(8)$$

Putting $a = 1$, $b = -2$

$$f(1, -2) = 1^2.(-2) + 3(-2) - 2 = -10$$

$f_x(x,y) = 2xy$	$f_x(1,-2) = -4$
$f_{xx}(x,y) = 2y$	$f_{xx}(1,-2) = -4$
$f_y(x,y) = x^2 + 3$	$f_y(1,-2) = 4$
$f_{yy}(x,y) = 0$	$f_{yy}(1,-2) = 0$
$f_{xy}(x,y) = 2x$	$f_{xy}(1,-2) = 2$

$$f_{xxx}(x,y) = f_{yyy}(x,y) = f_{xyy}(x,y) = 0$$

$$f_{xxy}(x,y) = 2 \qquad \therefore f_{xxy}(1,-2) = 2$$

The partial derivatives of order 4 and higher orders are 0.

$\therefore$ Substituting in (8) we get,

$$x^2y + 3y - 2 = -10 + (x-1).(-4) + (y+2).4 +$$

$$\frac{1}{2!}\Big((x-1)^2.(-4) + 2(x-1)(y+2).2 + (y+2)^2.0\Big) + \frac{1}{3!}\Big(3.(x-1)^2.(y+2).2\Big)$$

$$= -10 - 4(x-1) + 4(y+2) - 2(x-1)^2 + 2.(x-1)(y+2) + (x-1)^2.(y+2)$$

Example 4.3.5 If $f(x, y) = \tan^{-1}(xy)$, compute $f(0.9, -1.2)$ approximately

Solution:

$$f(0.9, -1.2) = f(1 - 0.1, -1 - 0.2)$$

$$= f(1 + h, -1 + k) \text{ where } h = -0.1, k = -0.2$$

$$= f(1, -1) + h..f_x(1,-1) + k.f_y(1,-1) +$$

$$(9)$$

$$\frac{1}{2!}\Big(h^2..f_{xx}(1,-1) + 2hk..f_{xy}(1,-1) + k^2.f_{yy}(1,-1)\Big) + \cdots$$

$$f(x, y) = \tan^{-1}(xy) \qquad \therefore f(1, -1) = \tan^{-1}(-1) = -\pi/4$$

$$f_x(x,y) = \frac{y}{1+x^2y^2} \qquad\qquad f_x(1,-1) = -\frac{1}{2}$$

$$f_y(x,y) = \frac{x}{1+x^2y^2} \qquad\qquad f_y(1,-1) = \frac{1}{2}$$

$$f_{xx}(x,y) = \frac{-y.2xy^2}{\left(1+x^2y^2\right)^2} \qquad\qquad f_{xx}(1,-1) = \frac{1}{2}$$

$$f_{yy}(x,y) = \frac{-x.2x^2y}{\left(1+x^2y^2\right)^2} \qquad\qquad f_{yy}(1,-1) = \frac{1}{2}$$

$$f_{xy}(x,y) = \frac{1-x^2y^2}{\left(1+x^2y^2\right)^2} \qquad\qquad f_{xy}(1,-1) = 0$$

Substituting in (9) we get,

$$f(0.9, -1.2) = -\frac{\pi}{4} + (-0.1).\left(-\frac{1}{2}\right) + (-0.2).\left(\frac{1}{2}\right) +$$

$$\frac{1}{2}\left((0.2)^2\left(\frac{1}{2}\right) + 2(-0.1)(-0.2).(0) + (-0.2)^2.\left(\frac{1}{2}\right)\right)$$

$$= -0.7854 + 0.05 - 0.1 + 0.5[0.005 + 0.02]$$

$$= -0.7854 + 0.05 - 0.1 + 0.0125 = -0.8229$$

Example 4.3.6 Expand e^{xy} in the neighbourhood of $(1, 1)$ up to second degree terms.

Solution:

$$f(x, y) = e^{xy} \qquad\qquad f(1, 1) = e$$
$$f_x(x, y) = e^{xy}.y \qquad\qquad f_x(1,1) = e$$
$$f_y(x, y) = e^{xy}.x \qquad\qquad f_y(1,1) = e$$
$$f_{xx}(x, y) = e^{xy}.y^2 \qquad\qquad f_{xx}(1,1) = e$$
$$f_{yy}(x, y) = e^{xy}.x^2 \qquad\qquad f_{yy}(1,1) = e$$
$$f_{xy}(x, y) = e^{xy} + x.e^{xy}.y \qquad\qquad f_{xy}(1,1) = 2e$$

Now, $f(x,y) = f(1,1) + (x-1).f_x(1,1) + (y-1).f_y(1,1) +$

$$\frac{1}{2!}\left((x-1)^2.f_{xx}(1,1) + 2(x-1)(y-1).f_{xy}(1,1) + (y-1)^2.f_{yy}(1,1)\right) + \cdots$$

$$= e + (x-1).e + (y-1).e + \frac{1}{2}(x-1)^2.e + (x-1).(y-1).2e + \frac{1}{2}(y-1)^2.e + \cdots$$

$$= e\left[1 + (x-1) + (y-1) + \frac{1}{2}(x-1)^2 + 2(x-1).(y-1) + \frac{1}{2}(y-1)^2 + \cdots\right]$$

Example 4.3.7 Find the percentage error in the area of an ellipse if one percent error is made in measuring the major and minor axes.

Solution:

Area of the ellipse is $A = \pi ab$ where $2a$ and $2b$ are the lengths of the major and minor axes of the ellipse.

Let $\quad 2a = d_1$ and $2b = d_2$.

Given $\dfrac{\Delta d_1}{d_1} \times 100 = 1$ and $\dfrac{\Delta d_2}{d_2} \times 100 = 1$

Now $A = \pi \cdot \dfrac{d_1}{2} \cdot \dfrac{d_2}{2}$

Method 1. $\quad A = \dfrac{\pi}{4} d_1 \cdot d_2 = f(d_1, d_2)$ $\hspace{3cm}$ (1)

$\therefore \quad \dfrac{\partial f}{\partial d_1} = \dfrac{\pi}{4} \cdot d_2$ and $\dfrac{\partial f}{\partial d_2} = \dfrac{\pi}{4} \cdot d_1$.

Now $\Delta A = \dfrac{\partial f}{\partial d_1} \Delta d_1 + \dfrac{\partial f}{\partial d_2} \Delta d_2$

$\therefore \quad \Delta A = \dfrac{\pi}{4} \cdot d_2 \cdot \Delta d_1 + \dfrac{\pi}{4} \cdot d_1 \cdot \Delta d_2$

Dividing by A, we get,

$\dfrac{\Delta A}{A} = \dfrac{\Delta d_1}{d_1} + \dfrac{\Delta d_2}{d_2}$ $\hspace{2cm}$ (Using (1))

$\therefore \quad \dfrac{\Delta A}{A} \times 100 = \dfrac{\Delta d_1}{d_1} \times 100 + \dfrac{\Delta d_2}{d_2} \times 100$

$\hspace{3cm} = 1 + 1 = 2$

$\therefore$ Percentage error in the area of the ellipse is equal to 2%.

Method 2. $\quad \log A = \log \dfrac{\pi}{4} + \log d_1 + \log d_2$ $\quad$ (Using (1))

Taking differentials on both sides, we get,

$\dfrac{dA}{A} = \dfrac{dd_1}{d_1} + \dfrac{dd_2}{d_2}$

$\therefore$ The error relation is $\dfrac{\Delta A}{A} = \dfrac{\Delta d_1}{d_1} + \dfrac{\Delta d_2}{d_2}$

$\therefore \quad \dfrac{\Delta A}{A} \times 100 = \dfrac{\Delta d_1}{d_1} \times 100 + \dfrac{\Delta d_2}{d_2} \times 100$

$\hspace{3cm} = 1 + 1 = 2$

$\therefore$ Percentage error in the area of the ellipse is equal to 2%.

Example 4.3.8 If the H.P required to propel a steamer varies as the cube of the velocity and square of the length, prove that a 3% increase in velocity and 4% increase in length will require an increase of about 17% in H.P.

Solution:
Let $H = cv^3l^2$ where H is H.P, v is velocity, l is length and c is an arbitrary constant.

Given $\dfrac{\Delta v}{v} \times 100 = 3$, $\dfrac{\Delta l}{l} \times 100 = 4$.

Now $\Delta H = \dfrac{\partial H}{\partial v} \Delta v + \dfrac{\partial H}{\partial l} \Delta l$

$\qquad = c.3v^2l^2\,\Delta v + cv^3.2l.\Delta l$

Dividing by H, we get,

$\qquad \dfrac{\Delta H}{H} = 3.\dfrac{\Delta v}{v} + 2.\dfrac{\Delta l}{l}$

$\therefore \qquad \dfrac{\Delta H}{H} \times 100 = \dfrac{\Delta v}{v} \times 100 + \dfrac{\Delta l}{l} \times 100$

$\qquad\qquad\qquad = 3.3 + 2.4 = 9 + 8 = 17$

$\therefore$ Increase in H.P = 17%.

Example 4.3.9 The focal length of a mirror is given by the formula $\dfrac{1}{f} = \dfrac{1}{v} - \dfrac{1}{u}$. If equal errors k is made in the determination of u and v show that the percentage error in f is $100k\left(\dfrac{1}{u} + \dfrac{1}{v}\right)$.

Solution:

$\qquad \dfrac{1}{f} = \dfrac{1}{v} - \dfrac{1}{u}$ $\qquad\qquad\qquad\qquad\qquad\qquad$ (1)

Given $\Delta u = \Delta v = k$ $\qquad\qquad\qquad\qquad\qquad\qquad$ (2)

Differentiating (1) and writing the error relation,

we get, $\dfrac{-1}{f^2}.\Delta f = \dfrac{-1}{v^2}.\Delta v + \dfrac{1}{u^2}.\Delta u$

i.e., $\qquad \dfrac{-1}{f^2}.\Delta f = \dfrac{-1}{v^2}.k + \dfrac{1}{u^2}.k$ $\qquad\qquad$ (Using (2))

i.e., $\qquad \dfrac{-1}{f^2}.\Delta f = \left(\dfrac{1}{u^2} - \dfrac{1}{v^2}\right).k$

i.e., $\qquad \dfrac{1}{f^2}.\Delta f = \left(\dfrac{1}{v} + \dfrac{1}{u}\right)\left(\dfrac{1}{v} - \dfrac{1}{u}\right).k$

i.e., $\dfrac{1}{f^2}.\Delta f = \left(\dfrac{1}{v}+\dfrac{1}{u}\right).\dfrac{1}{f}.k$

i.e., $\dfrac{1}{f}.\Delta f = k.\left(\dfrac{1}{v}+\dfrac{1}{u}\right)$

i.e., $\dfrac{\Delta f}{f}\times 100 = 100k.\left(\dfrac{1}{v}+\dfrac{1}{u}\right)$

$\therefore$ Percentage error in $f = 100k.\left(\dfrac{1}{v}+\dfrac{1}{u}\right)$.

EXERCISE 4.3

PART-A

1. Give the Taylor's series expansion of $f(x + h, y + k)$.
2. Expand $f(x + h, y + k)$ into a series of powers of h and k.
3. Give the Taylor's series expansion of $f(x, y)$ about the point (a, b).
4. Give the Taylor's series expansion of $f(x, y)$ in the neighbourhood of $(0, 0)$.
5. Write down the Maclaurin's expansion of $f(x, y)$.
6. Give the Taylor's expansion of e^{x+y} about the origin.
7. Expand $\dfrac{1}{1+x-y}$ into a series of powers of x and y.
8. Write down the Maclaurin's series for $\sin(x + y)$.
9. Write down the Maclaurin's series for $\cos(x + y)$.
10. Give the Taylor's series expansion of $e^x + e^y$ about the origin.
11. Define relative and percentage errors.
12. If an error of 1% is made in measuring the length and breadth of a rectangle, find the percentage error in the area of the rectangle.

PART-B

13. Expand $\tan^{-1}(y/x)$ in the neighbourhood of $(1, 1)$.
14. Expand $f(x, y) = x^2 y + \sin y + e^x$ into Taylor's series about $(1, \pi)$.
15. Expand e^{xy} about $(1, 1)$ up to three terms.
16. Expand $\cos x \cos y$ in powers of x and y.
17. Expand $e^x \cos y$ in powers of x and y up to the second degree terms.
18. Expand $e^x \cos y$ in the neighbourhood of $(1, \pi/4)$.
19. Expand $\dfrac{(x+h)(y+k)}{x+y+h+k}$ in powers of h and k up to the second degree terms.
20. Expand x^y about the point $(1, 1)$ up to the second degree terms.
21. Find the Taylor's expansion of $e^x \sin y$ near the point $(-1, \pi/4)$ up to third degree terms.

22. Use Taylor's series expansion to prove that $\cos(x+y) = 1 - \dfrac{(x+y)^2}{2!} + \dfrac{(x+y)^4}{4!} - \cdots$

23. Use Taylor's series expansion to prove that $\tan^{-1}(x+y) = (x+y) + \dfrac{(x+y)^3}{3} + \cdots$

24. Expand $x^2 y^2 + 2x^2 y + 3xy^2$ in powers of $(x+2)$ and $(y-1)$ up to third degree terms.

25. Prove that the Taylor's series expansion of $\log(1+x+y)$ is
$$(x+y) - \frac{1}{2}(x+y)^2 + \frac{1}{3}(x+y)^3 - \cdots$$

26. Expand $xy^2 + 2x - 3y$ in powers of $x+2$ and $y-1$ up to third degree terms.

27. Find the Taylors's series expansion of y^x about the point $(1, 1)$ up to second degree terms.

28. The range R of a projectile which starts with a velocity v at an elevation α is given by
$$R = \frac{v^2 \sin 2\alpha}{g}.$$ Find the percentage error in R due to an error of 1% in v and an error of 0.5% in α.

29. The torsional rigidity of a length of wire is obtained from the formula $N = \dfrac{8\pi I.l}{t^2 r^4}$. If l is decreased by 2%, t is increased by 1.5% and r is increased by 2%, show that the value of N is decreased by 13% approximately.

30. If $pv^2 = k$ and if the maximum relative error in p is not greater than 0.05 and that in v not greater than 0.025, show that error in k may range upto 10%.

31. If q is calculated from the formula $q = kr^2 \sqrt{h}$ where k alone is constant, show that a small percentage error in r is four times as serious as the same percentage error in h.

4.4   MAXIMA AND MINIMA OF FUNCTIONS OF TWO VARIABLES

Definition 4.4.1 A function $f(x, y)$ is said to have a **relative maximum** (or **local maximum**) at a point (a, b) if $f(a, b) \geq f(a+h, b+k)$ for all small values of h and k.
i.e., $f(a, b) \geq f(a+h, b+k)$ for all points $(a+h, b+k)$ in the neighbourhood of (a, b).
A function $f(x, y)$ is said to have a **relative minimum** (or **local minimum**) at a point (a, b) if $f(a, b) \leq f(a+h, b+k)$ for all small values of h and k.

Definition 4.4.2 A maximum or minimum value of a function is called an **extreme value** of the function.

Let $\Delta = f(a+h, b+k) - f(a, b)$.
If Δ has the same sign for all small values of h and k and if this sign is negative, then $f(a, b)$ is a maximum. If this sign is positive, then $f(a, b)$ is a minimum.
By Taylor's theorem

$$\Delta = \left(h\frac{\partial f}{\partial x} + k.\frac{\partial f}{\partial y} \right) + \frac{1}{2!}\left(h^2\frac{\partial^2 f}{\partial x^2} + 2hk\frac{\partial^2 f}{\partial x \partial y} + k^2.\frac{\partial^2 f}{\partial y^2} \right) + \cdots \qquad (1)$$

For small values of h and k, the second and higher order terms are negligible.

∴ Sign of Δ = Sign of $[h.f_x (a, b) + k.f_y (a, b)]$

Taking h = 0, we see that sign of Δ changes when sign of k changes. Hence Δ has the same sign only if $f_y(a, b) = 0$. Similarly taking k = 0, we see that Δ has the same sign only if $f_x(a, b) = 0$. Thus the **necessary conditions** for f(x, y) to have a maximum or minimum at (a, b) are that $f_x(a, b) = 0$ and $f_y(a, b) = 0$. When these conditions are satisfied for small values of h and k (1) gives,

Sign of Δ = Sign of $\dfrac{1}{2!}[h^2 r + 2hks + k^2.t]$,

where $r = \dfrac{\partial^2 f(a,b)}{\partial x^2}$, $s = \dfrac{\partial^2 f(a,b)}{\partial x \partial y}$, $t = \dfrac{\partial^2 f(a,b)}{\partial y^2}$

Now sign of Δ = Sign of $\dfrac{1}{2r}[(hr + ks)^2 + k^2(rt - s^2)]$ \qquad (2)

If $rt - s^2 > 0$, then R.H.S of (2) has the same sign as that of r for all values of h and k. Hence if $rt - s^2 > 0$, f(x, y) has a maximum or minimum at (a, b) according as r < 0 or r > 0. If $rt - s^2 < 0$, then sign of Δ will change with h and k and hence f(x, y) has no maximum or minimum at (a, b). If $rt - s^2 = 0$ or r = 0, further investigations are required to find whether there is a maximum or minimum at (a, b) or not.

Definition 4.4.3 A point (a, b) where f(x, y) has neither a maximum nor a minimum is called a **saddle point** of f(x, y).

Definition 4.4.4 For a function f(x, y), f(a, b) is said to be a **stationary value** if $f_x(a, b) = 0$ and $f_y(a, b) = 0$. The point (a, b) is called a **stationary point** of f(x, y).

Note:
(i)   A given function can have more than one maximum or minimum value.
(ii)   Every extreme value of a function f(x, y) is a stationary value. The converse need not be true.

4.4.1   PROCEDURE TO FIND MAXIMUM AND MINIMUM VALUES OF A FUNCTION f(x, y)

(1)   Solve the simultaneous equations $\dfrac{\partial f}{\partial x} = 0, \dfrac{\partial f}{\partial x} = 0$ .Let the solutions be (a, b), (c, d)... These solutions are **stationary points** of f(x, y).

(2)   At each of these stationary points, compute $r = \dfrac{\partial^2 f}{\partial x^2}$, $s = \dfrac{\partial^2 f}{\partial x \partial y}$, $t = \dfrac{\partial^2 f}{\partial y^2}$ and then $rt - s^2$.

(3)   (i) If $rt - s^2 > 0$ and $r < 0$ at (a, b), then $f(a, b)$ is a maximum value.

        (ii) If $rt - s^2 > 0$ and $r > 0$ at (a, b), then $f(a, b)$ is a minimum value.

        (iii) If $rt - s^2 < 0$ at (a, b) then $f(a, b)$ is neither a maximum nor a minimum.
             (a, b) is a saddle point of $f(x, y)$.

        (iv) If $rt - s^2 = 0$ or $r = 0$ at (a, b) further investigations are required to
             decide the nature of $f(x, y)$ at (a, b).

Note: In conditions (3) we can also use 't' in the place of 'r'.

4.4.2   CONSTRAINED MAXIMA AND MINIMA

We come across situations where we may require finding the extreme values of a function of several variables which are not all independent, but are connected by some given relation. The extrème values of the function in such a situation are called **constrained extreme values.**

   Let $f(x, y, z)$ be a function of three variables x, y and z which are not independent, but are connected by the relation $\varphi(x, y, z) = 0$. This relation can be used to eliminate one of the variables, say z from $f(x, y, z)$ and thus convert it into a function of only two variables. The unconstrained extreme values of the converted function of two independent variables can be computed. These are the constrained extreme values of the given function $f(x, y, z)$. When this method is very tedious or impracticable, we use the Lagrange's method of undetermined multipliers given below.

4.4.3   LAGRANGE'S METHOD OF UNDETERMINED MULTIPLIERS

Let $u = f(x, y, z)$ (1)
be a function of three variables connected by the relation $\varphi(x, y, z) = 0$ (2)
The necessary conditions for u to have stationary values are

$$\frac{\partial u}{\partial x} = 0, \ \frac{\partial u}{\partial y} = 0, \ \frac{\partial u}{\partial z} = 0$$

$\therefore$ $du = \dfrac{\partial u}{\partial x}.dx + \dfrac{\partial u}{\partial y}.dy + \dfrac{\partial u}{\partial z}.dz = 0$ (3)

Differentiating (2) we get $d\varphi = 0$

i.e., $d\varphi = \dfrac{\partial \varphi}{\partial x}.dx + \dfrac{\partial \varphi}{\partial y}.dy + \dfrac{\partial \varphi}{\partial z}.dz = 0$ (4)

Multiplying (4) by λ and adding with (3)

We get $\left(\dfrac{\partial u}{\partial x} + \lambda \dfrac{\partial \varphi}{\partial x}\right)dx + \left(\dfrac{\partial u}{\partial y} + \lambda \dfrac{\partial \varphi}{\partial y}\right)dy + \left(\dfrac{\partial u}{\partial z} + \lambda \dfrac{\partial \varphi}{\partial z}\right)dz = 0$

This equation will be satisfied if

$$\frac{\partial u}{\partial x} + \lambda \frac{\partial \varphi}{\partial x} = 0 \quad \frac{\partial u}{\partial y} + \lambda \frac{\partial \varphi}{\partial y} = 0 \quad \frac{\partial u}{\partial z} + \lambda \frac{\partial \varphi}{\partial z} = 0$$

i.e.,
$$\frac{\partial f}{\partial x} + \lambda \frac{\partial \varphi}{\partial x} = 0 \tag{5}$$

$$\frac{\partial f}{\partial y} + \lambda \frac{\partial \varphi}{\partial y} = 0 \tag{6}$$

$$\frac{\partial f}{\partial z} + \lambda \frac{\partial \varphi}{\partial z} = 0 \tag{7}$$

Solving the four equations (2), (5), (6) and (7) we get the values of x, y, z and λ which determine the stationary points and hence the stationary values of f(x, y, z).

Note:

(i) Lagrange's method gives only the stationary values of f(x, y, z). Further investigations are required to determine the nature of the stationary points.

(ii) Let F(x, y, z) = f(x, y, z) + $\lambda \varphi$(x, y, z). Then the four equations

$$\frac{\partial F}{\partial x} = 0, \ \frac{\partial F}{\partial y} = 0, \ \frac{\partial F}{\partial z} = 0 \text{ and } \frac{\partial F}{\partial \lambda} = 0 \text{ are same as equations (5), (6), (7) and (2).}$$

Hence the solutions of these equations give the stationary points and hence the stationary values of f(x, y, z). F(x, y, z) is called the **Auxiliary function** or **Lagrange's function.**

(iii) The parameter λ is an unknown multiplier called **Lagrange multiplier.**

(iv) If there are two constraints φ_1(x, y, z) = 0 and φ_2(x, y, z) = 0, then the Auxiliary function is F(x, y, z) = f(x, y, z) + $\lambda_1 \ \varphi_1$(x, y, z) + $\lambda_2 \ \varphi_2$(x, y, z) where λ_1 and λ_2 are Lagrange multipliers. The stationary values are obtained by solving the five equation $F_x = 0$, $F_y = 0$, $F_z = 0$, $F_{\lambda_1} = 0$ and $F_{\lambda_2} = 0$

Example 4.4.1 Find the maximum and minimum values of f(x, y) = $x^3 + y^3$ - 3axy.

Solution:

$$f(x, y) = x^3 + y^3 - 3axy$$
$$f_x = 3x^2 - 3ay$$
$$f_y = 3y^2 - 3ax$$

At the stationary points $f_x = 0$ and $f_y = 0$

i.e., $3x^2 - 3ay = 0$

and $3y^2 - 3ax = 0$

i.e., $x^2 = ay$ and $y^2 = ax$

i.e., $x^4 = a^2 y^2$

i.e., $x^4 = a^3 x$

i.e., $x(x^3 - a^3) = 0$

i.e., $x = 0$ or $x = a$

When $x = 0$, we get, $y = 0$ and when $x = a$, we get, $y = a$

$\therefore$ The stationary points are $(0,0)$ and (a, a)

Now $r = f_{xx} = 6x$

$s = f_{xy} = -3a$

$t = f_{yy} = 6y$

$\therefore rt - s^2 = 6x.6y - (-3a)^2$

$= 36xy - 9a^2$

(i)   At $(0, 0)$, $rt - s^2 = -9a^2 < 0$

$\therefore$ At $(0, 0)$, $f(x, y)$ has neither a maximum nor a minimum.

i.e., $(0, 0)$ is a saddle point.

(ii)   At (a, a), $rt - s^2 = 36a^2 - 9a^2 = 27a^2 > 0$

Also at (a, a), $r = 6a$.

If $a > 0$, then $r > 0$ and hence $f(a, a)$ is a minimum value.

If $a < 0$, then $r < 0$ and hence $f(a, a)$ is a maximum value. The maximum or minimum value at (a, a) is $f(a, a) = -a^3$.

Example 4.4.2 Find the maxima or minima of $f(x, y) = 2(x - y)^2 - x^4 - y^4$

Solution:

$f(x,y) = 2(x - y)^2 - x^4 - y^4$

$f_x = 4(x - y) - 4x^3$

$f_y = -4(x - y) - 4y^3$

$r = f_{xx} = 4 - 12x^2$

$s = f_{xy} = -4$

$t = f_{yy} = 4 - 12y^2$

Stationary points are obtained by solving $f_x = 0$ and $f_y = 0$.

i.e., $x - y - x^3 = 0$ (1)

and $-(x - y) - y^3 = 0$ (2)

Adding (1) and (2) $x^3 + y^3 = 0$

i.e., $(x + y)(x^2 - xy + y^2) = 0$

$\therefore$ $x = -y$ or $x^2 - xy + y^2 = 0$ (Check: $x^2 - xy + y^2 > 0$, always)

Putting in (1) $x = -y$, we get,

$-2y + y^3 = 0$

i.e., $y(y^2 - 2) = 0$

i.e., $y = 0, \sqrt{2}, -\sqrt{2}$

The corresponding x values are $0, -\sqrt{2}, \sqrt{2}$

$\therefore$ The stationary points are $(0, 0)$, $(\sqrt{2}, -\sqrt{2})$ and $(-\sqrt{2}, \sqrt{2})$

Now $rt - s^2 = (4 - 12x^2)(4 - 12y^2) - 16$ and $r = 4 - 12x^2$

(i) At $(0, 0)$, $rt - s^2 = 16 - 16 = 0$. This case needs further information.

Now $f(0, 0) = 0$

For points along the x-axis, where $y = 0$ we have $f(x, y) = 2x^2 - x^4 = x^2(2 - x^2)$ which is positive for points in the neighbourhood of $(0, 0)$.

Again for points along the line $y = x$ we have $f(x, y) = -2x^4$ which is negative. Thus in the neighbourhood of $(0, 0)$ there are points where $f(x, y) > f(0, 0)$ and there are points where $f(x, y) < f(0, 0)$. Hence $f(0, 0)$ is not an extreme value.

(i) At $(\sqrt{2}, -\sqrt{2})$ and $(-\sqrt{2}, \sqrt{2})$

We get, $rt - s^2 = (4 - 24)(4 - 24) - 16 = 400 - 16 = 384$.

i.e., $rt - s^2 > 0$ Also $r = 4 - 24 = -20 < 0$

Hence $f(x, y)$ has maximum values at $(\sqrt{2}, -\sqrt{2})$ and $(-\sqrt{2}, \sqrt{2})$

The maximum value is $f(\sqrt{2}, -\sqrt{2})$ or $f(-\sqrt{2}, \sqrt{2})$

i.e., $f(\sqrt{2}, -\sqrt{2}) = 2(2.\sqrt{2})^2 - 4 - 4 = 8$ or $f(-\sqrt{2}, \sqrt{2}) = 2(-2.\sqrt{2})^2 - 4 - 4 = 8$

Example 4.4.3 Find the maximum and minimum values of

$$f(x, y) = x^2 + xy + y^2 + \frac{1}{x} + \frac{1}{y}$$

Solution: Given $f(x, y) = x^2 + xy + y^2 + \frac{1}{x} + \frac{1}{y}$

$$f_x = 2x + y - \frac{1}{x^2}$$

$$f_y = 2y + x - \frac{1}{y^2}$$

$$r = f_{xx} = 2 + \frac{2}{x^3}$$

$$s = f_{xy} = 1$$

$$t = f_{yy} = 2 + \frac{2}{y^3}$$

At the stationary points $f_x = f_y = 0$

$\therefore$ $2x + y - \frac{1}{x^2} = 0$ i.e., $2x^3 + x^2y - 1 = 0$ (1)

and $x + 2y - \frac{1}{y^2} = 0$ i.e., $2y^3 + xy^2 - 1 = 0$ (2)

$(1) \times y - (2) \times x$ gives,

$$2x^3y - 2xy^3 - y + x = 0$$
$$2xy(x^2 - y^2) + (x - y) = 0$$
$$(x - y)[2xy(x + y) + 1] = 0$$

i.e., $x = y$ or $2xy(x + y) + 1 = 0$ (Investigate the case when $2xy(x + y) + 1 = 0$)

When $x = y$, (1) becomes $2x^3 + x^3 - 1 = 0$

i.e., $3x^3 = 1$

$$x = \sqrt[3]{\tfrac{1}{3}} \quad \text{and} \quad y = \sqrt[3]{\tfrac{1}{3}}$$

At $\left(\sqrt[3]{\tfrac{1}{3}}, \sqrt[3]{\tfrac{1}{3}} \right)$, $rt - s^2 = \left(2 + \tfrac{2}{x^3} \right)\left(2 + \tfrac{2}{y^3} \right) - 1$

$$= 64 - 1 = 63 > 0$$

$$r = 2 + \tfrac{2}{x^3} = 2 + 6 = 8 > 0$$

∴ $f(x, y)$ has a minimum value at $\left(\sqrt[3]{\tfrac{1}{3}}, \sqrt[3]{\tfrac{1}{3}} \right)$

The minimum value of $f(x, y) = \left(\tfrac{1}{3} \right)^{2/3} + \left(\tfrac{1}{3} \right)^{1/3} \left(\tfrac{1}{3} \right)^{1/3} + \left(\tfrac{1}{3} \right)^{2/3} + 3^{1/3} + 3^{1/3}$

$$= 3.3^{-2/3} + 2.3^{1/3}$$
$$= 3^{1/3} + 2.3^{1/3}$$
$$= 3.3^{1/3}$$
$$= 3^{4/3}$$

Example 4.4.4 Find the maxima and minima of $u(x, y) = \sin x \sin y \sin (x + y)$ if $0 < x, y < \pi$.

Solution: We have $u_x = \sin y [\sin x . \cos (x + y) + \cos x . \sin (x + y)]$
$$= \sin y \sin (2x + y)$$

Similarly, $u_y = \sin x . \sin (x + 2y)$.

$r = u_{xx} = 2 \sin y \cos (2x + y)$

$s = u_{xy} = \cos x . \sin (x + 2y) + \sin x . \cos (x + 2y)$
$$= \sin (2x + 2y)$$

$t = u_{yy} = 2 \sin x . \cos (x + 2y)$.

The stationary points are given by $u_x = 0$ and $u_y = 0$

i.e., $\sin y . \sin (2x + y) = 0$ (1)

and $\sin x . \sin (x + 2y) = 0$ (2)

From (1), $y \neq 0$, $y \neq \pi$ implies $\sin y \neq 0$ ∴ $\sin (2x + y) = 0$.

Similarly from (2), we get, $\sin (x + 2y) = 0$

Hence $2x + y = \pi$ and $x + 2y = \pi$ (∵ $x \neq 0$, $y \neq 0$)

Solving, we get, $x = \pi/3$, $y = \pi/3$

∴ $(\pi/3, \pi/3)$ is a stationary point in $0 < x, y < \pi$.

At $(\pi/3, \pi/3)$, $\quad r = 2 \sin \pi/3 . \cos \pi = -\sqrt{3}$

$$s = \sin(4\pi/3) = -\sin \pi/3 = -\sqrt{3}\big/_2$$

$$t = 2 \sin \pi/3 . \cos \pi = -\sqrt{3}$$

$\therefore rt - s^2 = 3 - (3/4) = 9/4 > 0$. Also $r = -\sqrt{3}$, < 0

$\therefore u(x, y)$ is maximum at $(\pi/3, \pi/3)$

The maximum value $= u(\pi/3, \pi/3)$

$$= \sin \frac{\pi}{3} . \sin \frac{\pi}{3} . \sin \frac{2\pi}{3} = \frac{\sqrt{3}}{2} . \frac{\sqrt{3}}{2} . \frac{\sqrt{3}}{2} = \frac{3\sqrt{3}}{8}$$

Example 4.4.5 Find the minimum value of $x^2 + y^2 + z^2$ when $xyz = a^3$.

Solution: Here we have to find the constrained minimum of $x^2 + y^2 + z^2$ subject to the condition $xyz = a^3$. i.e., $z = \dfrac{a^3}{xy}$

Let $f(x, y) = x^2 + y^2 + \dfrac{a^6}{x^2 y^2}$

$$f_x = 2x - \frac{2a^6}{x^3 y^2}$$

$$f_y = 2y - \frac{2a^6}{x^2 y^3}$$

$$r = f_{xx} = 2 + \frac{6.a^6}{x^4 y^2}$$

$$s = f_{xy} = \frac{4.a^6}{x^3 y^3}$$

$$t = f_{yy} = 2 + \frac{6.a^6}{x^2 y^4}$$

Stationary points are obtained by solving $f_x = 0$, $f_y = 0$.

i.e., $\quad x - \dfrac{a^6}{x^3 y^2} = 0$ $\hfill$ (1)

$\quad\quad y - \dfrac{a^6}{x^2 y^3} = 0$ $\hfill$ (2)

From (1) $\quad x^4 y^2 - a^6 = 0$

From (2) $\quad x^2 y^4 - a^6 = 0$

The stationary points are (a, a), $(-a, a)$, $(a, -a)$ and $(-a, -a)$.

At all these points $rt - s^2 = \left(2 + \dfrac{6a^6}{x^4 y^2}\right)\left(2 + \dfrac{6a^6}{x^2 y^4}\right) - \dfrac{16a^{12}}{x^6 y^6}$

$$= 8 \times 8 - 16 = 48 > 0.$$

Also $r = 8 > 0$

Hence $f(x, y)$ is minimum at the points (a, a), $(-a, a)$, $(a, -a)$ and $(-a, -a)$.

The minimum value of $f(x, y) = a^2 + a^2 + \dfrac{a^6}{a^2 a^2} = 3a^2$.

i.e., the minimum value of $x^2 + y^2 + z^2$ when $xyz = a^3$ is $3a^2$.

Example 4.4.6 Find the minimum value of $x^2 + y^2 + z^2$ where $ax + by + cz = p$.

Solution:

Method 1.   Let $f(x, y, z) = x^2 + y^2 + z^2$ (1)

To find the constrained minimum of $f(x, y, z)$ subject to the condition.

$$ax + by + cz = p \qquad\qquad\qquad (2)$$

i.e., $z = \dfrac{p - ax - by}{c}$

$\therefore$ $f(x, y, z) = x^2 + y^2 + \left[\dfrac{p - ax - by}{c}\right]^2$

$f_x = 2x + \dfrac{2}{c^2}(p - ax - by)(-a)$

$f_y = 2y + \dfrac{2}{c^2}(p - ax - by)(-b)$

$r = f_{xx} = 2 + \dfrac{2a^2}{c^2}$

$s = f_{xy} = \dfrac{2ab}{c^2}$

$t = f_{yy} = 2 + \dfrac{2b^2}{c^2}$

The stationary points are obtained by solving $f_x = 0$ and $f_y = 0$.

i.e., $c^2 x - a(p - ax - by) = 0$ (1)

and $c^2 y - b(p - ax - by) = 0$ (2)

$b \times (1) - a \times (2)$ gives $bc^2 x - ac^2 y = 0$

i.e., $c^2(bx - ay) = 0$

i.e., $bx = ay$

i.e., $x = \dfrac{a}{b}y$

Substituting in (1) $\dfrac{ac^2}{b}y - ap + a^2\dfrac{ay}{b} + aby = 0$

$$\dfrac{c^2}{b}y - p + \dfrac{a^2 y}{b} + by = 0$$

$$y(a^2 + b^2 + c^2) = bp$$

$$y = \dfrac{bp}{a^2 + b^2 + c^2}$$

$\therefore$ $x = \dfrac{a}{b}y = \dfrac{ap}{a^2 + b^2 + c^2}$ and $z = \dfrac{p - ax - by}{c} = \dfrac{cp}{a^2 + b^2 + c^2}$

The only stationary point is $\left(\dfrac{ap}{a^2 + b^2 + c^2}, \dfrac{bp}{a^2 + b^2 + c^2}, \dfrac{cp}{a^2 + b^2 + c^2}\right)$

$$rt - s^2 = \left(2 + 2\dfrac{a^2}{c^2}\right)\left(2 + 2\dfrac{b^2}{c^2}\right) - \dfrac{4a^2 b^2}{c^4}$$

$$= 4 + \dfrac{4b^2}{c^2} + \dfrac{4a^2}{c^2} > 0 \text{ at all points.}$$

Also r > 0 at all points.

$\therefore$ f(x, y, z) is minimum at the stationary point and the minimum value is

$$= \left(\dfrac{ap}{a^2 + b^2 + c^2}\right)^2 + \left(\dfrac{bp}{a^2 + b^2 + c^2}\right)^2 + \left(\dfrac{cp}{a^2 + b^2 + c^2}\right)^2$$

$$= \dfrac{p^2}{a^2 + b^2 + c^2}$$

Method 2.

We use Lagrange's method. Let $f(x, y, z) = x^2 + y^2 + z^2$.

$\varphi(x, y, z) = ax + by + cz - p$ and $F(x, y, z) = f(x, y, z) + \lambda\ \varphi(x, y, z)$ where λ is the Lagrange multiplier.

Then $F(x, y, z) = x^2 + y^2 + z^2 + \lambda(ax + by + cz - p)$

The stationary points are obtained by solving

$$F_x = 2x + a\lambda = 0 \tag{1}$$

$$F_y = 2y + b\lambda = 0 \tag{2}$$

$$F_z = 2z + c\lambda = 0 \tag{3}$$

and $F_\lambda = ax + by + cz - p$ $\tag{4}$

From (1), $x = -\dfrac{a\lambda}{2}$

From (2), $y = -\dfrac{b\lambda}{2}$

From (3), $z = -\dfrac{c\lambda}{2}$

From (4), $a\left(-\dfrac{a\lambda}{2}\right) + b\left(-\dfrac{b\lambda}{2}\right) + c\left(-\dfrac{c\lambda}{2}\right) = p$

$$\lambda = \frac{-2p}{a^2 + b^2 + c^2}$$

$$\therefore \quad x = \frac{ap}{a^2 + b^2 + c^2}, \quad y = \frac{bp}{a^2 + b^2 + c^2}, \quad z = \frac{cp}{a^2 + b^2 + c^2}$$

The only stationary point is $\left(\dfrac{ap}{a^2 + b^2 + c^2}, \dfrac{bp}{a^2 + b^2 + c^2}, \dfrac{cp}{a^2 + b^2 + c^2}\right)$

The minimum value of $f(x, y, z) = \left(\dfrac{ap}{a^2 + b^2 + c^2}\right)^2 + \left(\dfrac{bp}{a^2 + b^2 + c^2}\right)^2 + \left(\dfrac{cp}{a^2 + b^2 + c^2}\right)^2$

$$= \frac{p^2(a^2 + b^2 + c^2)}{\left(a^2 + b^2 + c^2\right)^2}$$

$$= \frac{p^2}{a^2 + b^2 + c^2}$$

Example 4.4.7 Find the volume of the greatest rectangular parallelepiped that can be inscribed in the ellipsoid $\dfrac{x^2}{a^2} + \dfrac{y^2}{b^2} + \dfrac{z^2}{c^2} = 1$.

Solution: The greatest rectangular parallelepiped that can be inscribed in the ellipsoid will have its corners on the ellipsoid and its sides parallel to the coordinate planes. Let the coordinates of the corners be $(\pm x, \pm y, \pm z)$.

Then the lengths of the edges of the parallelepiped are $2x, 2y, 2z$. Its volume is

$\quad V = 2x.2y.2z$

$\quad\quad = 8xyz.$

We have to find the maximum value of $V(x, y, z) = 8xyz$, subject to the condition

$$\varphi(x, y, z) = \frac{x^2}{a^2} + \frac{y^2}{b^2} + \frac{z^2}{c^2} - 1 = 0$$

The Auxiliary function $F(x, y, z)$ is given by $F(x, y, z) = V(x, y, z) + \lambda\, \varphi(x, y, z)$

i.e., $\quad F(x, y, z) = 8xyz + \lambda\left(\dfrac{x^2}{a^2} + \dfrac{y^2}{b^2} + \dfrac{z^2}{c^2} - 1\right)$

The stationary points are obtained by solving

$$F_x = 8yz + \lambda.\frac{2x}{a^2} = 0 \qquad\qquad\qquad (1)$$

$$F_y = 8xz + \lambda . \frac{2y}{b^2} = 0 \tag{2}$$

$$F_z = 8xy + \lambda . \frac{2z}{c^2} = 0 \tag{3}$$

$$F_\lambda = \frac{x^2}{a^2} + \frac{y^2}{b^2} + \frac{z^2}{c^2} - 1 = 0 \tag{4}$$

Multiplying (1), (2) and (3) by x, y and z respectively and adding we get.

$$24xyz + 2\lambda \left(\frac{x^2}{a^2} + \frac{y^2}{b^2} + \frac{z^2}{c^2} \right) = 0$$

i.e., $24xyz + 2\lambda (1) = 0$

i.e., $\lambda = -12xyz$

Substituting for λ in (1) we get

$$8yz - \frac{24x^2yz}{a^2} = 0$$

i.e., $a^2 = 3x^2$

i.e., $x = \dfrac{a}{\sqrt{3}}$ (as x can not be negative)

Similarly from (2) and (3) we get $y = \dfrac{b}{\sqrt{3}}$, $z = \dfrac{c}{\sqrt{3}}$

When x = 0, the volume of the parallelepiped V = 0.

$\therefore$ V must be maximum when $x = \dfrac{a}{\sqrt{3}}$, $y = \dfrac{b}{\sqrt{3}}$, $z = \dfrac{c}{\sqrt{3}}$

Hence the greatest volume = $8. \dfrac{a}{\sqrt{3}} . \dfrac{b}{\sqrt{3}} . \dfrac{c}{\sqrt{3}} = \dfrac{8abc}{3\sqrt{3}}$

Example 4.4.8 If $u = a^3x^2 + b^3y^2 + c^3z^2$, where $x^{-1} + y^{-1} + z^{-1} = 1$, show that stationary value of u is given by $x = \dfrac{\Sigma a}{a}$, $y = \dfrac{\Sigma a}{b}$, $z = \dfrac{\Sigma a}{c}$

Solution: Given $u = a^3x^2 + b^3y^2 + c^3z^2$ and $x^{-1} + y^{-1} + z^{-1} - 1 = 0$.

Let $F(x, y, z) = a^3x^2 + b^3y^2 + c^3z^2 + \lambda \left(\dfrac{1}{x} + \dfrac{1}{y} + \dfrac{1}{z} - 1 \right)$

The stationary points are given by,

$$F_x = 2a^3x + \lambda \left(\frac{-1}{x^2} \right) = 0 \tag{1}$$

$$F_y = 2b^3y + \lambda \left(\frac{-1}{y^2} \right) = 0 \tag{2}$$

$$F_y = 2c^3z + \lambda.\left(\frac{-1}{z^2}\right) = 0 \tag{3}$$

$$F_\lambda = \frac{1}{x} + \frac{1}{y} + \frac{1}{z} - 1 = 0 \tag{4}$$

From (1) $2a^3x^3 - \lambda = 0$

i.e., $x = \left(\dfrac{\lambda}{2}\right)^{\frac{1}{3}} \dfrac{1}{a}$

Similarly from (2) and (3) we get,

$$y = \left(\frac{\lambda}{2}\right)^{\frac{1}{3}} \frac{1}{b}, \ z = \left(\frac{\lambda}{2}\right)^{\frac{1}{3}} \frac{1}{c}$$

substituting for x, y, z in (4) we get

$$\left(\frac{2}{\lambda}\right)^{\frac{1}{3}} (a+b+c) - 1 = 0$$

i.e., $\left(\dfrac{2}{\lambda}\right)^{\frac{1}{3}} = \dfrac{1}{\Sigma a}$

i.e., $\left(\dfrac{\lambda}{2}\right)^{\frac{1}{3}} = \Sigma a$

$\therefore$ $x = \left(\dfrac{\lambda}{2}\right)^{\frac{1}{3}} \dfrac{1}{a} = \dfrac{\Sigma a}{a}$

Similarly $y = \dfrac{\Sigma a}{b}$ and $z = \dfrac{\Sigma a}{c}$.

Hence the stationary value of u is given by $x = \dfrac{\Sigma a}{a}, y = \dfrac{\Sigma a}{b}, z = \dfrac{\Sigma a}{c}$.

Example 4.4.9 The temperature T at any point (x, y, z) in space is $T = 400 \, xyz^2$. Find the highest temperature on the surface of the unit sphere $x^2 + y^2 + z^2 = 1$

Solution:
Given $T = 400 \, xyz^2$ and $x^2 + y^2 + z^2 - 1 = 0$.
Let $F(x, y, z) = 400 \, xyz^2 + \lambda(x^2 + y^2 + z^2 - 1)$
The stationary points are given by

$$F_x = 400 \, yz^2 + 2\lambda x = 0 \tag{1}$$
$$F_y = 400 \, xz^2 + 2\lambda y = 0 \tag{2}$$
$$F_z = 800 \, xyz + 2\lambda z = 0 \tag{3}$$
$$F_\lambda = x^2 + y^2 + z^2 - 1 = 0 \tag{4}$$

Multiplying (1), (2) and (3) by x, y and z respectively and adding we get.

$$1600\ xyz^2 + 2\lambda(x^2 + y^2 + z^2) = 0$$

i.e., $\quad 1600xyz^2 + 2\lambda(1) = 0$

i.e., $\quad \lambda = -800xyz^2$

Substituting for λ in (1) we get

$$400\ yz^2 - 1600\ xyz^2.x = 0$$

i.e., $\quad 1 - 4x^2 = 0$

i.e., $\quad x^2 = \frac{1}{4}$

i.e., $\quad x = \pm \frac{1}{2}$

Similarly from (2) and (3) we get $y = \pm \frac{1}{2}$ and $z = \pm \frac{1}{\sqrt{2}}$

Hence the stationary points of $T = 400\ xyz^2$ are $\left(\pm \frac{1}{2}, \pm \frac{1}{2}, \pm \frac{1}{\sqrt{2}}\right)$.

The temperature T is highest at the stationary points where T is positive.

$\therefore$ Highest temperature is given by $T = 400.\frac{1}{2}.\frac{1}{2}\left(\frac{1}{\sqrt{2}}\right)^2 = 50$

Example 4.4.10 Find the minimum distance from the point (3, 4, 15) to the cone $x^2 + y^2 = 4z^2$.

Solution: Let (x, y, z) be any point on the cone $x^2 + y^2 = 4z^2$. Then its distance from the point (3, 4, 15) is $d = \sqrt{(x-3)^2 + (y-4)^2 + (z-15)^2}$. First we find the minimum value of d^2 subject to the condition $x^2 + y^2 = 4z^2$.

Let $F(x, y, z) = (x-3)^2 + (y-4)^2 + (z-15)^2 + \lambda\ (x^2 + y^2 - 4z^2)$

The stationary points are given by,

$$F_x = 2(x-3) + 2\lambda x = 0 \tag{1}$$
$$F_y = 2(y-4) + 2\lambda y = 0 \tag{2}$$
$$F_z = 2(z-15) - 8\lambda z = 0 \tag{3}$$
$$F_\lambda = x^2 + y^2 - 4z^2 = 0 \tag{4}$$

From (1), $x = \dfrac{3}{1+\lambda}$

From (2), $y = \dfrac{4}{1+\lambda}$

From (3), $z = \dfrac{15}{1-4\lambda}$

Substituting in (4), $\left(\dfrac{3}{1+\lambda}\right)^2 + \left(\dfrac{4}{1+\lambda}\right)^2 = 4\left(\dfrac{15}{1-4\lambda}\right)^2$

i.e., $\quad 25(1-4\lambda)^2 = 4.225\ (1+\lambda)^2$

i.e., $\dfrac{1-4\lambda}{1+\lambda} = \pm 6$

From $\dfrac{1-4\lambda}{1+\lambda} = 6$ we get $\lambda = -\frac{1}{2}$

From $\dfrac{1-4\lambda}{1+\lambda} = -6$ we get $\lambda = -\frac{7}{2}$

When $\lambda = -\frac{1}{2}$, we get x = 6, y = 8, z = 5.

When $\lambda = -\frac{7}{2}$, we get x = -6/5, y = -8/5, z = 1.

Thus the stationary points are (6, 8, 5) and (-6/5, -8/5, 1).

Distance of (6, 8, 5) from (3, 4, 15) is $d = \sqrt{(6-3)^2 + (8-4)^2 + (5-15)^2}$

$$= \sqrt{125} \ = 5\sqrt{5}$$

Distance of (-6/5, -8/5, 1) from (3, 4, 15) is $d = \sqrt{\left(-\frac{6}{5}-3\right)^2 + \left(-\frac{8}{5}-4\right)^2 + (1-15)^2}$

$$= \sqrt{\dfrac{441}{25} + \dfrac{784}{25} + 196}$$

$$= \sqrt{49+196} \ = \sqrt{245} = 7\sqrt{5}$$

∴ The minimum distance from the point (3, 4, 15) to the cone $x^2 + y^2 = 4z^2$ is $5\sqrt{5}$.

Example 4.4.11 In a plane triangle, find the maximum value of cos A cos B cos C.

Solution: In any triangle A + B + C = 180°
              i.e., A + B + C = π radians.
Let f(A, B, C) = cos A cos B cos C and φ(A, B, C) = A + B + C - π.
Then the Auxiliary function is F(A, B, C) = f(A, B, C) + λ φ(A, B, C)
                                = cos A cos B cos C + λ (A + B + C - π)

The stationary points are given by
       F_A = -sin A cos B cos C + λ = 0   (1)
       F_B = -sin B cos A cos C + λ = 0   (2)
       F_C = -sin C cos A cos B + λ = 0   (3)
       F_λ = A + B + C - π = 0   (4)
 (1) – (2) gives –cos C [sin A cos B – cos A sin B] = 0
i.e.,   -cos C sin (A - B) = 0
i.e.,   cos C = 0, or sin (A - B) = 0
i.e.,   C = π/2 or A = B
Similarly from (2) and (3) we get, A = π/2 or B = C and
from (1) and (3) we get, B = π/2, A = C.
But cos A cos B cos C = 0 when A or B or C is π/2
∴ For cos A cos B cos C to be maximum, A = B = C.
Hence from (4) we get A = B = C = π/3.

i.e., cos A cos B cos C is maximum at $(\pi/3, \pi/3, \pi/3)$

The maximum value of cos A cos B cos C $= \cos\frac{\pi}{3}.\cos\frac{\pi}{3}.\cos\frac{\pi}{3}$

$$= \frac{1}{2}.\frac{1}{2}.\frac{1}{2} = \frac{1}{8}$$

Example 4.4.12 Find the points on the ellipse obtained as the curve of intersection of the surfaces $x + y = 1$ and $x^2 + 2y^2 + z^2 = 1$, which are nearest to and farthest from the origin.

Solution: Let (x, y, z) be any point on the ellipse.

Then $x + y = 1$ and $x^2 + 2y^2 + z^2 = 1$.

Distance of (x, y, z) from the origin is $d = \sqrt{x^2 + y^2 + z^2}$

Let $f(x, y, z) = x^2 + y^2 + z^2$

To find the points giving the minimum and maximum values of $f(x, y, z)$ subject to the constraints $x + y = 1$ and $x^2 + 2y^2 + z^2 = 1$.

The Auxiliary function is $F(x, y, z) = x^2 + y^2 + z^2 + \lambda_1(x + y - 1) + \lambda_2(x^2 + 2y^2 + z^2 - 1)$ where λ_1 and λ_2 are Lagrange multipliers.

The stationary points are given by

$$F_x = 2x + \lambda_1 + 2x\lambda_2 = 0 \tag{1}$$

$$F_y = 2y + \lambda_1 + 4y\lambda_2 = 0 \tag{2}$$

$$F_z = 2z + 2z\lambda_2 = 0 \tag{3}$$

$$F_{\lambda_1} = x + y - 1 = 0 \tag{4}$$

$$F_{\lambda_2} = x^2 + 2y^2 + z^2 - 1 = 0 \tag{5}$$

Eliminating λ_1 and λ_2 from (1), (2) and (3), we get,

$$\begin{vmatrix} 2x & 1 & 2x \\ 2y & 1 & 4y \\ 2z & 0 & 2z \end{vmatrix} = 0$$

i.e.,
$$\begin{vmatrix} x & 1 & x \\ y & 1 & 2y \\ z & 0 & z \end{vmatrix} = 0$$

i.e., $xz - (-yz) + x(-z) = 0$

i.e., $yz = 0$

∴ $y = 0$ or $z = 0$

Case 1. $y = 0$

From (4) we get $x = 1$

From (5) we get $z = 0$

∴ $(1, 0, 0)$ is a stationary point.

Case 2. $z = 0$

From (5) we get $x^2 + 2y^2 = 1$

From (4) we get $y = 1 - x$

$\therefore$ $x^2 + 2(1 - x)^2 = 1$

i.e., $3x^2 - 4x + 1 = 0$

          $x = 1, 1/3$

When x = 1, we get, y = 0

When x = 1/3, we get, y = 2/3

$\therefore$ The stationary points are (1, 0, 0) and (1/3, 2/3, 0)

Distance of these points from the origin are $\sqrt{1^2 + 0^2 + 0^2}$ and $\sqrt{(\frac{1}{3})^2 + (\frac{2}{3})^2 + 0^2}$

i.e., 1 and $\sqrt{5}/3$

$\therefore$ The nearest point is (1/3, 2/3 , 0) and the farthest point is (1, 0, 0).

Example 4.4.13 A tent having the form of a cylinder surmounted by a cone is to contain a given volume. If the canvas required is a minimum, show that the altitude of the cone is twice that of the cylinder.

Solution: Let x and y be the base radius and altitude of the cylinder and z be the altitude of the cone.

Then the base radius and slant height of the cone are x and $l = \sqrt{x^2 + z^2}$

Volume of the tent = $\pi x^2 y + \dfrac{1}{3}\pi x^2 z$

Given $\pi x^2 y + \dfrac{1}{3}\pi x^2 z = c$, a constant

The area of canvas required = $2\pi xy + \pi xl = 2\pi xy + \pi x \sqrt{x^2 + z^2}$

Let $f(x, y, z) = 2\pi xy + \pi x \sqrt{x^2 + z^2}$

To find minimum value of f(x, y, z) subject to the condition, $\pi x^2 y + \dfrac{1}{3}\pi x^2 z = c$

Let $F(x, y, z) = 2\pi xy + \pi x.\sqrt{x^2 + z^2} +\lambda\,[\,\pi x^2 y + \dfrac{1}{3}\pi x^2 z - c\,]$

The stationary points are given by

$$F_x = 2\pi y + \pi.\sqrt{x^2 + z^2} + \pi x.\frac{1}{\sqrt{x^2 + z^2}}.x + \lambda\left[2\pi xy + \frac{2}{3}\pi xz\right] = 0 \qquad (1)$$

$$F_y = 2\pi x + \lambda\pi x^2 = 0 \qquad\qquad\qquad (2)$$

$$F_z = \pi x.\frac{1}{\sqrt{x^2 + z^2}}.x + \lambda.\frac{1}{3}\pi x^2 = 0 \qquad\qquad (3)$$

$$F_\lambda = \pi x^2 y + \frac{1}{3}\pi x^2 z - c = 0 \tag{4}$$

From (2), we get, $2\pi x + \lambda \pi x^2 = 0$

i.e., $\pi x (2 + \lambda x) = 0$

i.e., $x = 0, -2/\lambda$

$x = 0$ is not possible.

$\therefore \ x = -2/\lambda$

From (3), we get, $\pi x z + \dfrac{\pi \lambda}{3}x^2 .\sqrt{x^2 + z^2} = 0$

i.e., $\quad \dfrac{-2}{\lambda}z + \dfrac{\lambda}{3}.\dfrac{4}{\lambda^2}\sqrt{\dfrac{4}{\lambda^2} + z^2} = 0$

i.e., $\quad \dfrac{4}{3}\sqrt{\dfrac{4}{\lambda^2} + z^2} = 2z$

i.e., $\quad \dfrac{16}{9}\left(\dfrac{4}{\lambda^2} + z^2\right) = 4z^2$

i.e., $\quad \dfrac{64}{\lambda^2} + 16z^2 = 36z^2$

i.e., $\quad 5z^2 = \dfrac{16}{\lambda^2}$

$\therefore \quad z = \pm \dfrac{4}{\sqrt{5}\lambda}$

Choosing $z = \dfrac{4}{\sqrt{5}\lambda}$, from (1) we get

$$2\pi y + \pi.\frac{6}{\sqrt{5}\lambda} + \pi.\frac{4}{\lambda^2}.\frac{\sqrt{5}\lambda}{6} + \lambda.2\pi\left(\frac{-2}{\lambda}\right)y + \frac{2}{3}\pi\lambda\left(\frac{-2}{\lambda}\right)\left(\frac{4}{\sqrt{5}\lambda}\right) = 0$$

i.e., $2\pi y + \dfrac{6\pi}{\sqrt{5}\lambda} + \dfrac{2\sqrt{5}.\pi}{3\lambda} - 4\pi y - \dfrac{16\pi}{3\sqrt{5}\lambda} = 0$

i.e., $-2y + \dfrac{18 + 10 - 16}{3\sqrt{5}.\lambda} = 0$

$\therefore \quad y = \dfrac{2}{\sqrt{5}\lambda}$

Similarly choosing $z = \dfrac{-4}{\sqrt{5}\lambda}$, we get $y = \dfrac{22}{3\sqrt{5}\lambda}$

Area of the canvas used will be minimum when $y = \dfrac{2}{\sqrt{5}\lambda}$ and $z = \dfrac{4}{\sqrt{5}\lambda}$. i.e., $z = 2y$

i.e., altitude of the cone is twice that of the cylinder.

EXERCISE 4.4

PART-A
1. Define 'relative maximum' and 'relative minimum' for a function of two variables.
2. Define 'extreme values' of a function of two variables.
3. Define 'stationary points' of f(x, y).
4. Define 'saddle point' of a function f(x, y).
5. What are the conditions for the point (a, b) to be a stationary point of f(a, b)?
6. State the conditions for the stationary point (a, b) of f(x, y) to be a maximum point.
7. State the conditions for the stationary point (a, b) of f(x, y) to be a minimum point.
8. State the conditions for the stationary point (a, b) to be a saddle point of f(x, y).
9. What are constrained extreme values?
10. Find the stationary point of $x^2 - xy + y^2 - 2x + y$.
11. Find the minimum point of $x^2 + y^2 + 6x + 12$.
12. Find the stationary points of $x^3 + y^3 - 3xy$.
13. Find the maximum point of $2 + 2x + 2y - x^2 - y^2$.
14. For the function f(x, y), $f_{xx} = 12x^2 - 4$, $f_{xy} = 4$, $f_{yy} = 12y^2 - 4$. Find the nature of the stationary points $\left(\sqrt{2}, -\sqrt{2}\right)$.
15. For the function f(x, y), $f_{xx} = 2$, $f_{xy} = 0$, $f_{yy} = 2$, find the nature of the stationary point (-3, 0).
16. Find the nature of the stationary point (1, 1) of the function f(x, y) if $f_{xx} = 6xy^3$, $f_{xy} = 9x^2y^2$, $f_{yy} = 6x^3y$.

PART-B
Find the maxima and minima of the following functions.
17. $x^3 + 3xy^2 - 15x^2 - 15y^2 + 72x$.
18. $x^3 + y^3 + 3xy$.
19. $x^4 + y^4 - 2x^2 + 4xy - 2y^2$.
20. $2(x^2 - y^2) - x^4 - y^4$.
21. $xy + \dfrac{a^3}{x} + \dfrac{a^3}{y}$.
22. $x^2y + xy^2 - axy$.
23. $x^2y^2 - 5x^2 - 8xy - 5y^2$.
24. $x^2 + y^2 + \dfrac{2}{x} + \dfrac{2}{y}$.
25. $x^2y^2 - x^2 - y^2$.
26. $xy\, e^{-x-y}$, $x > 0$, $y \geq 0$.
27. $x^3y^2(a - x - y)$.
28. $x^4 + 2x^2y - x^2 + 3y^2$.

29. Find the minimum value of $x^2 + y^2 + z^2$ when $x + y + z = 3a$.

30. Find the maximum product of three numbers whose sum is $3a$.

31. Find the maximum value of $x^2y^2z^2$ when $x^2 + y^2 + z^2 = a^2$.

32. Show that the maximum value of $xy + yz + zx$ when $x^2 + y^2 + z^2 = r^2$ is $-r^2/2$

33. Find the minimum value of $x^2 + y^2 + z^2$ subject to the condition $\dfrac{1}{x} + \dfrac{1}{y} + \dfrac{1}{z} = 1$

34. Find the maximum value of $x^m.y^n.z^p$ when $x + y + z = a$.

35. Find the maximum of xy^2z^3 given $x + y + z = b$ and $x > 0, y > 0, z > 0$.

36. Show the maximum of $x^p y^q z^r$ subject to $ax + by + cz = p + q + r$ is $\left(\dfrac{p}{a}\right)^p \left(\dfrac{q}{b}\right)^q \left(\dfrac{r}{c}\right)^r$.

37. Show that the minimum value of $a^3x^2 + b^3y^2 + c^3z^2$, when $\dfrac{1}{x} + \dfrac{1}{y} + \dfrac{1}{z} = \dfrac{1}{k}$ is
    $k^2(a + b + c)^3$.

38. A rectangular box, open at the top, is to have a volume of 32cc. Find the dimensions of the box that requires the least material for its construction.

39. Show that the rectangle of given perimeter which has maximum area is a square.

40. Show that the rectangular solid of maximum volume that can be inscribed in a sphere is a cube.

41. A thin closed rectangular box is to have one edge equal to twice the other, and a constant volume 72 m^3. Find the least surface area of the box.

42. Show that, of all rectangular parallelopipeds of given volume, the cube has the least surface.

43. Find the shortest and the longest distance from the point $(1, 2, -1)$ to the sphere $x^2 + y^2 + z^2 = 24$.

44. Find the points on the surface $z = x^2 + y^2$, that is nearest to the point $(3, -6, 4)$.

45. Find the points on the surface $z^2 = xy + 1$ whose distance from the origin is minimum.

46. Prove that $\sin A \sin B \sin C$ is maximum when $A = B = C = \pi/3$, in a $\triangle ABC$.

47. Find the shortest distance of the point $(2, 1, -3)$ from the plane $2x + y = 2z + 4$.

48. Find the point on the surface $x^2 + y^2 + z^2 = 1$ which is nearest to the point $(2, 0, 0)$.

49. Find the minimum distance from $(0, 0, 0)$ to the surface $z^2 = (x - y)^2 - 1$.

50. Find the maximum and minimum distances from the origin to the curve
    $5x^2 + 6xy + 5y^2 - 8 = 0$

51. Find the point on the curve of intersection of the surfaces $z = xy + 5$ and $x + y + z = 1$ which is nearest to the origin.

52. Find the greatest and least values of z, where (x, y, z) lies on the ellipse formed by the intersection of the plane $x + y + z = 1$ and the ellipsoid
    $16x^2 + 4y^2 + z^2 = 16$

53. Find the stationary values of $x^2 + y^2 + z^2$ subject to $x^2 + 2y^2 + 2z^2 = 1$ and
    $x + y + z = 0$

4.5   JACOBIANS

Definition 4.5.1 Let u and v be functions of two independent variables x and y. The

determinant $\begin{vmatrix} \dfrac{\partial u}{\partial x} & \dfrac{\partial u}{\partial y} \\ \dfrac{\partial v}{\partial x} & \dfrac{\partial v}{\partial y} \end{vmatrix}$ is called the **Jacobian** of u, v with respect to x, y and is written as

$\dfrac{\partial(u, v)}{\partial(x, y)}$ or $J\left(\dfrac{u, v}{x, y}\right)$.

The Jacobian is also called the functional determinant of u, v with respect to x, y.
Similarly the Jacobian of u, v, w with respect to x, y, z is defined as

$$\frac{\partial(u, v, w)}{\partial(x, y, z)} = \begin{vmatrix} \dfrac{\partial u}{\partial x} & \dfrac{\partial u}{\partial y} & \dfrac{\partial u}{\partial z} \\ \dfrac{\partial v}{\partial x} & \dfrac{\partial v}{\partial y} & \dfrac{\partial v}{\partial z} \\ \dfrac{\partial w}{\partial x} & \dfrac{\partial w}{\partial y} & \dfrac{\partial w}{\partial z} \end{vmatrix} \text{ or } \begin{vmatrix} u_x & u_y & u_z \\ v_x & v_y & v_z \\ w_x & w_y & w_z \end{vmatrix}$$

$\dfrac{\partial(u, v, w)}{\partial(x, y, z)}$ is also denoted by $J\left(\dfrac{u, v, w}{x, y, z}\right)$.

Note: An important application of Jacobians is in connection with the change of variables in multiple integrals. Computation of surface area and volume of three dimensional objects are made simpler when a transformation is applied on the variables of integration.

4.5.1   PROPERTIES OF JACOBIANS

(1) If $J = \dfrac{\partial(u, v)}{\partial(x, y)}$ and $J' = \dfrac{\partial(x, y)}{\partial(u, v)}$ then $JJ' = 1$.

Let $u = f(x, y)$ and $v = g(x, y)$

Suppose, solving for x and y, we get $x = \varphi(u, v)$ and $y = \chi(u, v)$.
Then,

$$\left.\begin{array}{l} \dfrac{\partial u}{\partial x}.\dfrac{\partial x}{\partial u} + \dfrac{\partial u}{\partial y}.\dfrac{\partial y}{\partial u} = \dfrac{\partial u}{\partial u} = 1 \\[3mm] \dfrac{\partial u}{\partial x}.\dfrac{\partial x}{\partial v} + \dfrac{\partial u}{\partial y}.\dfrac{\partial y}{\partial v} = \dfrac{\partial u}{\partial v} = 0 \\[3mm] \dfrac{\partial v}{\partial x}.\dfrac{\partial x}{\partial u} + \dfrac{\partial v}{\partial y}.\dfrac{\partial y}{\partial u} = \dfrac{\partial v}{\partial u} = 0 \\[3mm] \dfrac{\partial v}{\partial x}.\dfrac{\partial x}{\partial v} + \dfrac{\partial v}{\partial y}.\dfrac{\partial y}{\partial v} = \dfrac{\partial v}{\partial v} = 1 \end{array}\right\} \quad (1)$$

and

$$\therefore JJ' = \begin{vmatrix} \dfrac{\partial u}{\partial x} & \dfrac{\partial u}{\partial y} \\[2mm] \dfrac{\partial v}{\partial x} & \dfrac{\partial v}{\partial y} \end{vmatrix} \cdot \begin{vmatrix} \dfrac{\partial x}{\partial u} & \dfrac{\partial x}{\partial v} \\[2mm] \dfrac{\partial y}{\partial u} & \dfrac{\partial y}{\partial v} \end{vmatrix}$$

$$= \begin{vmatrix} \dfrac{\partial u}{\partial x} \cdot \dfrac{\partial x}{\partial u} + \dfrac{\partial u}{\partial y} \cdot \dfrac{\partial y}{\partial u} & \dfrac{\partial u}{\partial x} \cdot \dfrac{\partial x}{\partial v} + \dfrac{\partial u}{\partial y} \cdot \dfrac{\partial y}{\partial v} \\[3mm] \dfrac{\partial v}{\partial x} \cdot \dfrac{\partial x}{\partial u} + \dfrac{\partial v}{\partial y} \cdot \dfrac{\partial y}{\partial u} & \dfrac{\partial v}{\partial x} \cdot \dfrac{\partial x}{\partial v} + \dfrac{\partial v}{\partial y} \cdot \dfrac{\partial y}{\partial v} \end{vmatrix}$$

$$= \begin{vmatrix} 1 & 0 \\ 0 & 1 \end{vmatrix} = 1 \qquad\qquad \text{(Using (1))}$$

(2) Jacobian of Composite Functions: If u and v are functions of r and s, where r and s are functions of x and y, then,

$$\frac{\partial(u,v)}{\partial(x,y)} = \frac{\partial(u,v)}{\partial(r,s)} \cdot \frac{\partial(r,s)}{\partial(x,y)}$$

$$\frac{\partial(u,v)}{\partial(r,s)} \cdot \frac{\partial(r,s)}{\partial(x,y)} = \begin{vmatrix} \dfrac{\partial u}{\partial r} & \dfrac{\partial u}{\partial s} \\[2mm] \dfrac{\partial v}{\partial r} & \dfrac{\partial v}{\partial s} \end{vmatrix} \cdot \begin{vmatrix} \dfrac{\partial r}{\partial x} & \dfrac{\partial r}{\partial y} \\[2mm] \dfrac{\partial s}{\partial x} & \dfrac{\partial s}{\partial y} \end{vmatrix}$$

$$= \begin{vmatrix} \dfrac{\partial u}{\partial r} \cdot \dfrac{\partial r}{\partial x} + \dfrac{\partial u}{\partial s} \cdot \dfrac{\partial s}{\partial x} & \dfrac{\partial u}{\partial r} \cdot \dfrac{\partial r}{\partial y} + \dfrac{\partial u}{\partial s} \cdot \dfrac{\partial s}{\partial y} \\[3mm] \dfrac{\partial v}{\partial r} \cdot \dfrac{\partial r}{\partial x} + \dfrac{\partial v}{\partial s} \cdot \dfrac{\partial s}{\partial x} & \dfrac{\partial v}{\partial r} \cdot \dfrac{\partial r}{\partial y} + \dfrac{\partial v}{\partial s} \cdot \dfrac{\partial s}{\partial y} \end{vmatrix}$$

$$= \begin{vmatrix} \dfrac{\partial u}{\partial x} & \dfrac{\partial u}{\partial y} \\[2mm] \dfrac{\partial v}{\partial x} & \dfrac{\partial v}{\partial y} \end{vmatrix}$$

$$= \frac{\partial(u,v)}{\partial(x,y)}$$

(3) If u, v, w are functionally dependent functions of three independent variables x, y, z

then $\dfrac{\partial(u,v,w)}{\partial(x,y,z)} = 0$

As u, v, w are functionally dependent, f(u, v, w) = 0. Differentiating partially with respect to x, y and z, we get

$$\frac{\partial f}{\partial u} \cdot \frac{\partial u}{\partial x} + \frac{\partial f}{\partial v} \cdot \frac{\partial v}{\partial x} + \frac{\partial f}{\partial w} \cdot \frac{\partial w}{\partial x} = 0 \tag{1}$$

$$\frac{\partial f}{\partial u} \cdot \frac{\partial u}{\partial y} + \frac{\partial f}{\partial v} \cdot \frac{\partial v}{\partial y} + \frac{\partial f}{\partial w} \cdot \frac{\partial w}{\partial y} = 0 \tag{2}$$

$$\frac{\partial f}{\partial u} \cdot \frac{\partial u}{\partial z} + \frac{\partial f}{\partial v} \cdot \frac{\partial v}{\partial z} + \frac{\partial f}{\partial w} \cdot \frac{\partial w}{\partial z} = 0 \tag{3}$$

Eliminating $\dfrac{\partial f}{\partial u}$, $\dfrac{\partial f}{\partial v}$ and $\dfrac{\partial f}{\partial w}$ from equations (1), (2) and (3), we get

$$\begin{vmatrix} \dfrac{\partial u}{\partial x} & \dfrac{\partial v}{\partial x} & \dfrac{\partial w}{\partial x} \\[2mm] \dfrac{\partial u}{\partial y} & \dfrac{\partial v}{\partial y} & \dfrac{\partial w}{\partial y} \\[2mm] \dfrac{\partial u}{\partial z} & \dfrac{\partial v}{\partial z} & \dfrac{\partial w}{\partial z} \end{vmatrix} = 0$$

Interchanging the rows and columns,

$$\begin{vmatrix} \dfrac{\partial u}{\partial x} & \dfrac{\partial u}{\partial y} & \dfrac{\partial u}{\partial z} \\[2mm] \dfrac{\partial v}{\partial x} & \dfrac{\partial v}{\partial y} & \dfrac{\partial v}{\partial z} \\[2mm] \dfrac{\partial w}{\partial x} & \dfrac{\partial w}{\partial y} & \dfrac{\partial w}{\partial z} \end{vmatrix} = 0$$

i.e., $\dfrac{\partial(u, v, w)}{\partial(x, y, z)} = 0$

Note:

(i)   The converse of property (3) is also true. i.e., if $\dfrac{\partial(u, v, w)}{\partial(x, y, z)} = 0$, then

u, v and w are functionally dependent.

(ii) Under the transformations $x = x(u, v)$, $y = y(u, v)$, the double integral

$\displaystyle\iint f(x, y)\, dx\, dy$ is transformed into the double integral

$$\iint F(u, v)\, |\, J\, |\, du\, dv \quad \text{where } J = \dfrac{\partial(x, y)}{\partial(u, v)}$$

i.e., dx dy = | J | du dv

Similarly if x, y, z are functions of u, v, w then the volume element

$$dx\, dy\, dz = |\, J\, |\, du\, dv\, dw \quad \text{where } J = \dfrac{\partial(x, y, z)}{\partial(u, v, w)}$$

Example 4.5.1 If $x = r \cos\theta$, $y = r \sin\theta$, evaluate $J = \dfrac{\partial(x, y)}{\partial(r, \theta)}$ and $J' = \dfrac{\partial(r, \theta)}{\partial(x, y)}$ and prove that JJ' = 1.

Solution: We have $r = \sqrt{x^2 + y^2}$ and $\theta = \tan^{-1}\left(\dfrac{y}{x}\right)$

From $x = r \cos\theta$ and $y = r \sin\theta$, we have $\dfrac{\partial x}{\partial r} = \cos\theta$, $\dfrac{\partial x}{\partial \theta} = -r \sin\theta$

$\dfrac{\partial y}{\partial r} = \sin\theta$, $\dfrac{\partial y}{\partial r} = r \cos\theta$

$$\therefore \quad J = \begin{vmatrix} \dfrac{\partial x}{\partial r} & \dfrac{\partial x}{\partial \theta} \\ \dfrac{\partial y}{\partial r} & \dfrac{\partial y}{\partial \theta} \end{vmatrix} = \begin{vmatrix} \cos\theta & -r\sin\theta \\ \sin\theta & r\cos\theta \end{vmatrix}$$

$$= r(\cos^2\theta + \sin^2\theta) = r$$

From $r = \sqrt{x^2 + y^2}$ and $\theta = \tan^{-1}\left(\dfrac{y}{x}\right)$, we have

$$\frac{\partial r}{\partial x} = \frac{x}{\sqrt{x^2 + y^2}} = \frac{x}{r} \quad \text{and} \quad \frac{\partial r}{\partial y} = \frac{y}{\sqrt{x^2 + y^2}} = \frac{y}{r}$$

$$\frac{\partial \theta}{\partial x} = \frac{1}{1 + \left(\frac{y}{x}\right)^2} \cdot y \cdot \left(\frac{-1}{x^2}\right) = \frac{-y}{x^2 + y^2} = \frac{-y}{r^2}$$

$$\frac{\partial \theta}{\partial y} = \frac{1}{1 + \left(\frac{y}{x}\right)^2} \left(\frac{1}{x}\right) = \frac{x}{x^2 + y^2} = \frac{x}{r^2}$$

$$\therefore \quad J' = \begin{vmatrix} \dfrac{\partial r}{\partial x} & \dfrac{\partial r}{\partial y} \\ \dfrac{\partial \theta}{\partial x} & \dfrac{\partial \theta}{\partial y} \end{vmatrix} = \begin{vmatrix} \dfrac{x}{r} & \dfrac{y}{r} \\ \dfrac{-y}{r^2} & \dfrac{x}{r^2} \end{vmatrix}$$

$$= \frac{x^2 + y^2}{r^3} = \frac{r^2}{r^3} = \frac{1}{r} \qquad \therefore \quad JJ' = r.\frac{1}{r} = 1$$

Example 4.5.2 If $x = r\sin\theta\cos\varphi$, $y = r\sin\theta\sin\varphi$, $z = r\cos\theta$,

show that $\dfrac{\partial(x, y, z)}{\partial(r, \theta, \varphi)} = r^2\sin\theta$.

Solution:

$$x = r\sin\theta\cos\varphi$$

$$\therefore \quad \frac{\partial x}{\partial r} = \sin\theta\cos\varphi, \quad \frac{\partial x}{\partial \theta} = r\cos\theta.\cos\varphi, \quad \frac{\partial x}{\partial \varphi} = -r\sin\theta.\sin\varphi$$

$$y = r\sin\theta\sin\varphi$$

$$\therefore \quad \frac{\partial y}{\partial r} = \sin\theta\sin\varphi, \quad \frac{\partial y}{\partial \theta} = r\cos\theta.\sin\varphi, \quad \frac{\partial y}{\partial \varphi} = r\sin\theta.\cos\varphi$$

$$z = r\cos\theta$$

$$\therefore \quad \frac{\partial z}{\partial r} = \cos\theta, \quad \frac{\partial z}{\partial \theta} = -r\sin\theta, \quad \frac{\partial z}{\partial \varphi} = 0.$$

$$\frac{\partial(x, y, z)}{\partial(r, \theta, \varphi)} = \begin{vmatrix} \dfrac{\partial x}{\partial r} & \dfrac{\partial x}{\partial \theta} & \dfrac{\partial x}{\partial \varphi} \\ \dfrac{\partial y}{\partial r} & \dfrac{\partial y}{\partial \theta} & \dfrac{\partial y}{\partial \varphi} \\ \dfrac{\partial z}{\partial r} & \dfrac{\partial z}{\partial \theta} & \dfrac{\partial z}{\partial \varphi} \end{vmatrix}$$

$$= \begin{vmatrix} \sin\theta\cos\varphi & r\cos\theta\cos\varphi & -r\sin\theta\sin\varphi \\ \sin\theta\sin\varphi & r\cos\theta\sin\varphi & r\sin\theta\cos\varphi \\ \cos\theta & -r\sin\theta & 0 \end{vmatrix}$$

$= \sin\theta\cos\varphi \left[r^2\sin^2\theta\cos\varphi \right] - r\cos\theta\cos\varphi \left[-r\sin\theta\cos\theta\cos\varphi \right] +$
$\qquad\qquad\qquad (-r\sin\theta\sin\varphi)[-r\sin^2\theta\sin\varphi - r\cos^2\theta\sin\varphi]$

$= r^2\sin^3\theta\cos^2\varphi + r^2\sin^3\theta\sin^2\varphi + r^2\sin\theta\cos^2\theta\cos^2\varphi$
$\qquad\qquad\qquad\qquad + r^2\sin\theta\cos^2\theta\sin^2\varphi$

$= r^2\sin^3\theta + r^2\sin\theta\cos^2\theta$

$= r^2\sin\theta\,(\sin^2\theta + \cos^2\theta)$

$= r^2\sin\theta$

Example 4.5.3 If $u = xyz$, $v = xy + yz + zx$ and $w = x + y + z$, find $\dfrac{\partial(u, v, w)}{\partial(x, y, z)}$.

Solution: $u = xyz$

$\dfrac{\partial u}{\partial x} = yz, \quad \dfrac{\partial u}{\partial y} = xz, \quad \dfrac{\partial u}{\partial z} = xy$

$v = xy + yz + zx$

$\dfrac{\partial v}{\partial x} = y + z, \quad \dfrac{\partial v}{\partial y} = x + z, \quad \dfrac{\partial v}{\partial z} = x + y$

$w = x + y + z$

$\dfrac{\partial w}{\partial x} = 1, \quad \dfrac{\partial w}{\partial y} = 1, \quad \dfrac{\partial w}{\partial z} = 1$

$$\dfrac{\partial(u, v, w)}{\partial(x, y, z)} = \begin{vmatrix} yz & xz & xy \\ y+z & x+z & x+y \\ 1 & 1 & 1 \end{vmatrix}$$

$= yz\,(x + z - x - y) - xz\,(y + z - x - y) + xy\,(y + z - x - z)$

$= yz^2 - y^2z - xz^2 + x^2z + xy^2 - x^2y$

$= (x - y)\,(y - z)\,(z - x)$

Example 4.5.4 If $u = x + y + z$, $uv = y + z$, $uvw = z$, show that $\dfrac{\partial(x, y, z)}{\partial(u, v, w)} = u^2v$

Solution: Given $u = x + y + z$ (1)

$\qquad\qquad uv = y + z$ (2)

$\qquad\qquad uvw = z$ (3)

Using (2) in (1), we get,

$$x = u - (y + z)$$

i.e., $\quad x = u - uv$

i.e., $\quad x = u (1 - v)$ $\hspace{4cm}$ (4)

Using (3) in (2) we get,

$$y = uv - z$$

i.e., $\quad y = uv - uvw$

$$y = uv(1 - w)$$ $\hspace{4cm}$ (5)

From (4) $\quad \dfrac{\partial x}{\partial u} = 1 - v, \qquad \dfrac{\partial x}{\partial v} = -u, \qquad \dfrac{\partial x}{\partial w} = 0$

From (5) $\quad \dfrac{\partial y}{\partial u} = v.(1 - w), \qquad \dfrac{\partial y}{\partial v} = u.(1 - w), \qquad \dfrac{\partial y}{\partial w} = -uv$

From (3) $\quad \dfrac{\partial z}{\partial u} = vw, \qquad \dfrac{\partial z}{\partial v} = uw, \qquad \dfrac{\partial z}{\partial w} = uv$

$$\frac{\partial(x, y, z)}{\partial(u, v, w)} = \begin{vmatrix} 1 - v & -u & 0 \\ v(1 - w) & u(1 - w) & -uv \\ vw & wu & uv \end{vmatrix}$$

$$= (1 - v) [u^2v (1 - w) + u^2vw] + u [uv^2(1 - w) + uv^2w]$$

$$= (1 - v) (u^2v) + u^2v^2$$

$$= u^2v$$

Example 4.5.5 Examine the functional dependence of the functions $u = \dfrac{x + y}{x - y}$ and

$v = \dfrac{xy}{(x - y)^2}$. If they are dependent, find the relation between them.

Solution: $u = \dfrac{x + y}{x - y}$

$\therefore \qquad \dfrac{\partial u}{\partial x} = \dfrac{(x - y) - (x + y)}{(x - y)^2} = \dfrac{-2y}{(x - y)^2}$

$\dfrac{\partial u}{\partial y} = \dfrac{(x - y) + (x + y)}{(x - y)^2} = \dfrac{2x}{(x - y)^2}$

$v = \dfrac{xy}{(x - y)^2}$

$\therefore \qquad \dfrac{\partial v}{\partial x} = \dfrac{(x - y)^2.y - xy.2.(x - y)}{(x - y)^4} = \dfrac{-y(x + y)}{(x - y)^3}$

$\dfrac{\partial v}{\partial y} = \dfrac{(x - y)^2.x + xy.2(x - y)}{(x - y)^4}$

$$= \frac{x(x+y)}{(x-y)^3}$$

$$\therefore \quad \frac{\partial(u,v)}{\partial(x,y)} = \begin{vmatrix} \frac{\partial u}{\partial x} & \frac{\partial u}{\partial y} \\ \frac{\partial v}{\partial x} & \frac{\partial v}{\partial y} \end{vmatrix} = \begin{vmatrix} \dfrac{-2y}{(x-y)^2} & \dfrac{2x}{(x-y)^2} \\ \dfrac{-y(x+y)}{(x-y)^3} & \dfrac{x(x+y)}{(x-y)^3} \end{vmatrix}$$

$$= 0$$

$\therefore$ u and v are functionally dependent.

Now $u^2 - 4v = \dfrac{(x+y)^2}{(x-y)^2} - \dfrac{4xy}{(x-y)^2}$

$$= \frac{(x+y)^2 - 4xy}{(x-y)^2}$$

$$= \frac{(x-y)^2}{(x-y)^2}$$

$$= 1$$

$\therefore$ The relation between u and v is $u^2 - 4v = 1$.

Example 4.5.6 Are the functions $f_1 = x + y + z$, $f_2 = x^2 + y^2 + z^2$ and $f_3 = xy + yz + zx$ functionally dependent? If so, find the relation between them.

Solution: $f_1 = x + y + z$

$\therefore \quad \dfrac{\partial f_1}{\partial x} = 1, \ \dfrac{\partial f_1}{\partial y} = 1, \ \dfrac{\partial f_1}{\partial z} = 1$

$f_2 = x^2 + y^2 + z^2$

$\therefore \quad \dfrac{\partial f_2}{\partial x} = 2x, \ \dfrac{\partial f_2}{\partial y} = 2y, \ \dfrac{\partial f_2}{\partial z} = 2z$

$f_3 = xy + yz + zx$

$\therefore \quad \dfrac{\partial f_3}{\partial x} = y + z, \ \dfrac{\partial f_3}{\partial y} = z + x, \ \dfrac{\partial f_3}{\partial z} = x + y$

Hence, $\dfrac{\partial(f_1, f_2, f_3)}{\partial(x, y, z)} = \begin{vmatrix} 1 & 1 & 1 \\ 2x & 2y & 2z \\ y+z & z+x & x+y \end{vmatrix}$

$$= \begin{vmatrix} 1 & 0 & 0 \\ 2x & 2y-2x & 2z-2x \\ y+z & x-y & x-z \end{vmatrix} \quad \begin{array}{l} \text{Subtracting column1 from} \\ \text{column 2 and column 3} \end{array}$$

$$= 2(y - x)(x - z) - 2(x - y)(z - x)$$
$$= 0$$

∴ The functions f_1, f_2 and f_3 are functionally dependent.

$$f_1^2 = (x + y + z)^2$$
$$= x^2 + y^2 + z^2 + 2(xy + yz + zx)$$
$$= f_2 + 2f_3$$

i.e., $f_1^2 - f_2 - 2f_3 = 0$ is the relation between f_1, f_2 and f_3.

EXERCISE 4.5

PART- A

1. Define 'Jacobian'.
2. State any property of Jacobians.
3. If $\dfrac{\partial(x, y)}{\partial(u, v)} = 3$, what is $\dfrac{\partial(u, v)}{\partial(x, y)}$?.
4. If $u = x^2$, $v = y^2$; verify $\dfrac{\partial(x, y)}{\partial(u, v)} \cdot \dfrac{\partial(u, v)}{\partial(x, y)} = 1$.
5. If $x = uv$, $y = u + v$; verify $\dfrac{\partial(x, y)}{\partial(u, v)} \cdot \dfrac{\partial(u, v)}{\partial(x, y)} = 1$.
6. If $x = u(1 - v)$, $y = uv$; verify $\dfrac{\partial(x, y)}{\partial(u, v)} \cdot \dfrac{\partial(u, v)}{\partial(x; y)} = 1$.
7. If $x = 4(1 - v)$, $y = uv$; verify $JJ' = 1$.
8. If $u = x(1 + y)$, $v = y(1 + x)$ prove $\dfrac{\partial(u, v)}{\partial(x, y)} = 1 + u + v$.
9. If $u = x^2 - y^2$, $v = xy$ prove $\dfrac{\partial(u, v)}{\partial(x, y)} = 2(x^2 + y^2)$.
10. If $x = r \cos \theta$, $y = r \sin \theta$, find $\dfrac{\partial(x, y)}{\partial(r, \theta)}$.
11. Show that $\iint f(x, y)\, dx\, dy = \iint f(u(1 - v), uv).u\, du\, dv$.
12. If $x = u(1 + v)$ and $y = v(1 + u)$, what is $\dfrac{\partial(x, y)}{\partial(u, v)}$?
13. If $x = u^2 - v^2$, $y = 2uv$, find the Jacobian of x and y with respect to u and v.
14. If u, v, w are functions of x, y, z, what is the condition for the functional dependence of u, v, w?
15. Verify whether $u = \dfrac{x - y}{x + y}$, $v = \dfrac{x + y}{x}$ are functionally dependent.

PART- B

16. If $u = \dfrac{x+y}{1-xy}$, $v = \tan^{-1}x + \tan^{-1}y$, find $\dfrac{\partial(u,v)}{\partial(x,y)}$.

17. If $x = a \cosh\theta \cos\varphi$, $y = a \sinh\theta \sin\varphi$, show that $\dfrac{\partial(x,y)}{\partial(\theta,\varphi)} = \dfrac{a^2}{2}[\cosh 2\theta - \cos 2\varphi]$.

18. If $x = r\cos\theta$, $y = r\sin\theta$, $z = z$, find $\dfrac{\partial(x,y,z)}{\partial(r,\theta,z)}$.

19. If $u = yz/x$, $v = zx/y$, $w = xy/z$ find $\dfrac{\partial(u,v,w)}{\partial(x,y,z)}$.

20. If $u = xyz$, $v = xy + yz + zx$, $w = x + y + z$, find $\dfrac{\partial(u,v,w)}{\partial(x,y,z)}$.

21. If $u = x^2 - 2y$, $v = x + y + z$, $w = x - 2y + 3z$, find $\dfrac{\partial(u,v,w)}{\partial(x,y,z)}$.

22. If $F = xu + v - y$, $G = u^2 + vy + w$, $H = zu - v + vw$, compute $\dfrac{\partial(F,G,H)}{\partial(u,v,w)}$.

23. If $u = x + y + z$, $uv = y + z$, $uvw = z$, show that $\dfrac{\partial(x,y,z)}{\partial(u,v,w)} = u^2v$.

24. If $u = x + y + z$, $u^2v = y + z$, $u^3w = z$, show that $\dfrac{\partial(u,v,w)}{\partial(x,y,z)} = u^5$.

25. If $u = 1 - x$, $v = x(1-y)$, $w = xy(1-z)$, find the value of $\dfrac{\partial(u,v,w)}{\partial(x,y,z)}$.

26. If $x = a(u+v)$, $y = b(u-v)$ and $u = r^2\cos 2\theta$, $v = r^2\sin 2\theta$, find $\dfrac{\partial(x,y)}{\partial(r,\theta)}$.

27. If $u = 2xy$, $v = x^2 - y^2$ and $x = r\cos\theta$, $y = r\sin\theta$, prove $\dfrac{\partial(u,v)}{\partial(r,\theta)} = -4r^3$.

28. Verify whether $u = \dfrac{x+y}{1-xy}$, $v = \tan^{-1}x + \tan^{-1}y$, are functionally dependent, and if so, find the relation between them.

29. Test whether $u = x + y - z$, $v = x - y + z$, $x^2 + y^2 + z^2 - 2yz$ are functionally dependent? If so, find the relation between them.

30. Verify whether $u = \dfrac{x-y}{x+y}$, $v = \dfrac{x+y}{x}$ are functionally dependent. If so, find the relation between them.

31. If $u = x + y + z$, $v = x^2 + y^2 + z^2$, $w = x^3 + y^3 + z^3 - 3xyz$, prove that u, v, w are functionally dependent. Find the relation between them.

32. If $x^2 + y^2 + r^2 - \theta^2 = 0$ and $xy + r\theta = 0$, show that $\dfrac{\partial(r,\theta)}{\partial(x,y)} = \dfrac{x^2 - y^2}{r^2 + \theta^2}$.

33. If $x = r^2 - \theta^2$, $y = 2r\theta$, find $\dfrac{\partial r}{\partial x}, \dfrac{\partial \theta}{\partial x}$.

34. If $u = xyz$, $v = x^2 + y^2 + z^2$, $w = x + y + z$, show that $\dfrac{\partial x}{\partial u} = \dfrac{1}{(x-y)(x-z)}$.

35. If $p = x + y^2$, $q = y + z^2$, $r = z + x^2$, show that $\dfrac{\partial x}{\partial p} = \dfrac{1}{1 + 8xyz}$.

4.6 DIFFERENTIATION UNDER THE INTEGRAL SIGN

Consider the function $f(x, \alpha)$ where α is a parameter.

Let $F(\alpha) = \displaystyle\int_a^b f(x, \alpha)\,dx$

In order to find the derivative $F'(\alpha)$, when it exists, it is not always possible to first evaluate the integral to find $F(\alpha)$ and then to find the derivative, on the other hand $F'(\alpha)$ can be computed directly using Leibnitz's rules.

4.6.1 LEIBNITZ'S RULE 1 (for constant limits of integration)

If $f(x, \alpha)$ and $\dfrac{\partial f(x, \alpha)}{\partial \alpha}$ are continuous functions of x and α, then

$$\frac{d}{d\alpha}\left[\int_a^b f(x, \alpha)\,dx\right] = \int_a^b \frac{\partial f(x, \alpha)}{\partial \alpha}\,dx \quad \text{where a and b are constants independent of } \alpha.$$

Proof: Let $F(\alpha) = \displaystyle\int_a^b f(x, \alpha)\,dx$

$$F(\alpha + \Delta\alpha) - F(\alpha) = \int_a^b f(x, \alpha + \Delta\alpha)\,dx - \int_a^b f(x, \alpha)\,dx$$

$$= \int_a^b \left[f(x, \alpha + \Delta\alpha) - f(x, \alpha)\right]dx$$

By Mean Value Theorem, we have, $f(\alpha + h) - f(\alpha) = h.\,f'(\alpha + \theta h)$, $0 < \theta < 1$

$$\therefore \quad f(x, \alpha + \Delta\alpha) - f(x, \alpha) = \Delta\alpha.\frac{\partial f(x, \alpha + \theta\Delta\alpha)}{\partial \alpha}, \; 0 < \theta < 1$$

Hence, $F(\alpha + \Delta\alpha) - F(\alpha) = \Delta\alpha \int\limits_{a}^{b} \dfrac{\partial f(x, \alpha + \theta\Delta\alpha)}{\partial\alpha} dx$

$\therefore \qquad \dfrac{F(\alpha + \Delta\alpha) - F(\alpha)}{\Delta\alpha} = \int\limits_{a}^{b} \dfrac{\partial f(x, \alpha + \theta\Delta\alpha)}{\partial\alpha} dx$

Taking limits on both sides as $\Delta\alpha \to 0$, we get,

$$F'(\alpha) = \int\limits_{a}^{b} \dfrac{\partial f(x, \alpha)}{\partial\alpha} dx$$

i.e., $\qquad \dfrac{d}{d\alpha}\left[\int\limits_{a}^{b} f(x, \alpha)\, dx \right] = \int\limits_{a}^{b} \dfrac{\partial f(x, \alpha)}{\partial\alpha} dx$

4.6.2   LEIBNITZ'S RULE 2 (for variable limits of integration)

If $f(x, \alpha)$ and $\dfrac{\partial f(x, \alpha)}{\partial\alpha}$ are continuous functions of x and α, then

$$\dfrac{d}{d\alpha}\left[\int\limits_{\varphi(\alpha)}^{\chi(\alpha)} f(x, \alpha)\, dx \right] = \int\limits_{\varphi(\alpha)}^{\chi(\alpha)} \dfrac{\partial f(x, \alpha)}{\partial\alpha} dx + \dfrac{d\chi}{d\alpha}.f(\chi(\alpha), \alpha) - \dfrac{d\varphi}{d\alpha}.f(\varphi(\alpha), \alpha),$$

provided $\varphi(\alpha)$ and $\chi(\alpha)$ posses continuous first order derivatives.

Note:

(1) $\dfrac{d}{d\alpha}\left[\int\limits_{a}^{\chi(\alpha)} f(x, \alpha)\, dx \right] = \int\limits_{a}^{\chi(\alpha)} \dfrac{\partial f(x, \alpha)}{\partial\alpha} dx + \dfrac{d\chi}{d\alpha}.f(\chi(\alpha), \alpha)$

(2) $\dfrac{d}{d\alpha}\left[\int\limits_{\varphi(\alpha)}^{b} f(x, \alpha)\, dx \right] = \int\limits_{\varphi(\alpha)}^{b} \dfrac{\partial f(x, \alpha)}{\partial\alpha} dx - \dfrac{d\varphi}{d\alpha}.f(\varphi(\alpha), \alpha)$

Example 4.6.1 Differentiating under the integral sign $\int\limits_{0}^{x} \dfrac{dx}{x^2 + a^2} = \dfrac{1}{a}\tan^{-1}\dfrac{x}{a}$, find the

value of $\int\limits_{0}^{x} \dfrac{dx}{\left(x^2 + a^2\right)^2}$.

Solution: Let $F(a) = \int\limits_{0}^{x} \dfrac{dx}{x^2 + a^2}$

Then $\quad F'(a) = \int\limits_{0}^{x} \dfrac{\partial}{\partial a}\left(\dfrac{1}{x^2 + a^2}\right).dx \qquad\qquad$ (By Leibnitz's rule.1) i.e., $F'(a) = \int_0^x \dfrac{-1.2a}{(x^2+a^2)^2}.dx = -2a.\int_0^x \dfrac{dx}{(x^2+a^2)^2}$ (1)

But, $F(a) = \dfrac{1}{a}\tan^{-1}\dfrac{x}{a}$

$\therefore \quad F'(a) = \dfrac{1}{a}.\dfrac{1}{1+(x/a)^2}\left(\dfrac{-x}{a^2}\right)+\left(\dfrac{-1}{a^2}\right).\tan^{-1}\dfrac{x}{a}$

$\qquad = \dfrac{-x}{a(x^2+a^2)} - \dfrac{1}{a^2}\tan^{-1}\dfrac{x}{a}$ (2)

From (1) and (2) we get

$-2a.\int_0^x \dfrac{dx}{(x^2+a^2)^2} = \dfrac{-x}{a(x^2+a^2)} - \dfrac{1}{a^2}\tan^{-1}\dfrac{x}{a}$

i.e., $\int_0^x \dfrac{dx}{(x^2+a^2)^2} = \dfrac{x}{2a^2(x^2+a^2)} + \dfrac{1}{2a^3}\tan^{-1}\dfrac{x}{a}$

Example 4.6.2 Evaluate $\int_0^\pi \log(1+a\cos x)\,dx$ if $|a| < 1$

Solution: Let $F(a) = \int_0^\pi \log(1+a\cos x)\,dx$

Then $F'(a) = \int_0^\pi \dfrac{\partial}{\partial a}\log(1+a\cos x)\,dx$ (By Leibnitz's rule.1)

$\qquad = \int_0^\pi \dfrac{\cos x}{1+a\cos x}\,dx$

$\qquad = \dfrac{1}{a}\int_0^\pi \dfrac{-1+1+a\cos x}{1+a\cos x}\,dx$

$\qquad = \dfrac{-1}{a}\int_0^\pi \dfrac{dx}{1+a\cos x} + \dfrac{1}{a}\int_0^\pi dx$ (1)

Substituting $t = \tan x/2$ so that $dx = \dfrac{2dt}{1+t^2}$ $\cos x = \dfrac{1-t^2}{1+t^2}$, we get,

$\int_0^\pi \dfrac{dx}{1+a\cos x} = \int_0^\infty \dfrac{2dt}{(1+t^2)\left[1+a\left(\dfrac{1-t^2}{1+t^2}\right)\right]}$

$$= 2 \int_0^\infty \frac{dt}{(1+a)+(1-a)t^2}$$

$$= \frac{2}{1-a} \int_0^\infty \frac{dt}{\frac{1+a}{1-a}+t^2}$$

$$= \frac{2}{1-a} \cdot \frac{1}{\sqrt{\frac{1+a}{1-a}}} \cdot \tan^{-1} \frac{t}{\sqrt{\frac{1+a}{1-a}}} \Bigg]_0^\infty$$

i.e., $$\int_0^\pi \frac{dx}{1+a \cos x} = \frac{\pi}{\sqrt{1-a^2}} \tag{2}$$

Also, $$\int_0^\pi dx = \pi \tag{3}$$

Using (2) and (3) in (1) we get $F'(a) = \dfrac{-1}{a} \cdot \dfrac{\pi}{\sqrt{1-a^2}} + \dfrac{1}{a} \cdot \pi$

Integrating with respect to a

$$F(a) = -\pi \int \frac{da}{a\sqrt{1-a^2}} + \pi \int \frac{da}{a} + c \tag{4}$$

In the integral, $\displaystyle\int \frac{da}{a\sqrt{1-a^2}}$,   put $a = 1/t$.   i.e., $da = \dfrac{-1}{t^2} dt$

$\therefore$
$$\int \frac{da}{a\sqrt{1-a^2}} = \int \frac{\frac{-1}{t^2} da}{\frac{-1}{t} \sqrt{1-\left(\frac{1}{t}\right)^2}}$$

$$= -\int \frac{dt}{\sqrt{t^2-1}}$$

$$= -\log\left(t + \sqrt{t^2-1}\right)$$

$$= -\log\left(\frac{1}{a} + \frac{1}{a}\sqrt{1-a^2}\right)$$

$\therefore$ (4) becomes, $F(a) = \pi\log\left(\dfrac{1}{a} + \dfrac{1}{a}\sqrt{1-a^2}\right) + \pi \log a + c$

i.e., $$F(a) = \pi \log\left(1 + \sqrt{1-a^2}\right) + c \tag{5}$$

Now, $$F(0) = \int_0^\pi \log(1) \, dx = 0$$

∴ Substituting a = 0 in (5), we get, $0 = \pi \log 2 + c$

i.e., $c = -\pi \log 2$

∴ $F(a) = \pi \log \left(1 + \sqrt{1 - a^2}\right) - \pi \log 2$

i.e., $\int_0^{\pi} \log(1 + a \cos x)\, dx = \pi \log\left(\frac{1}{2} + \frac{1}{2}\sqrt{1 - a^2}\right)$

Example 4.6.3 Evaluate $\int_0^{\infty} e^{-x^2} \cos \alpha x\, dx$ and $\int_0^{\infty} x e^{-x^2} \sin \alpha x\, dx$

Solution: Let $F(\alpha) = \int_0^{\infty} e^{-x^2} \cos \alpha x\, dx$ \hfill (1)

Then, $F'(\alpha) = \int_0^{\infty} \frac{\partial}{\partial \alpha}\left(e^{-x^2} \cos \alpha x\right) dx$ \quad (By Leibnitz's rule.1)

$= \int_0^{\infty} -e^{-x^2} \sin \alpha x\, x\, dx$

$= \int_0^{\infty} \sin \alpha x \left(-x e^{-x^2}\right) dx$

$= \sin \alpha x . \dfrac{e^{-x^2}}{2}\Bigg]_0^{\infty} - \int_0^{\infty} \dfrac{e^{-x^2}}{2} \cos \alpha x\, \alpha\, dx$

$= \dfrac{-\alpha}{2} . F(\alpha)$

∴ $\dfrac{F'(\alpha)}{F(\alpha)} = \dfrac{-\alpha}{2}$

$\int \dfrac{F'(\alpha)}{F(\alpha)} d\alpha = - \int \dfrac{\alpha}{2} d\alpha$

i.e., $\log F(\alpha) = \dfrac{-\alpha^2}{4} + \log c$, where c is a constant.

i.e., $F(\alpha) = c.e^{\frac{-\alpha^2}{4}}$ \hfill (2)

∴ $F(0) = c$

From (1), $F(0) = \int_0^{\infty} e^{-x^2} dx = \dfrac{\sqrt{\pi}}{2}$

$$\therefore \qquad F(\alpha) = \int_0^\infty e^{-x^2} \cos \alpha x \, dx = \frac{\sqrt{\pi}}{2} . e^{\frac{-\alpha^2}{4}} \qquad \text{(Using (2))}$$

Now differentiating $F(\alpha)$ with respect to α, we get,

$$F'(\alpha) = \int_0^\infty e^{-x^2} (-\sin \alpha x).x \, dx = \frac{\sqrt{\pi}}{2} . e^{\frac{-\alpha^2}{4}} . \left(\frac{-2\alpha}{4} \right)$$

i.e., $\qquad \displaystyle\int_0^\infty xe^{-x^2} \sin \alpha x \, dx = \frac{\sqrt{\pi}}{4} \alpha. e^{\frac{-\alpha^2}{4}}$

Example 4.6.4 Prove that $\displaystyle\int_0^\infty \frac{e^{-x}}{x} \left(1 - e^{-\alpha x} \right) dx = \log (1+\alpha), \ \alpha > -1$

Solution: Let $F(\alpha) = \displaystyle\int_0^\infty \frac{e^{-x}}{x} \left(1 - e^{-\alpha x} \right) dx \qquad (1)$

Then $\quad F'(\alpha) = \displaystyle\int_0^\infty \frac{\partial}{\partial x} \left[\frac{e^{-x}}{x} \left(1 - e^{-\alpha x} \right) \right] dx$

$$= \int_0^\infty \frac{e^{-x}}{x} . e^{-\alpha x} .x \, dx$$

$$= \int_0^\infty e^{-(1+\alpha)x} . dx$$

$$= \frac{e^{-(1+\alpha)x}}{-(1+\alpha)} \Bigg]_0^\infty$$

$$= \frac{1}{1+\alpha} \quad \text{when } 1 + \alpha > 0 \text{ or } \alpha > -1$$

$\therefore \qquad F(\alpha) = \displaystyle\int \frac{1}{1+\alpha} \, d\alpha + c$

i.e., $\qquad F(\alpha) = \log (1+\alpha) + c \qquad (2)$

From (1), we get $F(0) = 0$

$\therefore$ (2) Gives, $0 = \log (1+0) + c$

i.e., $\quad c = 0$

$\therefore$ (2) Becomes, $F(\alpha) = \log (1+\alpha)$

i.e., $\qquad \displaystyle\int_0^\infty \frac{e^{-x}}{x} \left(1 - e^{-\alpha x} \right) dx = \log (1+\alpha)$

Example 4.6.5 Prove that $\int\limits_0^{\pi/2} \dfrac{\log(1 + y\sin^2 x)}{\sin^2 x}\, dx = \pi\left[\sqrt{1+y} - 1\right]$

Solution: Let $F(y) = \int\limits_0^{\pi/2} \dfrac{\log(1 + y\sin^2 x)}{\sin^2 x}\, dx$ (1)

Then $F'(y) = \int\limits_0^{\pi/2} \dfrac{\partial}{\partial y}\left[\dfrac{\log(1 + y\sin^2 x)}{\sin^2 x}\right] dx$ **(By Leibnitz's rule.1)**

$= \int\limits_0^{\pi/2} \dfrac{1}{1 + y\sin^2 x}\, dx$

$= \int\limits_0^{\pi/2} \dfrac{\sec^2 x}{\sec^2 x + y\tan^2 x}\, dx$ **(Substituting $t = \tan x$ so that $dt = \sec^2 x\, dx$)**

$= \int\limits_0^{\infty} \dfrac{dt}{1 + t^2 + y\cdot t^2}$

$= \int\limits_0^{\infty} \dfrac{dt}{1 + (1+y)t^2}$

$= \dfrac{1}{(1+y)} \int\limits_0^{\infty} \dfrac{dt}{(1+y)^{-1} + t^2}$

$= \dfrac{1\cdot\sqrt{1+y}}{(1+y)}\left. \tan^{-1}\left(t\cdot\sqrt{1+y}\right)\right]_0^{\infty}$

i.e., $F'(y) = \dfrac{\pi}{2\sqrt{1+y}}$

Integrating w.r.t y, we get

$F(y) = \pi\sqrt{1+y} + c$ (2)

From (1), $F(0) = 0$

$\therefore$ From (2) $F(0) = \pi + c$

i.e., $0 = \pi + c$

i.e., $c = -\pi$

$\therefore$ From (2), $F(y) = \pi\left[\sqrt{1+y} - 1\right]$

i.e., $\int\limits_0^{\pi/2} \dfrac{\log(1 + y\sin^2 x)}{\sin^2 x}\, dx = \pi\left[\sqrt{1+y} - 1\right]$

Example 4.6.6 Evalute $\int\limits_0^\alpha \dfrac{\log{(1+\alpha x)}}{1+x^2}\,dx$ and hence prove that $\int\limits_0^1 \dfrac{\log{(1+x)}}{1+x^2}\,dx = \dfrac{\pi}{8}.\log 2$

Solution: Let $F(\alpha) = \int\limits_0^\alpha \dfrac{\log{(1+\alpha x)}}{1+x^2}\,dx$ \hfill (1)

Using Leibnitz's rule.2(when the limits are variables)

$$F'(\alpha) = \int\limits_0^\alpha \frac{\partial}{\partial \alpha}\left(\frac{\log{(1+\alpha x)}}{1+x^2}\right)dx + \frac{d(\alpha)}{d\alpha}.\frac{\log{(1+\alpha^2)}}{1+\alpha^2} - 0$$

$$= \int\limits_0^\alpha \frac{x}{(1+\alpha x)(1+x^2)}dx + \frac{\log{(1+\alpha^2)}}{1+\alpha^2}$$ \hfill (2)

Now $\dfrac{x}{(1+\alpha x)(1+x^2)} = \dfrac{A}{(1+\alpha x)} + \dfrac{Bx+C}{1+x^2}$ (By partial fractions)

$x = A\,(1+x^2) + (1+\alpha x)\,(Bx+C)$

Equating the coefficients

$A + C = 0$

$B + C\alpha = 1$

$A + B\alpha = 0$

Solving for A, B, C we get $A = \dfrac{-\alpha}{1+\alpha^2}$, $\quad B = \dfrac{1}{1+\alpha^2}$, $\quad C = \dfrac{\alpha}{1+\alpha^2}$

$\therefore \quad \int\limits_0^\alpha \dfrac{x}{(1+\alpha x)(1+x^2)}dx = \dfrac{-\alpha}{1+\alpha^2}\int\limits_0^\alpha \dfrac{dx}{1+\alpha x} + \dfrac{1}{1+\alpha^2}\int\limits_0^\alpha \dfrac{x}{1+x^2}\,dx + \dfrac{\alpha}{1+\alpha^2}\int\limits_0^\alpha \dfrac{dx}{1+x^2}$

$$= \left[\dfrac{-1}{1+\alpha^2}\log{(1+\alpha x)} + \dfrac{1}{2(1+\alpha^2)}\log{(1+x^2)} + \dfrac{\alpha}{1+\alpha^2}\tan^{-1}x\right]_0^\alpha$$

$$= \dfrac{-1}{1+\alpha^2}\log{(1+\alpha^2)} + \dfrac{1}{2(1+\alpha^2)}\log{(1+\alpha^2)} + \dfrac{\alpha}{1+\alpha^2}\tan^{-1}\alpha$$

Using this in (2) we get,

$$F'(\alpha) = \dfrac{1}{2}.\dfrac{\log{(1+\alpha^2)}}{1+\alpha^2} + \dfrac{\alpha}{1+\alpha^2}\tan^{-1}\alpha$$

Integrating with respect to α

$$F(\alpha) = \dfrac{1}{2}.\int\dfrac{\log{(1+\alpha^2)}}{1+\alpha^2}\,d\alpha + \int\dfrac{\alpha}{1+\alpha^2}\tan^{-1}\alpha\,d\alpha$$

$$= \dfrac{1}{2}.\log{(1+\alpha^2)}.\tan^{-1}\alpha - \dfrac{1}{2}\int\tan^{-1}\alpha\dfrac{1}{1+\alpha^2}.2\alpha\,d\alpha + \int\dfrac{\alpha}{1+\alpha^2}\tan^{-1}\alpha\,d\alpha + c$$

i.e., $\quad F(\alpha) = \dfrac{1}{2}.\log{(1+\alpha^2)}.\tan^{-1}\alpha + c$ \hfill (3)

From (1), $\quad$ F(0) = 0
From (3), $\quad$ F(0) = c
∴ c = 0

Hence $\quad F(\alpha) = \dfrac{1}{2}.\log(1+\alpha^2).\tan^{-1}\alpha$ $\hspace{3cm}$ (4)

Putting $\alpha = 1$, $F(1) = \displaystyle\int_0^1 \dfrac{\log(1+x)}{1+x^2}dx$

$\hspace{3cm} = \dfrac{1}{2}\log(2).\tan^{-1}(1) = \dfrac{\pi}{8}\log 2$

Example 4.6.7 Prove that $\dfrac{d}{d\alpha}\displaystyle\int_0^{a^2}\tan^{-1}(x/a)dx = 2a\tan^{-1}a - \dfrac{1}{2}\log(a^2+1)$

Solution: Let $F(a) = \displaystyle\int_0^{a^2}\tan^{-1}(x/a)dx$

Then $\quad F'(a) = \displaystyle\int_0^{a^2}\dfrac{\partial}{\partial a}\left(\tan^{-1}(x/a)\right)dx + \dfrac{d}{da}(a^2).\tan^{-1}\left(\dfrac{a^2}{a}\right)$ $\hspace{1cm}$ (By Leibnitz's rule 2.)

$\hspace{1.5cm} = \displaystyle\int_0^{a^2}\dfrac{1}{1+(x/a)^2}.x.\left(\dfrac{-1}{a^2}\right)dx + 2a\tan^{-1}a$

$\hspace{1.5cm} = \displaystyle\int_0^{a^2}\dfrac{-x}{a^2+x^2}dx + 2a\tan^{-1}a$

$\hspace{1.5cm} = \dfrac{-1}{2}\log(a^2+x^2)\Bigg]_0^{a^2} + 2a\tan^{-1}a$

$\hspace{1.5cm} = \dfrac{-1}{2}\left[\log(a^2+a^4)-\log a^2\right] + 2a\tan^{-1}a$

i.e., $\quad \dfrac{d}{d\alpha}\displaystyle\int_0^{a^2}\tan^{-1}(x/a)dx = -\dfrac{1}{2}\log(a^2+1) + 2a\tan^{-1}a$

$\hspace{3cm} = 2a\tan^{-1}a - \dfrac{1}{2}\log(a^2+1)$

Example 4.6.8 If $y = \displaystyle\int_0^x f(t)\sin[k(x-t)]dt$, prove that y satisfies the differential

equation $\dfrac{d^2y}{dx^n} + k^2y = k.f(x)$.

Solution: Given $y = \int_0^x f(t) \sin[k(x-t)]\,dt$ (1)

$\therefore \dfrac{dy}{dx} = \int_0^x \dfrac{\partial}{\partial x}\left(f(t)\sin[k(x-t)]\right)dt + \dfrac{d(x)}{dx}\cdot f(x)\cdot\sin[k(x-x)]$ (By Leibnitz's rule 2.)

$= \int_0^x f(t)\cos[k(x-t)]\cdot k\,dt$

$= \int_0^x k\cdot f(t)\cos[k(x-t)]\,dt$

Similarly $\dfrac{d^2 y}{dx^2} = \int_0^x \dfrac{\partial}{\partial x}\left(k\cdot f(t)\cos[k(x-t)]\right)dt + \dfrac{d(x)}{dx}k\cdot f(x)\cos[k(x-x)]$

                                             (By Leibnitz's rule 2.)

$= -\int_0^x k^2\cdot f(t)\cdot\sin[k(x-t)]\,dt + k\cdot f(x)$

$= -k^2 y + k\cdot f(x)$ (Using (1))

i.e., $\dfrac{d^2 y}{dx^n} + k^2 y = k\cdot f(x)$

EXERCISE 4.6

PART- A

1. State Liebnitz's rule for $\dfrac{d}{d\alpha}\left[\int_a^b f(x,\alpha)\,dx\right]$.

2. State Liebnitz's rule for differentiation under the integral sign when both the limits are variables.

3. State Liebnitz's rule for $\dfrac{d}{d\alpha}\left[\int_{\varphi(\alpha)}^b f(x,\alpha)\,dx\right]$.

4. State Liebnitz's rule for $\dfrac{d}{d\alpha}\left[\int_a^{\chi(\alpha)} f(x,\alpha)\,dx\right]$.

5. Prove that $\dfrac{d}{d\alpha}\left[\int_0^1 \dfrac{x^\alpha - 1}{\log x}\,dx\right] = \dfrac{1}{1+\alpha}$ when $\alpha \ge 0$.

6. Prove $\dfrac{d}{dy}\left[\int_0^1 \log(x^2 + y^2)\,dx\right] = 2.\tan^{-1}\left(\dfrac{1}{y}\right)$.

PART-B

7. Evaluate $\int_0^1 \frac{x^\alpha - 1}{\log x} dx$, $\alpha \ge 0$

8. Evluate $\int_0^\alpha \frac{\log(1+\alpha x)}{1+x^2} dx$ and hence show that $\int_0^1 \frac{\log(1+x)}{1+x^2} dx = \frac{\pi}{8}.\log_e 2$.

9. Show that $\int_0^\infty \frac{\tan^{-1}(\alpha x)}{x.(1+x^2)} dx = \frac{\pi}{2}.\log(1+\alpha)$ where $\alpha \ge 0$.

10. By successive differentiation of $\int_0^1 x^m dx = \frac{1}{m+1}$ w.r.t m, evaluate $\int_0^1 x^m (\log x)^n dx$.

11. Prove that $\int_0^\pi \frac{\log(1+\sin \alpha.\cos x)}{\cos x} dx = \pi\alpha$.

12. If $\int_0^\pi \frac{dx}{a+b\cos x} = \frac{\pi}{\sqrt{a^2-b^2}}.(a > b)$, show that $\int_0^\pi \frac{dx}{(a+b\cos x)^2} = \frac{-\pi b}{(a^2-b^2)^{3/2}}$.

13. If $\int_0^\infty \frac{\cos \alpha x}{a^2+x^2} dx = \frac{\pi}{2a}.e^{-\alpha a}$, $a > 0$, prove $\int_0^\infty \frac{x.\sin \alpha x}{a^2+x^2} dx = \frac{\pi}{2}.e^{-\alpha a}$.

14. If a, b > 0 prove that $\int_0^\infty \frac{e^{-ax} - e^{-bx}}{x} dx = \log(b/a)$.

15. Prove $\int_0^\infty \frac{e^x - e^{-ax}}{x.\sec x} dx = \frac{1}{2} \log\left[\frac{1}{2}.(1+a^2)\right]$ if a > 0.

16. Prove $\int_0^{\pi/2} \frac{\log(1+\alpha.\sin x)}{\sin x} dx = \frac{1}{2}\left[\pi.\sin^{-1} \alpha - (\sin^{-1} \alpha)^2\right]$.

17. Prove $\int_0^\infty x^n.e^{-\alpha x} dx = \frac{n!}{\alpha^{n+1}}$, $\alpha > 0$.

18. $\int_0^\infty \frac{dx}{x^2+a^2} dx = \frac{\pi}{2a}$; show that $\int_0^\infty \frac{dx}{(x^2+a^2)^{n+1}} dx = \frac{(2n)!\pi}{2^{2n+1}.(n!)^2 a^{2n+1}}$.

19. Prove $\int_{-\pi/2}^{\pi/2} \log(1+a.\sin x) dx = \pi.\log\left(\frac{a}{2} + \frac{1}{2}.\sqrt{1-a^2}\right)$, if 0 < a < 1

20. Prove $\int_0^\pi \log(1-2a.\cos x + a^2) dx = 2\pi.\log a$ if $|a| > 1$

$$= 0 \qquad \text{if } |a| < 1.$$

ANSWERS

EXERCISE 4.1

(6) 3u (7) u   (8) tan u   (9) 2 tan u   (10) 0

(11) 0   (12) 2.nu   (13) 3u.log u   (14) 0   (20) 2u

EXERCISES 4.2

(2) $y^2\left(\dfrac{1}{t}+2x\right)\cos(xy^2)$ (3) $du = \cos(xy^2)\,(y^2\,dx + 2xy\,dy)$

(4) $y\,(1 + \log xy)\,dx + x\,(1 + \log xy)\,dy$ (5) $8a^5t^6\,(4t + 7)$

(6) $-2x(y - x)/y$ (7) $\log(xy) + 2 - \dfrac{ax}{y^2 - ax}$ (8) $-y/x$

(9) $-\dfrac{x^2 + 2xy + 2y^2}{x^2 + 4xy + y^2}$ (10) $\dfrac{2x}{a^2}.(a^2 - b^2).\cos(x^2 + y^2)$ (25) $\dfrac{2u^2 - v^2}{2uv}$ (28)

$\dfrac{\partial^2 z}{\partial u \partial v} = 0$ (29) $\dfrac{\partial^2 z}{\partial u^2} = 0$ (30) $\dfrac{\partial^2 u}{\partial z \partial \bar{z}} = 0$ (31)

$\dfrac{3}{\sqrt{1-t^2}}$ (32) $\dfrac{2(b^2 - a^2).x}{b^2}.\cos(x^2 + y^2)$ (33) $\dfrac{du}{dt} = 0$

(34) $2a^2xy\,(3a^4 + x^2y^2)\Big/(a^2x - y^3)^3$ (35) y/x (36) $\dfrac{b^2 - ac}{(ay + b)^3}$

EXERCISE 4.3

(6) $1 + x + y + \dfrac{x^2}{2} + xy + \dfrac{y^2}{2} + \cdots$ (7) $1 - x + y + x^2 - 2xy + y^2 + \ldots$

(8) $(x + y) - \dfrac{(x + y)^3}{3!} + \cdots$ (9) $1 - \dfrac{(x + y)^2}{2!} + \dfrac{(x + y)^4}{4!} - \cdots$

(10) $2 + x + y + \dfrac{x^2}{2!} + \dfrac{y^2}{2!} + \cdots$ (12) 2%

(13) $\dfrac{\pi}{4} - \dfrac{1}{2}(x - 1) + \dfrac{1}{2}(y - 1) + \dfrac{1}{4}(x - 1)^2 - \dfrac{1}{4}(y - 1)^2 + \cdots$

(14) $(\pi + e) + (2\pi + e)(x - 1) + \dfrac{1}{2!}(x - 1)^2(2\pi + e) + 2(x - 1)(y - \pi)^2 + \cdots$

(15) $e\left[1 + (x - 1) + (y - 1) + \dfrac{1}{2!}\left((x - 1)^2 + 4(x - 1)(y - 1) + (y - 1)^2\right) + \cdots\right]$

(16) $1 - \dfrac{x^2}{2} - \dfrac{y^2}{2} + \dfrac{x^4}{24} + \dfrac{x^2y^2}{4} + \dfrac{y^4}{24} + \cdots$

(17) $1 + x + \dfrac{x^2}{2} - \dfrac{y^2}{2} + \cdots$

(18) $\dfrac{e}{\sqrt{2}}\left[1 + (x-1) - \left(y - \dfrac{\pi}{4}\right) + \dfrac{(x-1)^2}{2!} - (x-1)\left(y - \dfrac{\pi}{4}\right) - \dfrac{\left(y - \dfrac{\pi}{4}\right)^2}{2!} + \cdots \right]$

(19) $\dfrac{xy}{x+y} + \dfrac{hy^2}{(x+y)^2} + \dfrac{k.x^2}{(x+y)^2} - \dfrac{h^2 y^2}{(x+y)^3} + \dfrac{2hkxy}{(x+y)^3} - \dfrac{k^2 x^2}{(x+y)^3} + \cdots$

(20) $1 + (x-1) + (x-1)(y-1) + \cdots$

(21) $\dfrac{1}{e\sqrt{2}}\left[1 + (x+1) + \left(y - \dfrac{\pi}{4}\right) + \dfrac{1}{2!}\left((x+1)^2 + 2(x+1)\left(y - \dfrac{\pi}{4}\right) - \left(y - \dfrac{\pi}{4}\right)^2 \right) + \right.$

$\left. \dfrac{1}{3!}\left((x+1)^3 + 3(x+1)^2\left(y - \dfrac{\pi}{4}\right) - 3(x+1)\left(y - \dfrac{\pi}{4}\right)^2 - \left(y - \dfrac{\pi}{4}\right)^3 \right) + \cdots \right]$

(24) $6 + \dfrac{1}{1!}\left(-9(x+2) + 4(y-1)\right) + \dfrac{1}{2!}\left(6(x+2)^2 - 20(x+2)(y-1) - 4(y-1)^2\right) +$

$\dfrac{1}{3!}\left(24(x+2)^2(y-1) - 6(x+2)(y-1)^2\right) + \cdots$

(26) $-9 + 3(x+2) - 7(y-1) + 2(x+2)(y-1) - 2(y-1)^2 + (x+2)(y-1)^2 + \cdots$

(27) $1 + (y-1) + (x-1)(y-1) + \cdots$

(28) $(28)\ 2 + \alpha.\cot 2\alpha$

EXERCISE 4.4

(10) (1, 0)   (11) (-3, 0)   (12) (0, 0), (1, 1)   (13) (1, 1)

(14) f(x, y) is minimum at $\left(\sqrt{2}, -\sqrt{2}\right)$ (15) f(x, y) is minimum at (-3, 0)

(16) Saddle point.

(17) Maximum value 112 at (4, 0), minimum value 106 at (6, 0)

(18) Maximum value -1 at (-1, -1)

(19) Minimum value is -8 at $\left(\sqrt{2}, -\sqrt{2}\right)$ and $\left(-\sqrt{2}, \sqrt{2}\right)$

(20) Maximum at (±1, 0), minimum at (0, ±1)

(21) Minimum at (a, a)

(22) Minimum at (a/3, a/3) if a > 0, maximum at (a/3, a/3) if a < 0

(23) Maximum at (0, 0)

(24) Minimum at (1, 1)

(25) Maximum at (0, 0), saddle points (±1, ±1).

(26) Maximum at (1, 1)

(27) Maximum at (a/2, a/3)

(28) Minimum at $\left(\pm \dfrac{\sqrt{3}}{2}, \dfrac{-1}{4} \right)$

(29) $3a^2$ (30) a^3 (31) $\left(\dfrac{a^2}{3} \right)^3$ (33) 27

(34) $a^{m+n+p} \cdot \dfrac{m^m \cdot n^n \cdot p^p}{(m+n+p)^{m+n+p}}$

(35) Maximum value = 108 at (1, 2, 3)

(38) 4cm, 4cm, 2cm (41) 108 (43) $\sqrt{6}, 3\sqrt{6}$

(44) (1, -2, 5) (45) (0, 0, 1) and (0, 0, -1)

(47) 7/3 (48) (1, 0, 0) (49) $\dfrac{1}{\sqrt{2}}$

(50) 2 and 1 (51) (2, -2, 1) and (-2, 2, 1) (52) 8/3, -8/7

(53) $\left(0, \dfrac{1}{2}, -\dfrac{1}{2} \right), \left(0, -\dfrac{1}{2}, \dfrac{1}{2} \right), \left(-\dfrac{1}{\sqrt{2}}, \dfrac{1}{2\sqrt{2}}, \dfrac{1}{2\sqrt{2}} \right), \left(\dfrac{1}{\sqrt{2}}, -\dfrac{1}{2\sqrt{2}}, -\dfrac{1}{2\sqrt{2}} \right)$

EXERCISE 4.5

(10) r (12) $u + v + 1$ (13) $4(u^2 + v^2)$

(16) 0 (18) r (19) 4

(20) $(x - y)(y - z)(z - x)$ (21) $10x + 4$

(22) $x(yv + 1 - w) + z - 2uv$ (25) $-x^2 y$ (26) $-8ab - r^3$

(28) $u = \tan v$ (29) $u^2 + v^2 = 2w$ (30) $uv = 2 - v$

(31) $2w = u(3v - u^2)$ (33) $\dfrac{r}{2(r^2 + \theta^2)}, \dfrac{-\theta}{2(r^2 + \theta^2)}.$

EXERCISE 4.6

(7) $\log (1 + \alpha)$ (8) $\dfrac{1}{2} \log(1 + \alpha^2) \cdot \tan^{-1} \alpha + c$ (10) $\dfrac{(-1)^n \cdot n!}{(m+1)^{n+1}}$

CHAPTER 5

ORDINARY DIFFERENTIAL EQUATIONS

5.0   INTRODUCTION

'Differential Equations' forms the basis of applied mathematics and is the most important branch of modern mathematics. Differential equations play an important role in solving problems in the different fields of engineering and physical sciences. Applications of differential equations can be found in such diverse areas as biology, medicine, statistics, sociology and economics.

Definition 5.0.1 An equation involving independent and dependent variables and the derivatives of one or more dependent variables with respect to one or more independent variables is called a **differential equation.**

The following equations are examples of differential equations.

$$\frac{dy}{dx} = \cos x \tag{1}$$

$$\frac{\partial^2 V}{\partial x^2} + \frac{\partial^2 V}{\partial y^2} + \frac{\partial^2 V}{\partial z^2} = 0 \tag{2}$$

$$p\frac{d^2 y}{dx^2} = \left[1 + \left(\frac{dy}{dx}\right)^2\right]^{3/2} \tag{3}$$

$$\frac{d^2 y}{dx^2} + \left(\frac{dy}{dx}\right)^3 = e^x \tag{4}$$

$$\frac{\partial^3 z}{\partial t^3} = k\left(\frac{\partial^2 z}{\partial x^2}\right)^2 \tag{5}$$

Definition 5.0.2 A differential equation which involves derivatives with respect to a single independent variable is known as an **ordinary differential equation**.

Equation (1), (3) and (4) are examples of ordinary differential equations.

Definition 5.0.3 A differential equation which contains two or more independent variables and partial derivatives with respect to them is called a **partial differential equation.**

Equations (2) and (5) are examples of partial differential equations.

Definition 5.0.4 The **order** of the highest order derivative involved in a differential equation is called the **order** of a differential equation. The **degree** of a differential equation is the degree of the highest order derivative present in the equation.

Note:

   The degree of a differential equation is an integer. Before computing the degree, the differential equation should be made free from fractional powers, as far as the derivatives are concerned. For example, equation (1) is first order, first degree. Equation (3) is second order; second degree where as equation (4) is second order, first degree equation. Equation (5) is third order, first degree in t where as it is second order, second degree in x.

5.0.1   SOLUTION OF A DIFFERENTIAL EQUATION

Definition 5.0.5 A **solution (or integral)** of a differential equation is a relation between the dependent and independent variables, not involving the derivatives such that this relation and the derivatives obtained from it satisfy the given differential equation.

For example,

(i) $y = \sin x$ is a solution of $\dfrac{dy}{dx} = \cos x$

(ii) $x = A \cos (nt + b)$ is a solution of $\dfrac{d^2x}{dt^2} + n^2 x = 0$.

In this solution A and b are arbitrary constants.

Definition 5.0.6 The **general (or complete) solution** of a differential equation is a solution in which the number of arbitrary constants is equal to the order of the differential equation. A **particular solution** is that which can be obtained from the general solution by giving particular values to the arbitrary constants.

For example,

(1) $y = \sin x + c$ is the general solution of $\dfrac{dy}{dx} = \cos x$, where as $y = \sin x$ is only a particular solution, obtained by putting $c = 0$ in the general solution.

(2) $x = A \cos (nt + b)$ is the general solution of $\dfrac{d^2x}{dt^2} + n^2 x = 0$, where as $x = \cos (nt + \pi/4)$ is a particular solution, obtained by putting $A = 1$ and $b = \pi/4$ in the general solution.

5.0.2   LINEAR AND NONLINEAR DIFFERENTIAL EQUATIONS

Definition 5.0.7 A differential equation in which the dependent variables and all its derivatives present occur in the first degree only is known as a **linear differential equation**. A differential equation which is not linear is called a **nonlinear differential equation**.

The most general linear differential equation of nth order is

$$a_0\frac{d^n y}{dx^n}+a_1\frac{d^{n-1}y}{dx^{n-1}}+a_2\frac{d^{n-2}y}{dx^{n-2}}+\cdots+a_{n-1}\frac{dy}{dx}+a_n\cdot y=\Phi \qquad (6)$$

where $a_0, a_1,..., a_n$ and Φ are functions of x or constants and $a_0 \neq 0$
If $\Phi = 0$, then equation (6) is called a **linear homogeneous** differential equation.
If $\Phi \neq 0$, then equation (6) is called a **linear non-homogeneous** differential equation.

Using the differential operator symbols $D \equiv \dfrac{d}{dx}, D^2 \equiv \dfrac{d^2}{dx^2}, \cdots, D^n \equiv \dfrac{d^n}{dx^n}$, equation (6)

becomes $(a_0 D^n + a_1 D^{n-1} + ... + a_{n-1}D + a_n)\, y = \Phi$
or $f(D).y = \Phi$ $\qquad\qquad (7)$
where $f(D) = a_0 D^n + a_1 D^{n-1} + ... + a_{n-1}D + a_n.$
The equation,
             $f(D).y = 0$ $\qquad\qquad\qquad (8)$
is the linear homogeneous differential equation corresponding to equation (7).

Note:
(i)   If $Y = u$ is the general solution of equation (8) that contains n arbitrary constants and $y = v$ is a particular solution of equation (7), that contains no arbitrary constants, then $y = u + v$ is the general solution of equation (7). u is called the **complementary function (C.F)** and v is called the **particular integral (P.I)** of the solution of the differential equation (7).

(ii)   The solutions $y_1, y_2,..., y_n$ of the linear homogenous differential equation (8) are said to be **linearly dependent** if there exists a set of n constants $c_1, c_2,..., c_n$ not all zero, such that $c_1 y_1 + c_2 y_2 + ... + c_n y_n = 0$.
        Otherwise, $y_1, y_2,..., y_n$ are said to be **linearly independent**.

(iii)   The necessary and sufficient condition for the n solutions $y_1, y_2,..., y_n$ of the linear homogeneous differential equation (8) to be linearly independent is that the determinant

$$\begin{vmatrix} y_1 & y_2 & \cdots & y_n \\ y_1' & y_2' & \cdots & y_n' \\ y_1'' & y_2'' & \cdots & y_n'' \\ \vdots & & & \\ y_1^{(n-1)} & y_2^{(n-1)} & \cdots & y_n^{(n-1)} \end{vmatrix} \text{ does not vanish identically.}$$

This determinant is called the **Wronskian** of the set of functions $y_1, y_2, ..., y_n$ and is denoted by $W(y_1, y_2 ... y_n)$.

Example 5.0.1 Show that $y_1(x) = \sin x$ and $y_2(x) = \sin x - \cos x$ are linearly independent solutions of $\dfrac{d^2y}{dx^2} + y = 0$

Solution: $\dfrac{d^2y}{dx^2} + y = 0$ (1)

$y_1(x) = \sin x$

$\therefore \quad y_1'(x) = \cos x, \quad y_1''(x) = -\sin x$

Now $y_1''(x) + y_1(x) = -\sin x + \sin x = 0$

$\therefore y_1(x)$ is a solution of equation (1).

$y_2(x) = \sin x - \cos x$

$\therefore \quad y_2'(x) = \cos x + \sin x$

$y_2''(x) = -\sin x + \cos x.$

Now $y_2''(x) + y_2(x) = -\sin x + \cos x + \sin x - \cos x = 0$

$\therefore y_2(x)$ is a solution of equation (1).

Also $W(y_1, y_2) = \begin{vmatrix} y_1 & y_2 \\ y_1' & y_2' \end{vmatrix} = \begin{vmatrix} \sin x & \sin x - \cos x \\ \cos x & \cos x + \sin x \end{vmatrix}$

$= \sin x (\cos x + \sin x) - \cos x (\sin x - \cos x)$

$= 1$

$\neq 0$

$\therefore y_1(x)$ and $y_2(x)$ are linearly independent solution of equation (1).

Example 5.0.2 Show that the functions x, x^2, x^3 are independent. Determine the differential equation for which x, x^2, x^3 form a set of linearly independent solutions.

Solution: Let $y_1(x) = x, y_2(x) = x^2, y_3(x) = x^3$

Then $W(y_1, y_2, y_3) = \begin{vmatrix} y_1 & y_2 & y_3 \\ y_1' & y_2' & y_3' \\ y_1'' & y_2'' & y_3'' \end{vmatrix}$

$= \begin{vmatrix} x & x^2 & x^3 \\ 1 & 2x & 3x^2 \\ 0 & 2 & 6x \end{vmatrix}$

$= x(12x^2 - 6x^2) - 1(6x^3 - 2x^3)$

$= 6x^3 - 4x^3 = 2x^3 \neq 0$

$\therefore y_1, y_2, y_3$ are linearly independent.

The general solution of the required differential equation is $y = c_1 y_1 + c_2 y_2 + c_3 y_3$ where c_1, c_2, c_3 are arbitrary constants.

i.e., $\quad y = c_1 x + c_2 x^2 + c_3 x^3$ $\hspace{3cm}$ (1)

Differentiating (1) thrice w.r.t x, we get

$$y' = c_1 + c_2.2x + c_3.3x^2 \hspace{3cm} (2)$$
$$y'' = 2c_2 + 3c_3.2x \hspace{3.5cm} (3)$$
$$y''' = 6c_3 \hspace{4.5cm} (4)$$

From (4), $\quad c_3 = \dfrac{y'''}{6}$

From (3), $\quad y'' = 2c_2 + 6\dfrac{y'''}{6}.x$

i.e., $\quad c_2 = \dfrac{y'' - xy'''}{2}$

From (2), $\quad y' = c_1 + (y'' - x y''').x + \dfrac{y'''}{6}.3x^2$

i.e., $\quad c_1 = y' - x.y'' + \dfrac{1}{2}.x^2 y'''$

Substituting the values of c_1, c_2, c_3 in (1), we get,

$$y = (y' - xy'' + \dfrac{1}{2}.x^2 y''').x + \dfrac{1}{2}(y'' - xy''').x^2 + \dfrac{1}{6}y'''.x^3$$

i.e., $\quad 6y = 6xy' - 6x^2 y'' + 3x^3 y''' + 3x^2 y'' - 3x^3 y''' + x^3 y'''$

i.e., $\quad x^3 y''' - 3x^2 y'' + 6xy' - 6y = 0$, which is the required differential equation.

5.1 LINEAR DIFFERENTIAL EQUATIONS WITH CONSTANT COEFFICIENTS

The general form of the linear differential equation of the nth order with constant coefficients is

$$a_0 \dfrac{d^n y}{dx^n} + a_1 \dfrac{d^{n-1}y}{dx^{n-1}} + a_2 \dfrac{d^{n-2}y}{dx^{n-2}} + \cdots + a_{n-1}\dfrac{dy}{dx} + a_n.y = \Phi \hspace{1cm} (1)$$

where $a_0, a_1 \ldots a_n$ are constants and $a_0 \neq 0$.

i.e., $\quad (a_0 D^n + a_1.D^{n-1} + \ldots + a_{n-1}.D + a_n) y = \Phi$ $\hspace{2cm}$ (2)

i.e., $\quad f(D).y = \Phi$ where $f(D) = a_0 D^n + a_1 D^{n-1} + \ldots + a_{n-1}D + a_n$

$\qquad f(D).y = 0$ $\hspace{6cm}$ (3)

is the linear homogenous differential equation corresponding to (1).

Replacing the operator D by m we get $f(m) = a_0 m^n + a_1 m^{n-1} + \ldots + a_{n-1}.m + a_n$.

The equation,

$$f(m) = a_0 m^n + a_1 m^{n-1} + \ldots + a_{n-1}.m + a_n = 0 \qquad (4)$$

is called the **auxiliary equation (A.E)** corresponding to equation (1).

5.1.1   COMPUTATION OF COMPLEMENTARY FUNCTION (C.F)

The complementary function (C.F) of equation (1) is the general solution of equation (3). The solutions of equation (3) depend on the nature of roots of the auxiliary equation (A.E) (4).

Case 1. The roots of the A.E are real and distinct.

Let the roots of the A.E. (4) be $m_1, m_2, \ldots, m_n$. Then $e^{m_1 x}, e^{m_2 x}, \ldots, e^{m_n x}$ are independent solutions of equation (3).

Then $y = c_1 e^{m_1 x} + c_2 e^{m_2 x} + \ldots + c_n e^{m_n x}$ where $c_1, c_2, \ldots, c_n$ are arbitrary constants, is the general solution of equation (3).

Hence the C.F of equation (1) is $u = c_1 e^{m_1 x} + c_2 e^{m_2 x} + \ldots + c_n e^{m_n x}$

Case 2. The roots of the A.E. are real, but not distinct.

Let the roots of the A.E (4) be $m_1, m_1, m_3, m_4, \ldots, m_n$. Then the C.F of equation (1) is

$$u = (c_1 + c_2 x)e^{m_1 x} + c_3 e^{m_3 x} + \ldots + c_n e^{m_n x}$$

In general, if the first k roots of the A.E. are equal, then the C.F is

$$u = (c_1 + c_2 x + \cdots + c_k x^{k-1})e^{m_1 x} + c_{k+1} e^{m_{k+1} x} + \ldots + c_n e^{m_n x}$$

and similarly for other equal roots.

Case 3. The A.E has complex roots

Let $m_1 = \alpha + i\beta$ and $m_2 = \alpha - i\beta$ be two of the roots of the A.E. Then the C.F is

$$u = e^{\alpha x}(c_1 \cos \beta x + c_2 \sin \beta x) + c_3 e^{m_3 x} + \ldots + c_n e^{m_n x} \text{ or equivalently}$$

$$u = e^{\alpha x} c_1 \cos(\beta x + c_2) + c_3 e^{m_3 x} + \ldots + c_n e^{m_n x}.$$

Also $\cos(\beta x + c_2)$ can be replaced by $\sin(\beta x + c_2)$.

Note:

In case the A.E has two pairs of complex roots equal, say $m_1 = m_3 = \alpha + i\beta$ and $m_2 = m_4 = \alpha - i\beta$ then the C.F is

$$u = e^{\alpha x}\left[(c_1 + c_2 x)\cos \beta x + (c_3 + c_4 x)\sin \beta x\right] + c_5 e^{m_5 x} + \ldots + c_n e^{m_n x} \text{ and similarly}$$

for others.

5.1.2   COMPUTATION OF PARTICULAR INTEGRAL (P.I)

The P.I of the solution of the equation $f(D).y = \Phi$ is a particular solution of $f(D).y = \Phi$.

i.e., $\quad P.I = \dfrac{1}{f(D)}\Phi$ where $\dfrac{1}{f(D)}$ is the inverse operator of f(D).

i.e., $\quad f(D).\left[\dfrac{1}{f(D)}\Phi\right] = \Phi.$ Thus $\dfrac{1}{f(D)}\Phi$

satisfies the equation $f(D).y = \Phi$ and is, therefore, its particular integral.

Note:

(i) $\quad \dfrac{1}{D}\Phi = \int\Phi dx$ $\hspace{5cm}$ (5)

$\quad$ Let $\dfrac{1}{D}.\Phi = y$

$\quad$ Then $D.\dfrac{1}{D}.\Phi = D.y$

$\quad$ i.e., $\quad \Phi = Dy$

$\quad$ i.e., $\quad \Phi = \dfrac{dy}{dx}$

$\quad$ Integrating both sides w.r.t x, we get, $y = \int\Phi dx$

$\quad$ i.e., $\dfrac{1}{D}.\Phi = \int\Phi.dx$

(ii) $\quad \dfrac{1}{D-a}.\Phi = e^{ax}\int\Phi.e^{-ax}dx$ $\hspace{4cm}$ (6)

$\quad$ Let $\dfrac{1}{D-a}.\Phi = y$

$\quad$ Then $(D-a).\dfrac{1}{D-a}.\Phi = (D-a).y$

$\quad$ i.e., $\Phi = (D-a).y$

$\quad$ i.e., $\dfrac{dy}{dx} - ay = \Phi$ $\hspace{5cm}$ (7)

$\quad$ i.e., $e^{-ax}\dfrac{dy}{dx} - a.e^{-ax}.y = e^{-ax}.\Phi$

$\quad \dfrac{d}{dx}(y.e^{-ax}) = e^{-ax}.\Phi$

$\quad$ Integrating both sides w.r.t x, we get, $y.e^{-ax} = \int e^{-ax}.\Phi dx$

$\quad \therefore\ y = e^{ax}\int\Phi.e^{-ax}dx$

$\quad$ i.e., $\dfrac{1}{D-a}.\Phi = e^{ax}\int\Phi.e^{-ax}dx$ $\hspace{4cm}$ (8)

(iii)   Equation (7) is a first order linear differential equation and e^{-ax} is called the **integrating factor (I.F)** of the differential equation (7). For the general linear differential equation of the first order, $\dfrac{dy}{dx} + Py = Q$, where P and Q are functions of x, the integrating factor (I.F) is $e^{\int P dx}$ and the general solution is

$$y.e^{\int p dx} = \int Q.e^{\int p dx}.dx + c \qquad\qquad (9)$$

5.1.3 PARTICULAR INTEGRALS OF SIMPLE FUNCTIONS

Case 1. $\Phi = e^{ax}$

Now, $De^{ax} = a.e^{ax}$

$\qquad D^2 e^{ax} = a^2.e^{ax}$

$$\vdots$$

$$D^n e^{ax} = a^n.e^{ax}$$

$\therefore\qquad f(D)e^{ax} = (a_0.D^n + a_1.D^{n-1} + \ldots + a_{n-1}.D + a_n)\,e^{ax}$

$\qquad\qquad\quad = (a_0.a^n + a_1.a^{n-1} + \ldots + a_{n-1}.a + a_n)\,e^{ax}$

$\qquad\qquad\quad = f(a).e^{ax}$

$\therefore\qquad \dfrac{1}{f(D)}.f(D).e^{ax} = \dfrac{1}{f(D)}.f(a).e^{ax}$

i.e., $e^{ax} = \dfrac{1}{f(D)}.f(a).e^{ax}$

Dividing by f(a), we get,

$$\dfrac{1}{f(D)}.e^{ax} = \dfrac{1}{f(a)}.e^{ax} \text{ , provided } f(a) \neq 0$$

When f(a) = 0, $(D-a)^r$ is a factor of f(D) for some positive integer r

i.e., $f(D) = (D-a)^r.\varphi(D)$ where $\varphi(a) \neq 0$.

When r = 1, $f(D) = (D-a).\varphi(D)$ where $\varphi(a) \neq 0$.

$$\dfrac{1}{f(D)}.e^{ax} = \dfrac{1}{(D-a).\varphi(D)}.e^{ax}$$

$$= \dfrac{1}{(D-a)}.\dfrac{1}{\varphi(a)}.e^{ax}$$

$$= \dfrac{1}{\varphi(a)}.\dfrac{1}{(D-a)}.e^{ax}$$

$$= \dfrac{1}{\varphi(a)}.e^{ax}\int e^{ax}.e^{-ax}dx \quad \text{ by(6)}$$

$$= \frac{1}{\varphi(a)}.e^{ax}.x$$

$$= \frac{1}{\varphi(a)}.\frac{x}{1!}.e^{ax}$$

When r = 2, f(D) = (D-a)^2.φ(D) where φ(a) ≠ 0.

$$\frac{1}{f(D)}.e^{ax} = \frac{1}{(D-a)^2.\varphi(D)}.e^{ax}$$

$$= \frac{1}{\varphi(a)}.\frac{1}{(D-a)}.x.e^{ax}$$

$$= \frac{1}{\varphi(a)}.e^{ax} \int x.e^{ax}.e^{-ax}dx$$

$$= \frac{1}{\varphi(a)}.\frac{x^2}{2!}.e^{ax}$$

In general, when f(D) = (D-a)^r.φ(D) where φ(a) ≠ 0,

$$\frac{1}{f(D)}.e^{ax} = \frac{1}{(D-a)^r.\varphi(D)}.e^{ax}$$

$$= \frac{1}{\varphi(a)}.\frac{x^r}{r!}.e^{ax}$$

Case 2. Φ = sin (ax + b) or cos (ax + b) where a and b are constants.

Now, D sin (ax + b) = a cos (ax + b)

D^2 sin (ax + b) = -a^2 sin (ax + b)

D^3 sin (ax + b) = -a^3 cos (ax + b)

D^4 sin (ax + b) = a^4 sin (ax + b)

i.e., $\left(D^2\right)^2$ sin (ax + b) = $\left(-a^2\right)^2$ sin (ax + b)

.

.

.

In general $\left(D^2\right)^r$ sin (ax + b) = $\left(-a^2\right)^r$ sin (ax + b) and

$\varphi\left(D^2\right)$sin (ax + b) = $\varphi\left(-a^2\right)$sin (ax + b)

∴ $\frac{1}{\varphi(D^2)}.\varphi\left(D^2\right)$sin (ax + b) = $\frac{1}{\varphi(D^2)}.\varphi\left(-a^2\right)$sin (ax + b)

i.e., sin (ax + b) = $\frac{1}{\varphi(D^2)}.\varphi\left(-a^2\right)$sin (ax + b)

Hence, if $\varphi\left(-a^2\right) \neq 0$, $\frac{1}{\varphi(D^2)}$sin (ax + b) = $\frac{1}{\varphi(-a^2)}$.sin (ax + b)

Similarly if $\varphi(-a^2) \neq 0$, $\dfrac{1}{\varphi(D^2)}.\cos(ax+b) = \dfrac{1}{\varphi(-a^2)}.\cos(ax+b)$

If $\varphi(-a^2) = 0$, $D^2 + a^2$ is a factor of $\varphi(D^2)$ and $\varphi(D^2) = (D^2 + a^2)\chi(D^2)$ where $\chi(-a^2) \neq 0$

$$\therefore \frac{1}{\varphi(D^2)}\sin(ax+b) = \frac{1}{(D^2+a^2).\chi(D^2)}\sin(ax+b)$$

$$= \frac{1}{\chi(-a^2)}.\frac{1}{(D^2+a^2)}\sin(ax+b) \qquad (10)$$

$$\frac{1}{(D^2+a^2)}\sin(ax+b) = \text{Imaginary Part of } \frac{1}{(D^2+a^2)}e^{i(ax+b)}$$

$$= \text{I.P. of } \frac{1}{(D+ia).(D-ia)}e^{i(ax+b)}$$

$$= \text{I.P. of } \frac{1}{2ia}.\frac{1}{(D-ia)}e^{i(ax+b)}$$

$$= \text{I.P. of } \frac{1}{2ia}.x.e^{i(ax+b)}$$

$$= \text{I.P. of } \frac{-ix.e^{i(ax+b)}}{2a}$$

$$= \frac{-x}{2a}.\cos(ax+b)$$

Similarly, $\dfrac{1}{(D^2+a^2)}\cos(ax+b) = \text{Real Part of } \dfrac{1}{(D^2+a^2)}e^{i(ax+b)}$

$$= \frac{x}{2a}.\sin(ax+b)$$

$\therefore$ From (10) $\dfrac{1}{\varphi(D^2)}\sin(ax+b) = \dfrac{1}{\chi(-a^2)}.\left(\dfrac{-x}{2a}\right)\cos(ax+b).$

Similarly, $\dfrac{1}{\varphi(D^2)}\cos(ax+b) = \dfrac{1}{\chi(-a^2)}.\left(\dfrac{x}{2a}\right)\sin(ax+b).$

While finding the P.I $= \dfrac{1}{f(D)}\sin(ax+b)$, D^2 is replaced by $-a^2$, D^3 by $-a^2D$, D^4 by a^4 etc.

After these substitutions, $f(D)$ will be of the form $\alpha D + \beta$.

Then $\dfrac{1}{\alpha D + \beta}.\sin(ax+b) = \dfrac{\alpha D - \beta}{\alpha^2 D^2 - \beta^2}.\sin(ax+b)$

$$= \frac{1}{-\alpha^2 a^2 - \beta^2}.(\alpha D - \beta).\sin(ax+b)$$

$$= \frac{-1}{\alpha^2 a^2 + \beta^2} . [a\alpha \cos(ax + b) - \beta \sin(ax + b)]$$

Similarly $\dfrac{1}{\alpha D + \beta} . \cos(ax + b) = \dfrac{1}{-\alpha^2 a^2 - \beta^2} . [-a\alpha \sin(ax + b) - \beta \cos(ax + b)]$

$$= \frac{1}{\alpha^2 a^2 + \beta^2} . [a\alpha \sin(ax + b) + \beta \cos(ax + b)]$$

Case 3. $\Phi = x^m$, where m is a positive integer

$$\text{P.I.} = \frac{1}{f(D)} . x^m$$
$$= [f(D)]^{-1} . x^m$$

Using the Binomial theorem, expand $[f(D)]^{-1}$ as far as the terms containing D^m and operate on x^m, term by term. Since $D^{m+1}(x^m) = 0$, $D^{m+2}(x^m) = 0$ and so on, we don't need the terms in the Binomial expansion of $[f(D)]^{-1}$ beyond D^m.

Note:

(i) $(1 - D)^{-1} = 1 + D + D^2 + D^3 + \ldots\ldots$

(ii) $(1 - D)^{-2} = 1 + 2D + 3D^2 + 4D^3 + \ldots\ldots$

(iii) $(1 - D)^{-n} = 1 + \dfrac{n}{1!}D + \dfrac{n(n+1)}{2!}D^2 + \dfrac{n(n+1)(n+2)}{3!}D^3 + \ldots\ldots$

Case 4. $\Phi = e^{ax} . V$ where V is a function of x.

Let u be a function of x. Then,

$$D(e^{ax} . u) = e^{ax} . Du + a . e^{ax} . u$$
$$= e^{ax} . (D + a) . u$$
$$D^2(e^{ax} . u) = D(e^{ax} . Du + a . e^{ax} . u)$$
$$= e^{ax} . D^2u + 2 . a . e^{ax} Du + a^2 e^{ax} . u$$
$$= e^{ax} . (D + a)^2 . u$$

In general, $D^n(e^{ax} . u) = e^{ax} . (D + a)^n . u$

$\therefore \quad f(D)(e^{ax} . u) = e^{ax} . f(D + a) . u$

$$\frac{1}{f(D)} . f(D)(e^{ax} . u) = \frac{1}{f(D)} . (e^{ax}) f(D + a) . u$$

i.e, $\quad e^{ax} . u = \dfrac{1}{f(D)} . (e^{ax}) f(D + a) . u \qquad\qquad (11)$

Put $f(D + a) . u = V$ in equation (11)

i.e., $\quad u = \dfrac{1}{f(D + a)} . V$

Then $e^{ax} \cdot \dfrac{1}{f(D+a)} \cdot V = \dfrac{1}{f(D)} e^{ax} \cdot V$

i.e., $\dfrac{1}{f(D)} e^{ax} \cdot V = e^{ax} \cdot \dfrac{1}{f(D+a)} \cdot V$ (12)

Given V, a function of x, $\dfrac{1}{f(D+a)} \cdot V$ can be computed.

Case 5. $\Phi = x^r.\sin(ax+b)$ or $x^r.\cos(ax+b)$

P.I. $= \dfrac{1}{f(D)} \cdot X = \dfrac{1}{f(D)} \cdot x^r \sin(ax+b)$

$\qquad = $ Imaginary part of $\dfrac{1}{f(D)} \cdot x^r e^{i(ax+b)}$

$\qquad = $ I.P. of $e^{i(ax+b)} \cdot \dfrac{1}{f(D+ia)} \cdot x^r$ (Using (12))

P.I. is obtained after computing $\dfrac{1}{f(D+ia)} \cdot x^r$

Similarly, $\dfrac{1}{f(D)} \cdot x^r \cos(ax+b) = $ R.P. of $e^{i(ax+b)} \cdot \dfrac{1}{f(D+ia)} \cdot x^r$

Case 6. Φ is any other function of x.

P.I. $= \dfrac{1}{f(D)} \cdot \Phi$

$\qquad = \dfrac{1}{a_0(D-m_1)((D-m_2)\cdots(D-m_n)} \cdot \Phi$

$\qquad = \left(\dfrac{A_1}{D-m_1} + \dfrac{A_2}{D-m_2} + \cdots + \dfrac{A_n}{D-m_n} \right).\Phi$ by splitting into partial fractions

$\qquad = \dfrac{A_1}{D-m_1} \cdot \Phi + \dfrac{A_2}{D-m_2} \cdot \Phi + \cdots + \dfrac{A_n}{D-m_n} \cdot \Phi$

$\qquad = A_1 e^{m_1 x} \int \Phi.e^{-m_1 x} dx + A_2 e^{m_2 x} \int \Phi.e^{-m_2 x} dx + \cdots + A_n e^{m_n x} \int \Phi.e^{-m_n x} dx$ (Using (8))

As we need only a P.I. arbitrary constants are not added in the above equation. P.I. is obtained after computing the integrals in the equation.

5.1.4   RULES FOR COMPUTING PARTICULAR INTEGRAL

Rule 1: $\dfrac{1}{f(D)}.e^{ax} = \dfrac{1}{f(a)}.e^{ax}$ provided $f(a) \neq 0$.

Rule 2: $\dfrac{1}{(D-a)^r}.e^{ax} = \dfrac{1}{\varphi(a)}.\dfrac{x^r}{r!}.e^{ax}$

Rule 3: $\dfrac{1}{f(D^2)}\sin(ax+b) = \dfrac{1}{f(-a^2)}.\sin(ax+b)$

and $\dfrac{1}{f(D^2)}.\cos(ax+b) = \dfrac{1}{f(-a^2)}.\cos(ax+b)$ when $f(-a^2) \neq 0$

Rule 4: $\dfrac{1}{(D^2+a^2)}\sin(ax+b) = \dfrac{-x}{2a}.\cos(ax+b)$

$\dfrac{1}{(D^2+a^2)}\cos(ax+b) = \dfrac{x}{2a}.\sin(ax+b)$

Rule 5: $\dfrac{1}{f(D)}.e^{ax}.V = e^{ax}\dfrac{1}{f(D+a)}.V$

Rule 6: $\dfrac{1}{f(D)}.x^m = [f(D)]^{-1}.x^m$

Rule 7: $\dfrac{1}{f(D)}.x^r \sin(ax+b) = \text{I.P. of } e^{i(ax+b)}.\dfrac{1}{f(D+ia)}.x^r$

and $\dfrac{1}{f(D)}.x^r \cos(ax+b) = \text{R.P. of } e^{i(ax+b)}.\dfrac{1}{f(D+ia)}.x^r$

Example 5.1.1 Solve $(3D^2 + D - 14)y = 13e^{2x}$

Solution: The A.E is $3m^2 + m - 14 = 0$

i.e., $3(m-2)(m+\tfrac{7}{3}) = 0$

$\therefore \quad m = 2, -\tfrac{7}{3}$

The C.F is $u = c_1e^{2x} + c_2e^{-\frac{7}{3}x}$

The P.I is $v = \dfrac{1}{3(D-2)(D+\frac{7}{3})}.13.e^{2x}$

$\qquad = \dfrac{13}{3(D-2)(2+\frac{7}{3})}.e^{2x}$ (By Rule 1)

$\qquad = \dfrac{1}{D-2}.e^{2x}$

$\qquad = \dfrac{x}{1!}.e^{2x}$ (By Rule 2)

$\therefore$ The general solution is $y = u + v$

i.e., $y = c_1e^{2x} + c_2e^{-\frac{7}{3}x} + x.e^{2x}$.

Example 5.1.2 Solve $(D^2 - 4D + 3)y = \cos 2x$

Solution: The A.E is $m^2 - 4m + 3 = 0$

i.e., $(m - 3)(m - 1) = 0$

$\therefore \qquad m = 3, 1$

The C.F is $u = c_1.e^{3x} + c_2.e^x$

The P.I is $v = \dfrac{1}{(D^2 - 4D + 3)}.\cos 2x$

$\qquad = \dfrac{1}{-2^2 - 4D + 3}.\cos 2x$ \hfill (By Rule 3)

$\qquad = \dfrac{-1}{4D + 1}.\cos 2x$

$\qquad = \dfrac{-(4D - 1)}{16D^2 - 1}.\cos 2x$

$\qquad = \dfrac{-(4D - 1)}{16(-2^2) - 1}.\cos 2x$ \hfill (By Rule 3)

$\qquad = \dfrac{1}{65}[4D \cos 2x - \cos 2x]$

$\qquad = \dfrac{1}{65}[-8 \sin 2x - \cos 2x]$

$\qquad = -\dfrac{1}{65}[8 \sin 2x + \cos 2x]$

The general solution is $y = c_1.e^{3x} + c_2.e^x - \dfrac{1}{65}[8 \sin 2x + \cos 2x]$

Example 5.1.3 Solve $(D^3 - 3D^2 + 3D - 1)y = 7e^{2x} \sin x$

Solution: The A.E is $m^3 - 3m^2 + 3m - 1 = 0$

i.e., $(m - 1)^3 = 0$

$\therefore \qquad m = 1, 1, 1.$

The C.F is $u = (c_1 + c_2x + c_3x^2)e^x$

The P.I is $v = \dfrac{1}{D^3 - 3D^2 + 3D - 1}.7e^{2x} \sin x$

$\qquad = 7e^{2x}.\dfrac{1}{(D + 2)^3 - 3(D + 2)^2 + 3(D + 2) - 1}.\sin x$ \hfill (By Rule 5)

$\qquad = 7e^{2x}.\dfrac{1}{D^3 + 3D^2 + 3D + 1}.\sin x$

$$= 7e^{2x}.\frac{1}{D(-1^2)+3(-1^2)+3D+1}.\sin x \qquad \text{(By Rule 3)}$$

$$= 7e^{2x}.\frac{1}{2D-2}.\sin x$$

$$= \frac{7}{2}e^{2x}\frac{D+1}{D^2-1}.\sin x$$

$$= \frac{7}{2}e^{2x}.\frac{D+1}{-1^2-1}.\sin x$$

$$= -\frac{7}{4}e^{2x}.(\cos x + \sin x)$$

The general solution is $y = u + v$

i.e., $\quad y = (c_1 + c_2x + c_3x^2)e^x - \frac{7}{4}e^{2x}.(\cos x + \sin x)$

Example 5.1.4 Solve $(D^3 - 7D - 6)y = (1+x)e^{2x}$

Solution: The A.E is $m^3 - 7m - 6 = 0$
$\qquad$ m = -1 is a root.

The other roots can be obtained as m = -2 and 3 (By synthetic division)
The C.F is $u = c_1e^{-x} + c_2e^{-2x} + c_3e^{3x}$

The P.I is $\quad v = \dfrac{1}{D^3 - 7D - 6}(1+x).e^{2x}$

$$= e^{2x}\frac{1}{(D+2)^3 - 7(D+2) - 6}(1+x) \qquad \text{(By Rule 5)}$$

$$= e^{2x}\frac{1}{D^3 + 6D^2 + 5D - 12}(1+x)$$

$$= \frac{e^{2x}}{-12}\left[1 - \frac{D^3 + 6D^2 + 5D}{12}\right]^{-1}(1+x) \qquad \text{(By Rule 6)}$$

$$= \frac{e^{2x}}{-12}\left[1 + \left(\frac{D^3 + 6D^2 + 5D}{12}\right) + \cdots\right](1+x)$$

$$= \frac{e^{2x}}{-12}\left[1 + x + \frac{5}{12}D(1+x)\right]$$

$$= \frac{e^{2x}}{-12}\left[1 + x + \frac{5}{12}\right]$$

$$= \frac{e^{2x}}{-12} \left[x + \frac{17}{12} \right]$$

The general solution is $y = u + v$

i.e., $y = c_1 e^{-x} + c_2 e^{-2x} + c_3 e^{3x} - \frac{e^{2x}}{12} \left[x + \frac{17}{12} \right]$

Example 5.1.5 Solve $(D^2 + 4)^2 y = \cos 2x$.

Solution: The A.E. is $(m^2 + 4)^2 = 0$

$\therefore$ $m = 2i, 2i, -2i, -2i$.

The C.F. is $u = (c_1 + c_2 x) \cos 2x + (c_3 + c_4 x) \sin 2x$

The P.I is $v = \dfrac{1}{\left(D^2 + 4\right)^2} . \cos 2x$

$$= \frac{1}{\left(D^2 + 4\right)\left(D^2 + 4\right)} . \cos 2x$$

$$= \frac{1}{\left(D^2 + 4\right)} . \frac{x}{2(2)} \sin 2x \qquad\qquad \text{(By Rule 4)}$$

$$= \frac{1}{4} . \frac{1}{\left(D^2 + 4\right)} . x . \sin 2x$$

$$= \text{I.P of } \frac{1}{4} e^{i2x} \frac{1}{\left(D + 2i\right)^2 + 4} . x \qquad\qquad \text{(By Rule 7)}$$

$$= \text{I.P of } \frac{1}{4} e^{i2x} \frac{1}{D^2 + 4iD - 4 + 4} . x$$

$$= \text{I.P of } \frac{1}{4} e^{i2x} \frac{1}{D(D + 4i)} . x$$

$$= \text{I.P of } \frac{1}{4} e^{i2x} \frac{1}{4iD} \left(1 + \frac{D}{4i}\right)^{-1} . x$$

$$= \text{I.P of } \frac{1}{4} e^{i2x} \frac{1}{4iD} \left(1 - \frac{D}{4i}\right) x$$

$$= \text{I.P of } \frac{1}{4} e^{i2x} \frac{1}{4iD} \left(x - \frac{1}{4i}\right)$$

$$= \text{I.P of } \frac{1}{4} e^{i2x} \frac{1}{4i} \int \left(x - \frac{1}{4i}\right) dx$$

$$= \text{I.P of } \frac{1}{4} e^{i2x} \frac{1}{4i} \left(\frac{x^2}{2} - \frac{x}{4i}\right)$$

$$= \text{I.P of } \frac{-1}{16} i.[\cos 2x + i \sin 2x] \left(\frac{x^2}{2} - \frac{x}{4i} \right)$$

$$= I.P \text{ of } \frac{-1}{16}.[\cos 2x + i \sin 2x] \left(i\frac{x^2}{2} - \frac{x}{4} \right)$$

$$= \frac{-1}{16} \left(\frac{x^2}{2} \cos 2x - \frac{x}{4} \sin 2x \right)$$

$$= \frac{-1}{64} \left(2x^2 \cos 2x - x \sin 2x \right)$$

The general solution is $y = u + v$

i.e., $y = (c_1 + c_2 x) \cos 2x + (c_3 + c_4 x) \sin 2x - \frac{1}{64}\left(2x^2 \cos 2x - x \sin 2x \right)$

Example 5.1.6 Solve $(D^2 - 4D + 4) y = 8x^2 e^{2x} \sin 2x$.

Solution: The A.E. is $m^2 - 4m + 4 = 0$

i.e., $(m - 2)^2 = 0$

$\therefore \quad m = 2, 2.$

The C.F is $u = (c_1 + c_2 x) e^{2x}$

The P.I is $\quad v = \dfrac{1}{D^2 - 4D + 4}.8x^2 e^{2x} \sin 2x$

$$= 8e^{2x} \frac{1}{(D+2)^2 - 4(D+2) + 4}.x^2 \sin 2x \qquad \text{(By Rule 5)}$$

$$= 8e^{2x} \frac{1}{D^2}.x^2 \sin 2x \qquad\qquad\qquad (1)$$

$$\frac{1}{D^2}.x^2 \sin 2x = \text{I.P of } e^{i2x} \frac{1}{(D+2i)^2}.x^2 \qquad \text{(By Rule 7)}$$

$$= \text{I.P of } e^{i2x} \frac{1}{(2i)^2 \left(1 + \frac{D}{2i}\right)^2}.x^2$$

$$= \text{I.P of } e^{i2x} \frac{1}{(-4)}.\left(1 + \frac{D}{2i}\right)^{-2}.x^2$$

$$= \text{I.P of } \frac{-1}{4}.e^{i2x}\left[1 - 2.\frac{D}{2i} + 3.\frac{D^2}{(2i)^2} \right]x^2$$

$$= \text{I.P of } \frac{-1}{4}.e^{i2x}\left[x^2 + iDx^2 - \frac{3}{4}D^2 x^2 \right]$$

$$= \text{I.P of } \frac{-1}{4}.(\cos 2x + i \sin 2x)\left[x^2 + i\,2x - \frac{3}{4}2 \right]$$

$$= \frac{-1}{4}\left[2x\cos 2x + x^2 \sin 2x - \tfrac{3}{2}\sin 2x\right]$$

∴ (1) becomes,

$$v = -8e^{2x}\frac{1}{4}\left[2x\cos 2x + x^2 \sin 2x - \tfrac{3}{2}\sin 2x\right]$$

$$= -e^{2x}\left[4x\cos 2x + 2x^2 \sin 2x - 3\sin 2x\right]$$

The general solution is $y = u + v$

i.e., $\quad y = (c_1 + c_2x)\,e^{2x} - e^{2x}\left[4x\cos 2x + 2x^2 \sin 2x - 3\sin 2x\right]$

Example 5.1.7 Solve $(D^3 - 6D^2 + 12D - 8)\,y = 16x^3 e^{4x}$

Solution: The A.E. is $m^3 - 6m^2 + 12m - 8 = 0$

i.e., $\quad (m-2)^3 = 0$

∴ $\quad m = 2, 2, 2$

The C.F is $u = (c_1 + c_2x + c_3x^2)e^{2x}$

The P.I is $v = \dfrac{1}{D^3 - 6D^2 + 12D - 8}.16x^3 e^{4x}$

$$= \frac{16.e^{4x}}{(D+4)^3 - 6(D+4)^2 + 12(D+4) - 8}.x^3 \qquad \text{(By Rule 5)}$$

$$= \frac{16.e^{4x}}{D^3 + 12D^2 + 48D + 64 - 6D^2 - 48D - 96 + 12D + 48 - 8}.x^3$$

$$= \frac{16.e^{4x}}{D^3 + 6D^2 + 12D + 8}.x^3$$

$$= \frac{16}{8}.e^{4x}\left[1 + \frac{D^3 + 6D^2 + 12D}{8}\right]^{-1}.x^3 \qquad \text{(By Rule 6)}$$

$$= 2.e^{4x}\left[1 - \left(\frac{D^3 + 6D^2 + 12D}{8}\right) + \left(\frac{D^3 + 6D^2 + 12D}{8}\right)^2 - \left(\frac{D^3 + 6D^2 + 12D}{8}\right)^3\right].x^3$$

$$= 2.e^{4x}\left[1 - \frac{D^3}{8} - \frac{6D^2}{8} - \frac{12D}{8} + \frac{144D^3}{64} + \frac{144D^2}{64} - \frac{12^3 D^3}{8^3}\right].x^3$$

$$= 2.e^{4x}\left[x^3 - \frac{6}{8} - \frac{6.6x}{8} - 12.\frac{3x^2}{8} + \frac{9}{4}.6 + \frac{9}{4}.6x - \frac{27}{8}.6\right]$$

$$= 2.e^{4x}\left[x^3 - \frac{9}{2}.x^2 + 9x - \frac{15}{2}\right]$$

$$= e^{4x}\left[2x^3 - 9x^2 + 18x - 15\right]$$

$\therefore$ The general solution is $y = u + v$

i.e., $\quad y = (c_1 + c_2 x + c_3 x^2)\, e^{2x} + e^{4x}\left[2x^3 - 9x^2 + 18x - 15\right]$

Example 5.1.8 Solve $(D^2 + 4)\, y = 4\tan 2x$

Solution: The A.E. is $m^2 + 4 = 0$

i.e., $\quad (m+2i)(m-2i) = 0$

$\therefore \qquad m = -2i,\ 2i$

The C.F is $u = c_1 \cos 2x + c_2 \sin 2x$

The P.I. is $v = \dfrac{1}{D^2 + 4}.4\tan 2x$

$$= \left[\frac{\tfrac{1}{4i}}{(D - 2i)} - \frac{\tfrac{1}{4i}}{(D + 2i)}\right].4\tan 2x$$

$$= \tfrac{1}{4i}.\frac{1}{(D - 2i)}.4\tan 2x - \tfrac{1}{4i}.\frac{1}{(D + 2i)}.4\tan 2x$$

$$= \frac{1}{i}.e^{2ix}\int e^{-2ix}\tan 2x\,dx - \frac{1}{i}.e^{-2ix}\int e^{2ix}\tan 2x\,dx \quad \left(\because \frac{1}{D-m}\Phi = e^{mx}\int \Phi.e^{-mx}dx\right)$$

$$= \frac{1}{i}.e^{2ix}\int (\cos 2x - i\sin 2x)\tan 2x\,dx - \frac{1}{i}.e^{-2ix}\int (\cos 2x + i\sin 2x)\tan 2x\,dx$$

$$= \frac{1}{i}.e^{2ix}\int (\sin 2x - i\sin^2 2x.\sec 2x)dx - \frac{1}{i}.e^{-2ix}\int (\sin 2x + i\sin^2 2x.\sec 2x)dx$$

$$= \frac{1}{i}.e^{2ix}\int (\sin 2x - i\sec 2x + i\cos 2x)dx - \frac{1}{i}.e^{-2ix}\int (\sin 2x + i\sec 2x - i\cos 2x)dx$$

$$\left(\because \sin^2 2x = 1 - \cos^2 2x\right)$$

$$= \frac{1}{i}.e^{2ix}\left[\frac{-1}{2}\cos 2x - \frac{i}{2}\log(\sec 2x + \tan 2x) + \frac{i}{2}\sin 2x\right]$$

$$\qquad\quad - \frac{1}{i}.e^{-2ix}\left[\frac{-1}{2}\cos 2x + \frac{i}{2}\log(\sec 2x + \tan 2x) - \frac{i}{2}\sin 2x\right]$$

$$= \left(\frac{e^{2ix} - e^{-2ix}}{2i}\right).(-\cos 2x) - \left(\frac{e^{2ix} + e^{-2ix}}{2}\right)\log(\sec 2x + \tan 2x)$$

$$\qquad\qquad + \left(\frac{e^{2ix} + e^{-2ix}}{2}\right).\sin 2x$$

$$= -\sin 2x \cos 2x - \cos 2x.\log(\sec 2x + \tan 2x) + \cos 2x.\sin 2x$$

$$= -\cos 2x.\log(\sec 2x + \tan 2x)$$

$\therefore$ The general solution is $y = u + v$

i.e., $y = c_1\cos 2x + c_2 \sin 2x - \cos 2x . \log (\sec 2x + \tan 2x)$

Example 5.1.9 Solve $(D^4 + 2D^2 + 1)y = x^2\cos x$

Solution: The A.E is $m^4 + 2m^2 + 1 = 0$

i.e., $(m^2 + 1)^2 = 0$

$\therefore$ $m = -i, i, -i, i$

The C.F is $u = (c_1 + c_2 x)\cos x + (c_3 + c_4 x) \sin x$

The P.I is $v = \dfrac{1}{D^4 + 2D^2 + 1} . x^2 \cos x$

$$= \text{R.P of } e^{ix} \frac{1}{(D+i)^4 + 2(D+i)^2 + 1} . x^2$$

$$= \text{R.P of } e^{ix} \frac{1}{D^4 + 4iD^3 - 4D^2} . x^2$$

$$= \text{R.P of } e^{ix} \left(\frac{-1}{4D^2}\right)\left(1 - \frac{D^2 + 4iD}{4}\right)^{-1} . x^2$$

$$= \text{R.P of } e^{ix} \left(\frac{-1}{4D^2}\right)\left[1 + \left(\frac{D^2 + 4iD}{4}\right) + \left(\frac{D^2 + 4iD}{4}\right)^2\right] . x^2$$

$$= \text{R.P of } e^{ix} \left(\frac{-1}{4D^2}\right)\left[x^2 + \frac{D^2}{4}x^2 + iDx^2 + i^2D^2x^2\right]$$

$$= \text{R.P of } e^{ix} \left[\frac{-1}{4D^2}x^2 - \frac{1}{16}x^2 - \frac{1}{4D}x^2 + \frac{1}{4}x^2\right]$$

$$= \text{R.P of } e^{ix} \left[\frac{-1}{4}\frac{x^4}{12} - \frac{1}{16}x^2 - \frac{i}{4}\frac{x^3}{3} + \frac{1}{4}x^2\right]$$

$$= \text{R.P of } (\cos x + i \sin x)\left[\frac{-1}{48}x^4 + \frac{3}{16}x^2 - \frac{i}{12}x^3\right]$$

$$= \cos x\left(\frac{-x^4}{48} + \frac{3}{16}x^2\right) + \sin x . \frac{x^3}{12}$$

$$= -\frac{1}{48}(x^4 - 9x^2).\cos x + \frac{x^3}{12}.\sin x$$

$\therefore$ The general solution is $y = u + v$

i.e., $y = (c_1 + c_2 x)\cos x + (c_3 + c_4 x) \sin x - \dfrac{1}{48}(x^4 - 9x^2).\cos x + \dfrac{x^3}{12}.\sin x$

Example 5.1.10 Solve $(D^4 - 1)y = \cos x . \cosh x$

Solution: The A.E is $m^4 - 1 = 0$

i.e., $(m^2 - 1)(m^2 + 1) = 0$

$\therefore$ $m = 1, -1, i, -i$

The C.F is $u = c_1 e^x + c_2 e^{-x} + c_3 \cos x + c_4 \sin x$

The P.I is $v = \dfrac{1}{D^4 - 1} \cos x . \cosh x$

$$= \frac{1}{D^4 - 1} \cos x . \left(\frac{e^x + e^{-x}}{2} \right)$$

$$= \frac{1}{2} \frac{1}{D^4 - 1} e^x \cos x + \frac{1}{2} \frac{1}{D^4 - 1} e^{-x} \cos x$$

$$= \frac{e^x}{2} \frac{1}{(D+1)^4 - 1} \cos x + \frac{e^{-x}}{2} \frac{1}{(D-1)^4 - 1} \cos x$$

$$= \frac{e^x}{2} \frac{1}{D^4 + 4D^3 + 6D^2 + 4D} \cos x + \frac{e^{-x}}{2} \frac{1}{D^4 - 4D^3 + 6D^2 - 4D} \cos x$$

$$= \frac{e^x}{2} \frac{1}{(-1)(-1) + 4D(-1) + 6(-1) + 4D} \cos x$$

$$\qquad + \frac{e^{-x}}{2} \frac{1}{(-1)(-1) - 4D(-1) + 6(-1) - 4D} \cos x$$

$$= \frac{e^x}{2} . \frac{1}{(-5)} . \cos x + \frac{e^{-x}}{2} . \frac{1}{(-5)} . \cos x$$

$$= \frac{-1}{5} \cos x . \left(\frac{e^x + e^{-x}}{2} \right)$$

$$= \frac{-1}{5} \cos x . \cosh x$$

$\therefore$ The complete solution is $y = u + v$

i.e., $y = c_1 e^x + c_2 e^{-x} + c_3 \cos x + c_4 \sin x - \dfrac{1}{5} \cos x . \cosh x$

EXERCISE 5.1

PART – A

1. Define order and degree of a differential equation.

2. Find the order and degree of $y . \dfrac{d^2 y}{dx^2} = x . \left(\dfrac{dy}{dx} \right)^{3/2}$

3. Define 'general solution' or 'complete solution' of a differential equation.

4. Write down the general solution of $\dfrac{dy}{dx} + Py = Q$.

5. What is the complete solution of $\dfrac{dy}{dx} - ay = 0$?

6. Give a particular solution of $\dfrac{dy}{dx} - ay = Q$

7. Find the C.F of $(D - 2)^2 y = \sin 2x$.

8. Find the C.F of $(D^4 + 2D^2 + 1)y = x^2$.

9. Find the C.F of $(D^3 + 8)^2 y = 2x$.

10. Find the general solution of $(D^4 - 2D^3 + D^2)y = 0$.

11. Find the complete solution of $(D^2 + 1)^2 y = 0$.

12. Find the complete solution of $(D^4 - 1)y = 0$.

13. Find the particular integral of $(D^2 + D + 1)y = x^2$.

14. Find a particular integral of $(D^2 - 4)y = x^3$.

15. Find the complete solution of $(D^2 + 4)y = 2 \sin 2x$.

PART – B
Solve the following differential equations.

16. $(D^3 - D)y = e^x + e^{-x}$.

17. $(D^4 - 2D^2 + 1)y = 40 \cosh x$.

18. $(D^2 - 3D + 2)y = \sin 3x$.

19. $(D^2 - 4D + 3)y = \sin 3x . \cos 2x$.

20. $(D^2 + 16)y = 2e^{-3x} + \cos 4x$.

21. $(D^3 + 6D^2 + 11D + 6)y = 2 \sin x$.

22. $(D^4 - 1)y = 4(\sin x + e^x)$.

23. $(D^2 + D + 1)y = x^2$.

24. $(D^2 + 3D - 4)y = x^2 - 2x$.

25. $(D^3 - D^2 - 6D)y = 1 + x^2$.

26. $D^2(D^2 + 4)y = 96x^2$

27. $(D^3 + 1)y = e^{2x} \sin x$.

28. $(D^3 - D^2 + 3D + 5)y = e^x \cos 2x$.

29. $(D^2 - 2D + 1)y = \sin x + x^2 e^x$.

30. $(D^4 - 2D^3 + D^2)y = x^3$.

31. $(D^4 + 8D^2 + 16)y = 16x + 10$.

32. $(D^2 + 4D + 4)y = e^{-x} \sin 2x$.

33. $(D^2 + 9)y = (x^2 + 1)e^{3x}$.

34. $(D^2 + 4D + 3)y = e^x \cos 2x - \cos 3x$.

35. $(D^2 - 2D + 1)y = x^2 e^{3x}$.

36. $(D^4 - 1)y = e^x \cos x$.

37. $(D^2 - 2D + 1)y = x \sin x$.

38. $(D^2 - 1)y = x^2 \cos x$.

39. $(D^2 - 2D + 1)y = x \, e^x \sin x$.

40. $(D^3 + 2D^2 + D)y = x^2 e^{2x} + \sin^2 x$.

41. $(D^2 + 1)y = x^2 \sin 2x$.

42. Solve $(D^2 - 7D + 6)y = e^{2x}$ given $y = y' = 0$ at $x = 0$.

43. Solve $(D^2 + 6D + 5)y = 16e^{3x} + 5e^{-2x}$ given $y = \dfrac{13}{3}$, $y' = \dfrac{157}{12}$ when $x = \log 2$

44. Solve $\ddot{x} + w^2 x = a \cos wt$.

45. Solve $\ddot{x} + n^2 x = E \cos pt$. Examine the case when $n = p$.

46. Solve $(D^2 + a^2)y = \sec ax$.

5.2 SIMULTANEOUS LINEAR DIFFERENTIAL EQUATIONS WITH CONSTANT COEFFICIENTS

In the last section we solved differential equations involving only two variables – one independent variable and one dependent variable. In case we have one independent variable, but more than one dependent variable, then we will need more than one differential equation to solve for the dependent variables in terms of the independent variable. We should have as many differential equations as the number of dependent variables.

For example, if x and y are two dependent variables and t is an independent variable, then two differential equations involving x, y, $\dfrac{dx}{dt}, \dfrac{dy}{dt}, \dfrac{d^2x}{dt^2}, \dfrac{d^2y}{dt^2}, \cdots$ form a pair of simultaneous differential equations. In this section we solve simultaneous differential equations of the form,

$$f_1(D).x + g_1(D).y = \varphi_1(t) \qquad (1)$$
$$f_2(D).x + g_2(D).y = \varphi_2(t) \qquad (2)$$

where f_1, f_2, g_1, g_2 are polynomials in the operator $D \equiv \dfrac{d}{dt}$ and φ_1, φ_2 are functions of t.

To solve equations (1) and (2), we proceed just like solving a pair of simultaneous linear algebraic equations in two variables x and y. Operating both sides of (1) by $g_2(D)$ and both sides of (2) by $g_1(D)$ and subtracting y is eliminated. i.e., we get the equation $[g_2(D).f_1(D) - g_1(D).f_2(D)]x = g_2(D).\varphi_1(t) - g_1(D).\varphi_2(t)$, which is of the form,

$$f(D).x = \varphi(t) \qquad (3)$$

(3) is a linear differential equation in x and t with constant coefficients and can be solved by the methods given in section 5.1. Substituting the value of x obtained from the solution of (3) in (1) or (2), the value y is obtained. The order of equation (3) is same as the degree of D in $f(D) = g_2(D).f_1(D) - g_1(D).f_2(D)$

$$= \begin{vmatrix} f_1(D) & g_1(D) \\ f_2(D) & g_2(D) \end{vmatrix} \qquad (4)$$

i.e., the order of equation (3) is the degree of the determinant (4).

Note:

The number of arbitrary constants that appear in the solutions of equations (1) and (2) should be equal to the order of equation (3). In case more arbitrary constants are introduced while solving the equations, the extra ones should be expressed in terms of the other constants. For example, if order of (3) is two, then the value of x will contain two arbitrary constants c_1 and c_2. In a similar way, if x is eliminated from (1) and (2) and we solve for y, then again the value of y will contain two arbitrary constants c_3 and c_4. But c_3 and c_4 can be expressed in terms of c_1 and c_2, thus resulting in only two independent arbitrary constants in the values of x and y.

Example 5.2.1 Solve $\dfrac{dx}{dt} + y = e^t, \dfrac{dy}{dt} - x = -t.$

Solution: Put $\dfrac{d}{dt} = D.$

Then $Dx + y = e^t$ (1)

   $x - Dy = t$ (2)

Operating (1) by D and adding with (2), we get,

   $D^2x + x = D.e^t + t.$

i.e., $(D^2 + 1)x = e^t + t$ (3)

Now A.E is $m^2 + 1 = 0.$ $\therefore m = \pm i.$

$C.F = c_1 \cos t + c_2 \sin t$

$P.I = \dfrac{1}{D^2 + 1}(e^t + t)$

$= \dfrac{1}{D^2 + 1}e^t + \dfrac{1}{D^2 + 1}t$

$= \dfrac{e^t}{1^2 + 1} + (1 + D^2)^{-1}t$

$= \dfrac{e^t}{2} + (1 - D^2)t = \dfrac{e^t}{2} + t$

Now, $x = C.F + P.I$

i.e., $x = c_1 \cos t + c_2 \sin t + \dfrac{e^t}{2} + t$

(Next we find y using one of the following methods)

Method 1. We eliminate x from equations (1) and (2).

Operating equation (2) by D and subtracting from (1), we get,

   $y + D^2y = e^t - Dt$

i.e., $(D^2 + 1)y = e^t - 1$ (4)

A.E is $m^2 + 1 = 0.$ $\therefore m = \pm i.$

C.F $= c_3 \cos t + c_4 \sin t$.

$$P.I = \frac{1}{D^2 + 1}(e^t - 1)$$

$$= \frac{1}{D^2 + 1}e^t - \frac{1}{D^2 + 1}$$

$$= \frac{e^t}{2} - (1 + D^2)^{-1}.1$$

$$= \frac{e^t}{2} - (1 - D^2).1$$

$$= \frac{e^t}{2} - 1$$

$\therefore$ Solution of equation (4) is

$$y = c_3 \cos t + c_4 \sin t + \frac{e^t}{2} - 1 \qquad (5)$$

The total number of arbitrary constants in the values of x and y should be equal to the degree of the determinant $\begin{vmatrix} D & 1 \\ 1 & -D \end{vmatrix} = -D^2 - 1$, which is equal to 2. Hence the extra arbitrary constants c_3 and c_4 can be expressed in terms of c_1 and c_2.

Substituting $x = c_1 \cos t + c_2 \sin t + \frac{e^t}{2} + t$

and $y = c_3 \cos t + c_4 \sin t + \frac{e^t}{2} - 1$ in (1), we get,

$$D(c_1 \cos t + c_2 \sin t + \frac{e^t}{2} + t) + c_3 \cos t + c_4 \sin t + \frac{e^t}{2} - 1 = e^t$$

i.e., $-c_1 \sin t + c_2 \cos t + \frac{e^t}{2} + 1 + c_3 \cos t + c_4 \sin t + \frac{e^t}{2} - 1 = e^t$

i.e., $(-c_1 + c_4) \sin t + (c_2 + c_3) \cos t = 0$.

This equation should be true for all values of t.

When $t = 0$, $\quad c_2 + c_3 = 0$

When $t = \pi/2$, $-c_1 + c_4 = 0$

i.e., $\quad c_3 = -c_2$ and $c_4 = c_1$.

Substituting these values in (5), we get $y = -c_2 \cos t + c_1 \sin t + \frac{e^t}{2} - 1$

$\therefore$ The general solution of the given simultaneous equations is

$$x = c_1 \cos t + c_2 \sin t + \frac{e^t}{2} + t$$

$$y = c_1 \sin t - c_2 \cos t + \frac{e^t}{2} - 1$$

Method 2. Substituting for x in equation (1) we get,
$$y = e^t - Dx$$

$$= e^t - D(c_1 \cos t + c_2 \sin t + \frac{e^t}{2} + t)$$

$$= e^t + c_1 \sin t - c_2 \cos t - \frac{e^t}{2} - 1$$

i.e., $y = c_1 \sin t - c_2 \cos t + \frac{e^t}{2} - 1$

∴ The general solution of the given simultaneous equations is

$$x = c_1 \cos t + c_2 \sin t + \frac{e^t}{2} + t$$

$$y = c_1 \sin t - c_2 \cos t + \frac{e^t}{2} - 1$$

Example 5.2.2 Solve $2(D - 2)x + (D - 1)y = e^t$, $(D + 3)x + y = 0$.

Solution: We have
$$2(D - 2).x + (D - 1)y = e^t \tag{1}$$
$$(D + 3)x + y = 0 \tag{2}$$

Operating the equation (2) by D -1 and subtracting from (1), we get
$$[2(D - 2) - (D - 1)(D + 3)]x = e^t$$
i.e., $(2D - 4 - D^2 - 2D + 3)x = e^t$
i.e., $(D^2 + 1)x = -e^t \tag{3}$
A.E is $m^2 + 1 = 0$ ∴ $m = \pm i$
C.F $= c_1 \cos t + c_2 \sin t$.

$$P.I = \frac{1}{D^2 + 1}(-e^t) = \frac{-e^t}{1^2 + 1} = \frac{-e^t}{2}$$

The solution of equation (3) is $x = c_1 \cos t + c_2 \sin t - \frac{e^t}{2}$

Substituting the value of x in (2), we get

$$(D + 3)(c_1 \cos t + c_2 \sin t - \frac{e^t}{2}) + y = 0$$

i.e., $c_1(-\sin t) + c_2 \cos t - \frac{e^t}{2} + 3c_1 \cos t + 3c_2 \sin t - \frac{3}{2}e^t + y = 0$

∴ $y = -(3c_1 + c_2) \cos t + (c_1 - 3c_2) \sin t + 2e^t$.

∴ The general solution of the given differential equation is

$$x = c_1 \cos t + c_2 \sin t - \frac{e^t}{2}$$

$$y = -(3c_1 + c_2) \cos t + (c_1 - 3c_2) \sin t + 2e^t.$$

Example 5.2.3 Solve $(2D + 1)x + (3D + 1)y = e^t$, $(D + 5)x + (D + 7)y = 2e^t$.

Solution: $(2D+1)x + (3D + 1)y = e^t$ (1)

$\qquad\qquad (D + 5)x + (D + 7)y = 2e^t$ (2)

Operating (1) by $(D + 7)$ and (2) by $(3D + 1)$ and subtracting, we get

$\qquad [(D + 7)(2D +1) - (3D +1)(D + 5)]x = (D + 7) e^t - (3D + 1) 2e^t$.

i.e., $(-D^2 - D + 2)x = e^t + 7e^t - 6e^t - 2e^t$.

i.e., $(D^2 + D - 2)x = 0$

A. E is $m^2 + m + 2 = 0$

i.e., $(m + 2)(m - 1) = 0$.

∴ $m = -2, 1$.

∴ Solution is $x = c_1 e^{-2t} + c_2 e^t$ (3)

Operating (1) by $(D + 5)$ and (2) by $(2D +1)$ and subtracting, we get

$\qquad [(D + 5)(3D +1) - (2D + 1)(D + 7)]y = (D + 5) e^t - (2D +1) 2e^t$.

i.e., $(D^2 + D - 2)y = e^t + 5e^t - 4e^t - 2e^t$.

i.e., $(D^2 + D - 2)y = 0$

A. E is $m^2 + m - 2 = 0$

i.e., $(m + 2)(m - 1) = 0$

∴ $m = -2, 1$.

∴ Solution is $y = c_3 e^{-2t} + c_4 e^t$ (4)

$$\begin{vmatrix} 2D+1 & 3D+1 \\ D+5 & D+7 \end{vmatrix} = -D^2 - D + 2, \text{ which is of degree 2 in D.}$$

Hence the general solution of the given differential equations should contain only two arbitrary constants.

Substituting in (1) the values of x and y from (3) and (4), we get

$\qquad (2D + 1)(c_1 e^{-2t} + c_2 e^t) + (3D + 1)(c_3 e^{-2t} + c_4 e^t) = e^t$.

i.e., $-4c_1 e^{-2t} + c_1 e^{-2t} + 2c_2 e^t + c_2 e^t + (-6) c_3 e^{-2t} + c_3 e^{-2t} + 3c_4 e^t + c_4 e^t = e^t$.

i.e., $(-3c_1 - 5c_3)e^{-2t} + (3c_2 + 4c_4 - 1)e^t = 0$.

This equation is true for all values of t.

∴ $-3c_1 - 5c_3 = 0$

$\qquad 3c_2 + 4c_4 - 1 = 0$

Solving we get, $c_3 = -\frac{3}{5}c_1$, $c_4 = -\frac{3}{4}c_2 + \frac{1}{4}$

$\therefore$ The general solution of the given differential equations is

$$x = c_1 e^{-2t} + c_2 e^t$$

$$y = -\frac{3}{5}c_1 e^{-2t} + \left(-\frac{3}{4}c_2 + \frac{1}{4}\right)e^t$$

Example 5.2.4 Solve $Dx + 2x + 3y = 0$, $Dy + 3x + 2y = 2e^{2t}$

Solution: $(D + 2)x + 3y = 0$ (1)

           $3x + (D + 2)y = 2e^{2t}$ (2)

Operating (1) by $(D + 2)$ and multiplying (2) by 3 and subtracting, we get,

      $((D + 2)^2 - 9)x = -6e^{2t}$.

i.e., $(D^2 + 4D - 5)x = -6e^{2t}$

A.E. is $m^2 + 4m - 5 = 0$

i.e., $(m + 5)(m - 1) = 0$.

$\therefore$ $m = -5, 1$.

$\therefore$ C.F. $= c_1 e^{-5t} + c_2 e^t$

$$P.I = \frac{1}{D^2 + 4D - 5}(-6)e^{2t}$$

$$= \frac{-6e^{2t}}{2^2 + 4.2 - 5}$$

$$= \frac{-6e^{2t}}{7}$$

$\therefore$ Solution is $x = c_1 e^{-5t} + c_2 e^t - \dfrac{6e^{2t}}{7}$ (3)

From (1), we have $3y = -D.x - 2x$

$$= -D(c_1 e^{-5t} + c_2 e^t - \frac{6e^{2t}}{7}) - 2(c_1 e^{-5t} + c_2 e^t - \frac{6e^{2t}}{7})$$

$$= 5c_1 e^{-5t} - c_2 e^t - \frac{12e^{2t}}{7} - 2c_1 e^{-5t} - 2c_2 e^t + \frac{12e^{2t}}{7}$$

$$= 3c_1 e^{-5t} - 3c_2 e^t + \frac{24e^{2t}}{7}$$

$\therefore$ $y = c_1 e^{-5t} - c_2 e^t + \dfrac{8e^{2t}}{7}$ (4)

$\therefore$ The general solutions of the differential equations are given by (3) and (4).

Example 5.2.5 Solve $Dx - Dy + 2y = \cos 2t$, $Dx + Dy - 2x = \sin 2t$

Solution: $Dx - (D - 2)y = \cos 2t$ (1)

$$(D - 2)x + Dy = \sin 2t \tag{2}$$

Operating (1) by D and (2) by (D - 2) and adding we get,

$$D^2x + (D - 2)^2x = D \cos 2t + (D-2) \sin 2t$$

i.e., $(2D^2 - 4D + 4)x = -2\sin 2t + 2\cos 2t - 2\sin 2t$

i.e., $(D^2 - 2D + 2)x = \cos 2t - 2\sin 2t$

A.E is $m^2 - 2m + 2 = 0$

$$m = 1 \pm i$$

$\therefore$ C.F $= e^t(c_1\cos t + c_2 \sin t)$

$$\text{P.I} = \frac{1}{D^2 - 2D + 2}(\cos 2t - 2\sin 2t)$$

$$= \frac{1}{-4 - 2D + 2}\cos 2t - 2.\frac{1}{-4 - 2D + 2}\sin 2t$$

$$= \frac{1}{-2(D + 1)}\cos 2t + \frac{1}{D + 1}\sin 2t$$

$$= \frac{-1}{2}\frac{(D - 1)}{(D^2 - 1)}\cos 2t + \frac{D - 1}{D^2 - 1}\sin 2t$$

$$= \frac{-1}{2(-5)}(-2\sin 2t - \cos 2t) + \frac{1}{(-5)}(2\cos 2t - \sin 2t)$$

$$= -\frac{1}{5}\sin 2t - \frac{1}{10}\cos 2t - \frac{2}{5}\cos 2t + \frac{1}{5}\sin 2t$$

$$= -\frac{1}{2}\cos 2t$$

$\therefore$ $x = e^t(c_1\cos t + c_2 \sin t) - \dfrac{1}{2}\cos 2t$ (3)

Adding (1) and (2) we get $2Dx - 2x + 2y = \cos 2t + \sin 2t$

$$2y = -2Dx + 2x + \cos 2t + \sin 2t$$

Substituting for x from (3), we get

$$2y = -2D(e^tc_1\cos t + e^tc_2 \sin t - \frac{1}{2}\cos 2t) + 2(e^tc_1\cos t + e^tc_2\sin t - \frac{1}{2}\cos 2t)$$

$$+ \cos 2t + \sin 2t$$

$$= -2(c_1 e^t \cos t - c_1 e^t\sin t + c_2 e^t \sin t + c_2 e^t \cos t + \sin 2t)$$

$$+2(c_1 e^t \cos t + c_2 e^t\sin t \frac{-1}{2} \cos 2t) + \cos 2t + \sin 2t$$

$$= e^t(-2c_2 \cos t + 2c_1 \sin t) - \sin 2t$$

$\therefore$ $y = e^t(c_1 \sin t + c_2 \cos t) - \dfrac{1}{2}\sin 2t$ (4)

$\therefore$ The general solutions of the differential equations are given by (3) and (4).

Example 5.2.6 Solve $\dfrac{d^2y}{dt^2} - x = 0$, $\dfrac{d^2x}{dt^2} - y = 0$

Solution: $x - D^2y = 0$ (1)

$\qquad\quad D^2x - y = 0$ (2)

$\begin{vmatrix} 1 & -D^2 \\ D^2 & -1 \end{vmatrix} = -1 + D^4$, which is of degree 4.

$\therefore$ The general solution should contain four arbitrary constants.

Operating (2) by D^2 and subtracting from (1), we get,

$\qquad x - D^4x = 0$

i.e., $\quad (D^4 - 1)x = 0$

A.E. is $m^4 - 1 = 0$

$\therefore \qquad m = 1, -1, i, -i$

$C.F = c_1e^x + c_2 e^{-x} + c_3 \cos t + c_4 \sin t$

$\therefore \qquad x = c_1e^x + c_2 e^{-x} + c_3 \cos t + c_4 \sin t$ (3)

From (2), $y = D^2x = D^2(c_1e^x + c_2 e^{-x} + c_3 \cos t + c_4 \sin t)$

$\qquad\qquad = D(c_1e^x - c_2 e^{-x} - c_3 \sin t - c_4 \cos t)$

i.e., $\quad y = c_1e^x + c_2 e^{-x} - c_3 \cos t - c_4 \sin t$ (4)

$\therefore$ The general solutions of the differential equations are given by (3) and (4).

Example 5.2.7 Solve $D^2x + 3x - 2y = 0$, $D^2x + D^2y - 3x + 5y = 0$

Solution: $\qquad (D^2 + 3)x - 2y = 0$ (1)

$\qquad\qquad (D^2 - 3)x + (D^2 + 5)y = 0$ (2)

Operating (1) by $D^2 - 3$ and (2) by $D^2 + 3$ and subtracting, we get,

$\qquad -(D^2 - 3)2y - (D^2 + 3)(D^2 + 5)y = 0$

i.e., $\quad (D^4 + 10D^2 + 9)y = 0$

A.E is $\quad m^4 + 10m^2 + 9 = 0$

i.e., $\quad (m^2 + 1)(m^2 + 9) = 0$

$\therefore \qquad m = +i, -i, +3i, -3i$

$\therefore \qquad y = c_1\cos t + c_2\sin t + c_3\cos 3t + c_4\sin 3t$ (3)

Operating (1) by $D^2 + 5$ and multiplying (2) by 2 and adding, we get,

$\qquad [(D^2 + 5)(D^2 + 3) + 2(D^2 - 3)]x = 0$

i.e., $\quad (D^4 + 10D^2 + 9)x = 0$

$\therefore \qquad x = d_1\cos t + d_2\sin t + d_3\cos 3t + d_4\sin 3t$ (4)

$\begin{vmatrix} D^2 + 3 & -2 \\ D^2 - 3 & D^2 + 5 \end{vmatrix} = D^4 + 10D^2 + 9$, which is of degree 4

∴ The general solution of the differential equations should contain only four **independent** arbitrary constants.

Substituting (3) and (4) in (1) we get

D^2 [$d_1\cos t + d_2\sin t + d_3\cos 3t + d_4\sin 3t$] + 3 [$d_1\cos t + d_2\sin t + d_3\cos 3t + d_4\sin 3t$]
$$-2 [c_1\cos t + c_2\sin t + c_3\cos 3t + c_4\sin 3t] = 0$$

i.e., $-d_1\cos t - d_2\sin t - 9 d_3\cos 3t - 9d_4\sin 3t + 3$ [$d_1\cos t + d_2\sin t + d_3\cos 3t$
$$+ d_4\sin 3t] -2 [c_1\cos t + c_2\sin t + c_3\cos 3t + c_4\sin 3t] = 0$$

i.e., $2(d_1 - c_1)\cos t + 2(d_2 - c_2)\sin t - 2(3d_3 + c_3)\cos 3t - 2(3d_4 + c_4)\sin 3t = 0$

This equation is true for all values of t

∴ $\quad d_1 - c_1 = 0 = d_2 - c_2 = 3d_3 + c_3 = 3d_4 + c_4$

i.e., $\quad d_1 = c_1, d_2 = c_2, d_3 = \dfrac{-1}{3}c_3, d_4 = \dfrac{-1}{3}c_4$

Hence $x = c_1\cos t + c_2\sin t - \dfrac{1}{3}(c_3\cos 3t + c_4\sin 3t)$ $\hspace{2cm}$ (5)

and the general solutions of the given differential equations are given by (3) and (5).

Example 5.2.8 Solve $\dfrac{d^2x}{dt^2} + y = \sin t, \quad \dfrac{d^2y}{dt^2} + x = \cos t$

Solution: $\quad D^2x + y = \sin t$ $\hspace{3cm}$ (1)
$\hspace{2.3cm} X + D^2y = \cos t$ $\hspace{3cm}$ (2)

Operating (1) by D^2 and subtracting from (2), we get,

$\quad x - D^4x = \cos t - D^2 \sin t$
$\hspace{2cm} = \cos t + \sin t$

i.e., $\quad (D^4 - 1)x = -\cos t - \sin t$

A.E. is $m^4 - 1 = 0$

$\hspace{1.5cm} m = 1, -1, i, -i$

C.F. $= c_1 e^t + c_2 e^{-t} + c_3 \cos t + c_4 \sin t$

P.I. $= \dfrac{1}{D^4 - 1}(-\cos t - \sin t) = \dfrac{1}{(D^2 + 1)(D^2 - 1)}(-\cos t - \sin t)$

$\hspace{2cm} = \dfrac{1}{(D^2 + 1)(-1 - 1)}(-\cos t - \sin t)$

$\hspace{2cm} = \dfrac{1}{2(D^2 + 1)}(\cos t + \sin t)$

$\hspace{2cm} = \dfrac{1}{2}\left(\dfrac{t}{2}\sin t - \dfrac{t}{2}\cos t\right)$

∴ $\quad x = c_1 e^t + c_2 e^{-t} + c_3 \cos t + c_4 \sin t + \dfrac{t}{4}\sin t - \dfrac{t}{4}\cos t$ $\hspace{1.5cm}$ (3)

From (1) we have $y = -D^2x + \sin t$

$$= -D^2 (c_1 e^t + c_2 e^{-t} + c_3 \cos t + c_4 \sin t + \frac{t}{4} \sin t - \frac{t}{4} \cos t) + \sin t$$

$$= -[c_1 e^t + c_2 e^{-t} - c_3 \cos t - c_4 \sin t] -$$

$$D[\frac{1}{4} \sin t + \frac{t}{4} \cos t - \frac{1}{4} \cos t + \frac{t}{4} \sin t] + \sin t$$

$$= -[c_1 e^t + c_2 e^{-t} - c_3 \cos t - c_4 \sin t] -$$

$$[\frac{1}{4} \cos t + \frac{1}{4} \cos t - \frac{t}{4} \sin t + \frac{1}{4} \sin t + \frac{1}{4} \sin t + \frac{t}{4} \cos t] + \sin t$$

i.e., $y = -c_1 e^t - c_2 e^{-t} + c_3 \cos t + c_4 \sin t - \frac{1}{2} \cos t + \frac{1}{2} \sin t + \frac{t}{4} \sin t + \frac{t}{4} \cos t$ (4)

(3) and (4) give the general solutions of the given differential equations.

Example 5.2.9 Solve $\dfrac{dx}{dt} + 2x - 3y = 5t$, $\dfrac{dy}{dt} - 3x + 2y = 2e^{2t}$

Solution: $(D + 2)x - 3y = 5t$ (1)

$-3x + (D + 2)y = 2e^{2t}$ (2)

Operating (1) by $D + 2$ and multiplying (2) by 3 and adding

$$[(D + 2)^2 - 9] x = (D + 2)5t + 6e^{2t}$$

$$[D^2 + 4D - 5]x = 5 + 10t + 6e^{2t}$$

A.E is $m^2 + 4m - 5 = 0$

i.e., $(m + 5)(m - 1) = 0$

$\therefore$ $m = -5, 1$.

$\therefore$ $C.F = c_1 e^{-5t} + c_2 e^t$

$$P.I = \frac{1}{D^2 + 4D - 5}(5 + 10t + 6e^{2t})$$

$$= \frac{1}{D^2 + 4D - 5}(5 + 10t) + \frac{6}{D^2 + 4D - 5}e^{2t}$$

$$= \frac{1}{-5}\left[1 - \frac{D^2 + 4D}{5}\right]^{-1}(5 + 10t) + \frac{6e^{2t}}{4 + 8 - 5}$$

$$= -\frac{1}{5}\left[1 + \frac{4D}{5}\right](5 + 10t) + \frac{6e^{2t}}{7}$$

$$= -\frac{1}{5}\left(5 + 10t + \frac{4}{5} \cdot 10\right) + \frac{6e^{2t}}{7}$$

$$= -\frac{1}{5}(13 + 10t) + \frac{6e^{2t}}{7}$$

$$\therefore \qquad x = c_1 e^{-5t} + c_2 e^t - \frac{1}{5}(13 + 10t) + \frac{6e^{2t}}{7} \qquad\qquad (3)$$

Substituting the value of x in (1), we get

$$(D + 2)\, [c_1 e^{-5t} + c_2 e^t - \frac{1}{5}(13 + 10t) + \frac{6e^{2t}}{7}] - 3y = 5t$$

i.e., $\quad -5c_1 e^{-5t} + c_2 e^t + \frac{12}{7} e^{2t} - 2 + 2c_1 e^{-5t} + 2c_2 e^t + \frac{12}{7} e^{2t} - 4t - \frac{26}{5} - 3y = 5t$

$$\therefore \qquad y = \frac{1}{3}[-3c_1 e^{-5t} + 3c_2 e^t + \frac{24}{7} e^{2t} - \frac{36}{5} - 9t - 2]$$

$$y = -c_1 e^{-5t} + c_2 e^t + \frac{8}{7} e^{2t} - \frac{12}{5} - 3t - \frac{2}{3} \qquad\qquad (4)$$

Hence the general solutions of the differential equations are given by (3) and (4).

Example 5.2.10 Solve $\dot{x} = 2y,\ \ \dot{y} = 2z,\ \ \dot{z} = 2x$

Solution: $\quad \dfrac{dx}{dt} = 2y,\ \ \dfrac{dy}{dt} = 2z,\ \ \dfrac{dz}{dt} = 2x$

Put $\quad \dfrac{d}{dt} = D$

Then $\quad$ Dx - 2y = 0 $\qquad\qquad\qquad\qquad\qquad\qquad$ (1)

$\qquad\qquad$ Dy - 2z = 0 $\qquad\qquad\qquad\qquad\qquad\qquad$ (2)

$\qquad\qquad$ Dz - 2x = 0 $\qquad\qquad\qquad\qquad\qquad\qquad$ (3)

Operating (1) by D, $D^2 x - 2Dy = 0$

i.e., $\quad D^2 x - 2(2z) = 0 \qquad\qquad\qquad$ (Using (2))

Again operating by D, we have

$$D^3 x - 4Dz = 0$$

i.e., $\quad D^3 x - 4(2x) = 0 \qquad\qquad\qquad$ (Using (3))

i.e., $\quad (D^3 - 8)x = 0$

A.E is $\ m^3 - 8 = 0$

i.e., $\quad (m - 2)(m^2 + 2m + 4) = 0$

$$\therefore \qquad m = 2,\ \ \frac{-2 \pm \sqrt{4 - 16}}{2}$$

i.e., $\quad m = 2,\ -1 + i\sqrt{3},\ -1 - i\sqrt{3}$

$$\therefore \qquad x = c_1 e^{2t} + e^{-t}(c_2 \cos \sqrt{3}\, t + c_3 \sin \sqrt{3}\, t) \qquad\qquad (4)$$

Now equation (1) gives $y = \dfrac{1}{2} Dx$

$$= \frac{1}{2} D [c_1 e^{2t} + e^{-t}(c_2 \cos \sqrt{3}\, t + c_3 \sin \sqrt{3}\, t)]$$

$$= \frac{1}{2} [2c_1 e^{2t} + e^{-t}(-c_2\sqrt{3} \sin \sqrt{3}\, t + c_3\sqrt{3} \cos \sqrt{3}\, t)$$

$$- e^{-t}(c_2 \cos \sqrt{3}\, t + c_3 \sin \sqrt{3}\, t)]$$

i.e., $y = c_1 e^{2t} + \dfrac{e^{-t}}{2}[(-\sqrt{3}\, c_2 - c_3) \sin \sqrt{3}\, t + (\sqrt{3}\, c_3 - c_2) \cos \sqrt{3}\, t]$ (5)

Equation (2) gives,

$$z = \frac{1}{2} Dy$$

$$= \frac{1}{2} D \left[c_1 e^{2t} + \frac{e^{-t}}{2}[(-\sqrt{3}\, c_2 - c_3) \sin \sqrt{3}\, t + (\sqrt{3}\, c_3 - c_2) \cos \sqrt{3}\, t] \right]$$

$$= \frac{1}{2} \left\{ 2c_1 e^{2t} - \frac{e^{-t}}{2}[(-\sqrt{3}\, c_2 - c_3) \sin \sqrt{3}\, t + (\sqrt{3}\, c_3 - c_2) \cos \sqrt{3}\, t] \right.$$

$$\left. + \frac{e^{-t}}{2}[\sqrt{3}\, (-\sqrt{3}\, c_2 - c_3) \cos \sqrt{3}\, t - \sqrt{3}\, (\sqrt{3}\, c_3 - c_2) \sin \sqrt{3}\, t] \right\}$$

i.e., $z = c_1 e^{2t} + \dfrac{e^{-t}}{2}[(\sqrt{3}\, c_2 + c_3 - 3c_3 + \sqrt{3}\, c_2) \sin \sqrt{3}\, t +$

$$(-\sqrt{3}\, c_3 + c_2 - 3c_2 - \sqrt{3}\, c_3) \cos \sqrt{3}\, t]$$

i.e., $z = c_1 e^{2t} + e^{-t}[(\sqrt{3}\, c_2 - c_3) \sin \sqrt{3}\, t - (c_2 + \sqrt{3}\, c_3) \cos \sqrt{3}\, t]$ (6)

The general solutions of the differential equations are given by (4), (5) and (6).

EXERCISE 5.2

PART-A

1. Find the number of arbitrary constants in the general solution of
   $Dx + y = \sin t$, $x + Dy = \cos t$.
2. Find the number of arbitrary constants in the general solution of $Dx - y = t$,
   $Dy + x = t^2$.
3. Find the number of arbitrary constants in the general solution of
   $D^2x + y = 3e^{2t}$, $Dx - Dy = 3e^{2t}$.
4. Solve for x from $Dx - y = t$, $x + Dy = 1$.
5. Find x if $D^2x - 3x - 4y = 0$, $D^2y + x + y = 0$.

6. Find the number of arbitrary constants in the general solution of
$(D-1)x + Dy = 2t + 1$, $(2D+1)x + 2Dy = t$.

PART-B
Solve the following simultaneous differential equations.

7. $Dx + 2x - 3y = 0$, $Dy - 3x + 2y = 0$.

8. $Dx + 2y + \sin t = 0$, $Dy - 2x - \cos t = 0$.

9. $Dx + 2x - 3y = t$, $Dy - 3x + 2y = e^{2t}$.

10. $\dfrac{dx}{dt} + 5x - 2y = t$, $\dfrac{dy}{dt} + 2x + y = 0$, given $x = y = 0$ when $t = 0$.

11. $\dfrac{dx}{dt} - \dfrac{dy}{dt} + 2y = \cos 2t$, $\dfrac{dx}{dt} + \dfrac{dy}{dt} - 2x = \sin 2t$.

12. $2Dx + 6x - y = 2 \sin 2t$, $Dy - 2x + 5y = 0$.

13. $4Dx - Dy + 3x = \sin t$, $Dx + y = \cos t$.

14. $\dfrac{dx}{dt} + y - 1 = \sin t$, $\dfrac{dy}{dt} + x = \cos t$

15. $Dx + Dy - 2y = 2 \cos t - 7 \sin t$, $Dx - Dy + 2x = 4 \cos t - 3 \sin t$.

16. $Dx = 5x + y$, $Dy = y - 4x$.

17. $Dx - 7x + y = 0$, $Dy - 2x - 5y = 0$.

18. $(D-1)x + Dy = 2t + 1$, $(2D+1)x + 2Dy = t$.

19. $(D+1)x + (2D+1)y = e^t$, $(D-1)x + (D+1)y = 1$.

20. $Dx - y = t$, $Dy + x = t^2$.

21. $(D-3)x + 2(D+2)y = 2 \sin t$, $2(D+1)x + (D-1)y = \cos t$

22. $2(2D+1)x + (9D+31)y = e^t$, $(3D+1)x + (7D+24)y = 3$.

23. $\dot{x} + 2y = \sin 2t$, $\dot{y} - 2x = \cos 2t$.

24. $\dot{x} + 5x + \dot{y} + 7y = 2e^t$, $2\dot{x} + x + 3\dot{y} + y = e^t$.

25. $\dfrac{dy}{dx} + y = z + e^x$, $\dfrac{dz}{dx} + z = y + e^x$.

26. $Dx - wy = a \cos pt$, $Dy + wx = a \sin pt$ given that $x = y = 0$ when $t = 0$.

27. $Dx + 2x = 2y$, $Dy + y = 3x$, given $x = 0$ and $Dy = \frac{1}{2}$ when $t = 0$.

28. $D^2x - 2Dy - x = e^t \cos t$, $D^2y + 2Dx - y = e^t \sin t$.

29. $2D^2x - Dy - 4x = 2t$, $2Dx + 4Dy - 3y = 0$.

30. $(D^2 - 5)x + 3y = \sin t$, $(D^2 + 5)y - 3x = t$.

5.3   EULER'S HOMOGENEOUS LINEAR DIFFERENTIAL EQUATIONS

The linear differential equation

$$a_0 x^n \frac{d^n y}{dx^n} + a_1 x^{n-1} \frac{d^{n-1}y}{dx^{n-1}} + \cdots + a_{n-1}.x.\frac{dy}{dx} + a_n.y = \Phi \qquad (1)$$

where $a_0, a_1, ..., a_n$ are constants and Φ is a function of x is called **Euler's linear differential equation**. This differential equation can be reduced to a linear differential equation with constant coefficients by changing the independent variable from x to t by means of the transformation $x = e^t$ or $t = \log x$. Then the linear differential equation with constant coefficients obtained from (1) can be solved by using the methods given in section 5.1.

When $\Phi = 0$, (1) is known as **Euler's homogeneous linear differential equation**.

In equation (1) the coefficient of $\dfrac{d^r y}{dx^r}$ is a multiple of x^r, for $r = 0, 1, 2, ..., n$.

Now put $x = e^t$ (2)

Differentiating w.r.t x

$$\frac{dx}{dt} = e^t \text{ or } \frac{dt}{dx} = \frac{1}{e^t} = \frac{1}{x}$$

$\therefore$

$$\frac{dy}{dx} = \frac{dy}{dt} \cdot \frac{dt}{dx} = \frac{1}{x}\frac{dy}{dt}$$

i.e.,

$$x\frac{dy}{dx} = \frac{dy}{dt}$$ (3)

Differentiating w.r.t x

$$x\frac{d^2y}{dx^2} + \frac{dy}{dx} = \frac{d^2y}{dt^2} \cdot \frac{dt}{dx}$$

i.e.,

$$x\frac{d^2y}{dx^2} + \frac{dy}{dx} = \frac{d^2y}{dt^2} \cdot \frac{1}{x}$$

$\therefore$

$$x^2\frac{d^2y}{dx^2} = \frac{d^2y}{dt^2} - x \cdot \frac{dy}{dx}$$

i.e.,

$$x^2\frac{d^2y}{dx^2} = \frac{d^2y}{dt^2} - \frac{dy}{dt}$$ (4)

We denote $\dfrac{d}{dx}$ by D and $\dfrac{d}{dt}$ by θ

From (3), $xD = \theta$

From (4), $x^2D^2 = \theta^2 - \theta = \theta(\theta - 1)$

Similarly $x^3D^3 = \theta(\theta - 1)(\theta - 2)$

$x^4D^4 = \theta(\theta - 1)(\theta - 2)(\theta - 3)$

In general $x^nD^n = \theta(\theta - 1)(\theta - 2)...(\theta - (n - 1))$.

$\therefore$ (1) Becomes, $a_0 x^n D^n y + a_1 x^{n-1} D^{n-1} y + ... + a_{n-1} xDy + a_n.y = \Phi$

i.e., $[a_0 \, \theta(\theta - 1)(\theta - 2)...(\theta - (n - 1)) + a_1. \, \theta(\theta - 1)(\theta - 2)...(\theta - (n - 2)) + ...$

$$+ a_{n-1}.\theta + a_n] \, y = \Phi(e^t) \qquad (5)$$

Equation (5) is a linear differential equation with constant coefficients and hence can be solved.

5.3.1 EQUATIONS REDUCIBLE TO EULER'S FORM (OR LEGENDRE'S LINEAR DIFFERENTIAL EQUATION)

The equations of the form

$$a_0(a+bx)^n \frac{d^n y}{dx^n} + a_1(a+bx)^{n-1} \frac{d^{n-1} y}{dx^{n-1}} + \cdots + a_{n-1}.(a+bx).\frac{dy}{dx} + a_n.y = \Phi \qquad (6)$$

where $a_0, a_1, \ldots, a_n$ are all constants and Φ is a function of x can be reduced to Euler's linear form by the substitution $a + bx = u$. The differential equations of the form (6) are called **Legendre's linear differential equations.**
Now put $a + bx = u$

$$\therefore \quad \frac{du}{dx} = b$$

$$\therefore \quad \frac{dy}{dx} = \frac{dy}{du}.\frac{du}{dx} = b.\frac{dy}{du}$$

$$\frac{d^2 y}{dx^2} = \frac{d}{dx}\left(\frac{dy}{dx}\right) = \frac{d}{dx}\left(\frac{dy}{du}\right)$$

i.e., $$\frac{d^2 y}{dx^2} = b\frac{d^2 y}{du^2}.\frac{du}{dx} = b^2 \frac{d^2 y}{du^2}$$

In general $$\frac{d^n y}{dx^n} = b^n \frac{d^n y}{du^n}$$

Substituting in (6), we get

$$a_0 b^n u^n \frac{d^n y}{du^n} + a_1 b^{n-1} u^{n-1} \frac{d^{n-1} y}{du^{n-1}} + \cdots + a_{n-1}.b.u.\frac{dy}{du} + a_n.y = \Phi\left(\frac{u-a}{b}\right)$$

i.e., $$A_0 u^n \frac{d^n y}{du^n} + A_1 u^{n-1} \frac{d^{n-1} y}{du^{n-1}} + \cdots + A_{n-1}.u.\frac{dy}{du} + A_n.y = \Phi\left(\frac{u-a}{b}\right) \qquad (7)$$

Equation (7) is Euler's linear differential equation and hence can be solved by putting $u = e^t$ and reducing it to a linear equation with constant coefficients.
Note: In equation (6) one can directly substitute $a + bx = e^t$, so that

$$(a+bx)^n \frac{d^n}{dx^n} = b^n \theta(\theta-1)(\theta-2)\ldots(\theta-(n-1)), \text{ where } \theta = \frac{d}{dt}$$

Example 5.3.1 Solve $x^2 \frac{d^2 y}{dx^2} - 2x\frac{dy}{dx} - 4y = x^4$.

Solution: Putting $x = e^t$, the equation reduces to $[\theta(\theta-1) - 2\theta - 4]y = e^{4t}$ where $\theta \equiv \frac{d}{dt}$

i.e., $(\theta^2 - 3\theta - 4)y = e^{4t}$

A.E is $m^2 - 3m - 4 = 0$

i.e., $(m+1)(m-4) = 0$

$\therefore \qquad m = -1, 4$

$C.F = c_1 e^{-t} + c_2 e^{4t}$

$$P.I = \frac{1}{(\theta+1)(\theta-4)} e^{4t}$$

$$= \frac{1}{(4+1)(\theta-4)} e^{4t}$$

$$= \frac{1}{5} e^{4t} \cdot \frac{1}{(\theta+4-4)} \,(1)$$

$$= \frac{e^{4t}}{5} \cdot \frac{1}{\theta} (1)$$

$$= \frac{e^{4t}}{5} \cdot t$$

$\therefore$ The complete solution is $y = c_1 e^{-t} + c_2 e^{4t} + \dfrac{e^{4t}}{5} \cdot t$

i.e., $y = \dfrac{c_1}{x} + c_2 x^4 + \dfrac{x^4}{5} \cdot \log x$ $(\because t = \log x)$

Example 5.3.2 Solve $\dfrac{d^2 y}{dx^2} + \dfrac{1}{x} \cdot \dfrac{dy}{dx} = \dfrac{12 \log x}{x^2}$.

Solution: The given differential equation is $x^2 \cdot \dfrac{d^2 y}{dx^2} + x \cdot \dfrac{dy}{dx} = 12 \log x.$ (1)

Putting $x = e^t$, (1) reduces to $[\theta(\theta-1) + \theta]y = 12t$ where $\theta \equiv \dfrac{d}{dt}$

i.e., $\theta^2 y = 12t$.

$C.F = c_1 + c_2 t$.

$$P.I = \frac{1}{\theta^2} 12t = \frac{1}{\theta} \cdot \int 12t \, dt$$

$$= \frac{1}{\theta} \cdot 6t^2$$

$$= \int 6t^2 \, dt$$

$$= 2t^3$$

$\therefore \qquad y = c_1 + c_2 t + 2t^3$

i.e., $y = c_1 + c_2 \log x + 2 (\log x)^3$

Example 5.3.3 Solve $(x^3 D^3 + 2x^2 D^2 - xD + 1)y = \log x$.

Solution: Put $x = e^t$ or $t = \log x$.

Then the equation becomes $[\theta(\theta - 1)(\theta - 2) + 2\theta(\theta - 1) - \theta + 1]y = t$ where $\theta \equiv \dfrac{d}{dt}$

i.e., $(\theta^3 - \theta^2 - \theta + 1)y = t$.

A.E is $m^3 - m^2 - m + 1 = 0$

i.e., $(m - 1)(m^2 - 1) = 0$. $\therefore m = 1, 1, -1$.

C.F. $= (c_1 + c_2 t)e^t + c_3 e^{-t}$.

$$
\begin{aligned}
\text{P.I.} &= \frac{1}{\theta^3 - \theta^2 - \theta + 1}.t \\
&= (1 + \theta^3 - \theta^2 - \theta)^{-1}.t \\
&= [1 - (\theta^3 - \theta^2 - \theta)].t \\
&= t + 1
\end{aligned}
$$

$\therefore \qquad y = (c_1 + c_2 t)e^t + c_3 e^{-t} + t + 1$

i.e., $\quad y = (c_1 + c_2 \log x)x + c_3 x^{-1} + \log x + 1$

Example 5.3.4 Solve $(x^2 D^2 + xD + 1)y = \log x . \sin(\log x)$.

Solution: Put $x = e^t$ or $t = \log x$.

Then the equation becomes

$$(\theta(\theta - 1) + \theta + 1)y = t \sin t \text{ where } \theta \equiv \frac{d}{dt}$$

$$(\theta^2 + 1)y = t \sin t.$$

A.E. is $m^2 + 1 = 0 \qquad \therefore m = +i, -i$

C.F. $= c_1 \cos t + c_2 \sin t$.

$$
\begin{aligned}
\text{P.E} &= \frac{1}{\theta^2 + 1} t \sin t \\[2mm]
&= \text{I.P. of } \frac{1}{\theta^2 + 1} t\, e^{it} \\[2mm]
&= \text{I.P. of } \frac{e^{it}}{(\theta + i)^2 + 1}.t \\[2mm]
&= \text{I.P. of } e^{it} \frac{1}{\theta^2 + 2i\theta}.t \\[2mm]
&= \text{I.P. of } \frac{e^{it}}{2i\theta}\left(1 + \frac{\theta}{2i}\right)^{-1}.t \\[2mm]
&= \text{I.P. of } \frac{e^{it}}{2i\theta}\left(1 - \frac{\theta}{2i}\right).t
\end{aligned}
$$

$$= \text{I.P. of } \frac{e^{it}}{2i\theta}\left(t - \frac{1}{2i}\right).t$$

$$= \text{I.P. of } \frac{e^{it}}{2i}\left(\frac{t^2}{2} - \frac{1}{2i}t\right)$$

$$= \text{I.P. of } \frac{-ie^{it}}{2}\left(\frac{t^2}{2} + \frac{it}{2}\right)$$

$$= \text{I.P. of } \frac{-i}{2}(\cos t + i\sin t)\left(\frac{t^2}{2} + \frac{it}{2}\right)$$

$$= \frac{-1}{2}\left(\frac{t^2}{2}\cos t - \frac{t}{2}\sin t\right)$$

$$= \frac{t}{4}\sin t - \frac{t^2}{4}\cos t$$

$\therefore$ The general solution is $y = c_1 \cos t + c_2 \sin t + \dfrac{t}{4}\sin t - \dfrac{t^2}{4}\cos t$

i.e., $\quad y = c_1 \cos(\log x) + c_2 \sin(\log x) + \dfrac{\log x}{4}\sin(\log x) - \dfrac{(\log x)^2}{4}\cos(\log x)$

Example 5.3.5 Solve $x^3 D^3 y - x^2 D^2 y + 2xDy - 2y = x^3 + 3x$.

Solution: $(x^3 D^3 - x^2 D^2 + 2xD - 2)y = x^3 + 3x$.
Put $\quad x = e^t$ or $t = \log x$.

The given equation reduces to $[\theta(\theta - 1)(\theta - 2) - \theta(\theta - 1) + 2\theta - 2]y = e^{3t} + 3e^t$ where $\theta \equiv \dfrac{d}{dt}$

i.e., $\quad (\theta^3 - 4\theta^2 + 5\theta - 2)y = e^{3t} + 3e^t$

A.E. is $m^3 - 4m^2 + 5m - 2 = 0$.

i.e., $\quad (m-1)(m-1)(m-2) = 0$. $\qquad \therefore m = 2, 1, 1$.

C.F. $= c_1 e^{2t} + (c_2 + c_3 t)e^t$

$$\text{P.I } = \frac{1}{(\theta - 2)(\theta - 1)^2}e^{3t} + \frac{1}{(\theta - 2)(\theta - 1)^2}3e^t$$

$$= \frac{e^{3t}}{(3 - 2)(3 - 1)^2} + 3.\frac{1}{(1 - 2)(\theta - 1)^2}e^t$$

$$= \frac{e^{3t}}{4} - 3e^t.\frac{1}{(\theta + 1 - 1)^2} \quad (1)$$

$$= \frac{e^{3t}}{4} - 3e^t.\frac{t^2}{2}$$

$$= \frac{1}{4}e^{3t} - \frac{3}{2}t^2 e^t$$

$\therefore$ The general solution is $y = c_1 e^{2t} + (c_2 + c_3 t)e^t + \frac{1}{4}e^{3t} - \frac{3}{2}t^2 e^t$

i.e., $y = c_1 x^2 + (c_2 + c_3 \log x)x + \frac{1}{4}x^3 - \frac{3}{2}(\log x)^2 x$.

Example 5.3.6 Solve $(x^2 D^2 - xD + 2)y = x \log x$

Solution: Put $x = e^t$ or $t = \log x$.

The given equation reduces to $[\theta(\theta -1) - \theta + 2]y = e^t.t$ where $\theta \equiv \dfrac{d}{dt}$

i.e., $(\theta^2 - 2\theta + 2)y = t.e^t$

A.E. is $m^2 - 2m + 2 = 0$. $\therefore m = 1 \pm i$.

C.F. $= e^t(c_1 \cos t + c_2 \sin t)$.

$$P.I = \frac{1}{\theta^2 - 2\theta + 2}t.e^t = \frac{e^t}{(\theta+1)^2 - 2(\theta+1) + 2}.t$$

$$= \frac{e^t}{\theta^2 + 1}.t$$
$$= e^t(1 + \theta^2)^{-1}.t$$
$$= e^t(1 - \theta^2).t$$
$$= e^t.t$$

Hence the general solution is $y = e^t(c_1 \cos t + c_2 \sin t) + t.e^t$

i.e., $y = x[c_1 \cos (\log x) + c_2 \sin (\log x)] + (\log x).x$

i.e., $y = x[c_1 \cos (\log x) + c_2 \sin (\log x)] + x.\log x$

Example 5.3.7 Solve $(x^2 D^2 - xD + 4)y = \cos (\log x) + x \sin (\log x)$.

Solution: Put $x = e^t$ or $t = \log x$.

The given equation reduces to $[\theta(\theta -1) - \theta + 4]y = \cos t + e^t \sin t$ where $\theta \equiv \dfrac{d}{dt}$

i.e., $(\theta^2 - 2\theta + 4)y = \cos t + e^t \sin t$

A.E. is $m^2 - 2m + 4 = 0$. $\therefore m = 1 \pm \sqrt{3} i$

C.F. $= e^t[c_1 \cos \sqrt{3}\,t + c_2 \sin \sqrt{3}\,t]$

$$P.I = \frac{1}{\theta^2 - 2\theta + 4}\cos t + \frac{1}{\theta^2 - 2\theta + 4}e^t \sin t$$

$$= \frac{1}{-1 - 2\theta + 4}\cos t + \frac{e^t}{(\theta+1)^2 - 2(\theta+1) + 4}\sin t$$

$$= \frac{3+2\theta}{9-4\theta^2}\cos t + \frac{e^t}{\theta^2+3}\sin t$$

$$= \frac{3+2\theta}{9-4(-1)}\cos t + \frac{e^t}{(-1)+3}\sin t$$

$$= \frac{1}{13}[3\cos t - 2\sin t] + \frac{1}{2}e^t.\sin t$$

∴ The general solution is $y = e^t[c_1\cos\sqrt{3}\,t + c_2\sin\sqrt{3}\,t] +$

$$\frac{1}{13}[3\cos t - 2\sin t] + \frac{1}{2}e^t.\sin t$$

i.e., $y = x[c_1\cos(\sqrt{3}.\log x) + c_2\sin(\sqrt{3}.\log x)] +$

$$\frac{1}{13}[3\cos(\log x) - 2\sin(\log x)] + \frac{1}{2}x.\sin(\log x)$$

Example 5.3.8 Solve $(x^2D^2 + 2xD - 20)y = (x+1)^2$

Solution: Put $x = e^t$ or $t = \log x$.

The given equation reduces to $[\theta(\theta-1) + 2\theta - 20]y = (e^t + 1)^2$ where $\theta \equiv \dfrac{d}{dt}$

i.e., $(\theta^2 + \theta - 20)y = (e^t + 1)^2$

The A.E. is $m^2 + m - 20 = 0$.

$\qquad (m+5)(m-4) = 0$

∴ $m = -5, 4$

C.F. $= c_1e^{-5t} + c_2e^{4t}$

$$\text{P.I} = \frac{1}{(\theta+5)(\theta-4)}(e^t+1)^2$$

$$= \frac{1}{(\theta+5)(\theta-4)}(e^{2t} + 2e^t + 1)$$

$$= \frac{e^{2t}}{(2+5)(2-4)} + 2\frac{e^t}{(1+5)(1-4)} + \frac{1}{(0+5)(0-4)}$$

$$= \frac{e^{2t}}{-14} + \frac{2e^t}{-18} + \frac{1}{-20}$$

∴ The complete solution is $y = c_1 e^{-5t} + c_2 e^{4t} - \dfrac{e^{2t}}{14} - \dfrac{2e^t}{18} - \dfrac{1}{20}$

i.e., $y = c_1 x^{-5} + c_2 x^4 - \dfrac{x^2}{14} - \dfrac{2x}{9} - \dfrac{1}{20}$

Example 5.3.9   Solve $(x^2D^2 + xD - 1)y = x^2e^{2x}$

Solution: Put $x = e^t$ or $t = \log x$.

The given equation reduces to $[\theta(\theta - 1) + \theta - 1]y = e^{2t}.e^{2e^t}$ where $\theta \equiv \dfrac{d}{dt}$

i.e., $(\theta^2 - 1)y = e^{2t}.e^{2e^t}$

A.E. is $m^2 - 1 = 0.$ $\therefore m = 1, -1$

C.F. $= c_1 e^t + c_2 e^{-t}$

$$P.I = \frac{1}{\theta^2 - 1} e^{2t}.e^{2e^t}$$

$$= \frac{1}{2}\left(\frac{1}{\theta - 1} - \frac{1}{\theta + 1}\right) e^{2t}.e^{2e^t}$$

$$= \frac{1}{2}\frac{1}{\theta - 1} e^{2t}.e^{2e^t} - \frac{1}{2}\frac{1}{\theta + 1} e^{2t}.e^{2e^t}$$

$$= \frac{1}{2} e^t \int e^{-t} e^{2t}.e^{2e^t} dt - \frac{1}{2} e^{-t} \int e^t e^{2t}.e^{2e^t} dt \qquad \left(\because \frac{1}{D - a} \Phi. = e^{at}\int e^{-at}.\Phi dt\right)$$

$$= \frac{1}{2} e^t \int e^{2e^t} e^t dt - \frac{1}{2} e^{-t} \int e^{2e^t} e^{3t} dt$$

$$= \frac{1}{2} e^t \int e^{2z} dz - \frac{1}{2} e^{-t} \int e^{2z} z^3 dz \qquad \text{(By substituting } e^t = z.)$$

$$= \frac{1}{2} e^t \frac{e^{2z}}{2} - \frac{1}{2} e^{-t}\left[z^3 \frac{e^{2z}}{2} - 3z^2 \frac{e^{2z}}{4} + 6z \frac{e^{2z}}{8} - 6\frac{e^{2z}}{16}\right] \text{(By Bernoulli's formula)}$$

$$= \frac{e^{2e^t}}{4}\left[e^t - e^{2t} + \frac{3e^t}{2} - \frac{3}{2} + \frac{3e^{-t}}{4}\right]$$

$$= \frac{e^{2e^t}}{4}\left[\frac{5e^t}{2} - e^{2t} - \frac{3}{2} + \frac{3e^{-t}}{4}\right]$$

$$\therefore \qquad y = c_1 e^t + c_2 e^{-t} + \frac{e^{2e^t}}{4}\left[\frac{5e^t}{2} - e^{2t} - \frac{3}{2} + \frac{3e^{-t}}{4}\right]$$

i.e., $\qquad y = c_1 x + \dfrac{c_2}{x} + \dfrac{e^{2e^t}}{4}\left[\dfrac{5e^t}{2} - e^{2t} - \dfrac{3}{2} + \dfrac{3e^{-t}}{4}\right]$

Example 5.3.10 Solve $(1 + 2x)^2 \dfrac{d^2 y}{dx^2} - 6(1 + 2x)\dfrac{dy}{dx} + 16y = 8(1 + 2x)^2$

Solution: Put $1 + 2x = e^t$ or $t = \log(1 + 2x)$

Then $(1+2x)^2 \dfrac{d^2}{dx^2} = 2^2\theta(\theta-1)$ and $(1+2x)\dfrac{d}{dx} = 2\theta$ where $\theta \equiv \dfrac{d}{dt}$

The equation reduces to $[2^2\theta(\theta-1) - 6.2\theta + 16]y = 8e^{2t}$

i.e., $(4\theta^2 - 16\theta + 16)y = 8e^{2t}$

i.e., $(\theta^2 - 4\theta + 4)y = 2e^{2t}$

A.E is $m^2 - 4m + 4 = 0$

$\therefore$ $m = 2, 2$

C.F $= (c_1 + c_2 t)e^{2t}$

P.I $= \dfrac{1}{\theta^2 - 4\theta + 4} 2e^{2t}$

$= \dfrac{2}{(\theta - 2)^2} e^{2t}$

$= 2e^{2t} \dfrac{1}{(\theta + 2 - 2)^2}(1)$

$= 2e^{2t} \dfrac{1}{\theta^2}(1)$

$= 2e^{2t} \dfrac{t^2}{2} = t^2 e^{2t}$

$\therefore$ $y = (c_1 + c_2 t)\,e^{2t} + t^2 e^{2t}$

i.e., $y = (c_1 + c_2 \log(1+2x))(1+2x)^2 + (\log(1+2x))^2 (1+2x)^2$

i.e., $y = (1+2x)^2 [(c_1 + c_2 \log(1+2x)) + (\log(1+2x))^2]$

Example 5.3.11 Solve $(x+1)^2 \dfrac{d^2 y}{dx^2} + (x+1)\dfrac{dy}{dx} = (2x+3)(2x+4)$

Solution: Put $x + 1 = e^t$ so that $(x+1)^2 \dfrac{d^2}{dx^2} = \theta(\theta-1)$ and $(x+1)\dfrac{d}{dx} = \theta$ where $\theta \equiv \dfrac{d}{dt}$

Then the equation becomes $[\theta(\theta-1) + \theta]y = (2e^t + 1)(2e^t + 2)$

i.e., $\theta^2 y = (2e^t + 1)(2e^t + 2)$

A.E is $m^2 = 0$ $\therefore m = 0, 0$

C.F $= c_1 + c_2 t$

P.I $= \dfrac{1}{\theta^2}(2e^t + 1)(2e^t + 2)$

$= \dfrac{1}{\theta^2}\left[4e^{2t} + 6e^t + 2\right]$

$= \dfrac{1}{\theta}\left[4\dfrac{e^{2t}}{2} + 6e^t + 2t\right]$

$$= 2\frac{e^{2t}}{2} + 6e^t + 2\frac{t^2}{2}$$
$$= e^{2t} + 6e^t + t^2$$

∴ The complete solution is $y = c_1 + c_2 t + e^{2t} + 6e^t + t^2$

i.e., $y = c_1 + c_2 \log(x + 1) + (x + 1)^2 + 6(x + 1) + [\log(x + 1)]^2$

Example 5.3.12 Solve $[x^3 D^2 + 3x^2 D + 5x]y = 2$

Solution: We have $[x^2 D^2 + 3xD + 5]y = 2x^{-1}$

Put $x = e^t$ so that $xD = \theta$, $x^2 D^2 = \theta(\theta - 1)$ where $\theta \equiv \dfrac{d}{dt}$

The equation reduces to $[\theta(\theta - 1) + 3\theta + 5]y = 2.e^{-t}$
$$(\theta^2 + 2\theta + 5)y = 2e^{-t}$$

The A.E is $m^2 + 2m + 5 = 0$ ∴ $m = -1 \pm 2i$

C.F $= e^{-t}[c_1 \cos 2t + c_2 \sin 2t]$

P.I $= \dfrac{1}{\theta^2 + 2\theta + 5} 2e^{-t} = \dfrac{2e^{-t}}{(-1)^2 + 2(-1) + 5} = \dfrac{e^{-t}}{2}$

∴ $y = e^{-t}[c_1 \cos 2t + c_2 \sin 2t] + \dfrac{e^{-t}}{2}$

i.e., $y = \dfrac{1}{x}[c_1 \cos (2\log x) + c_2 \sin (2\log x)] + \dfrac{1}{2x}$

EXERCISE 5.3

PART – A

1. State 'Euler's homogeneous linear differential equation' and give a transformation to convert it in to a linear equation with constant coefficients.
2. State the Legendre's linear differential equation and give a transformation to convert it into a linear equation with constant coefficients.
3. Transform $x^2 y'' + xy' + 4y = 0$ into a linear equation with constant coefficients.
4. Transform $\dfrac{d^2 y}{dx^2} - \dfrac{6y}{x^2} = x \log x$ into a linear equation with constant coefficients.
5. Convert $(5 + 2x)^2 y'' - 6(5 + 2x)y' + 8y = 0$ into a linear equation with constant coefficients.
6. Solve $x^2 y'' - xy' + y = 0$.
7. Solve $(x^2 D^2 + 7xD + 13)y = 0$.
8. Solve $(x^3 D^3 + 3x^2 D^2 - 2xD + 2)y = 0$.

PART-B
Solve the following differential equations.

9. $(x^2D^2 - 3xD + 4)y = 2x^2$.

10. $(x^2D^2 - 5xD + 8)y = 2x^3$, $y(-2) = 1$, $y'(-2) = 7$.

11. $(x^2D^2 - 2)y = x^2 + \dfrac{1}{x}$.

12. $(x^2D^2 - xD - 3)y = x^2 \log x$.

13. $(x^2D^2 - 3xD + 5)y = x^2 \sin(\log x)$.

14. $[x^2D^2 - (2m - 1)xD + (m^2 + n^2)]y = n^2x^m \log x$.

15. $(x^2D^2 - 2xD + 2)y = x + x^2 \log x + x^3$.

16. $(x^3D^3 - x^2D^2 + 2xD - 2)y = x^3 + 3x$.

17. $(x^2D^2 + 7xD + 13)y = \log x$.

18. $x^2 y'' + xy' + y = 4 \sin(\log x)$.

19. $(x^2D^2 - 4xD + 6)y = \dfrac{42}{x^4}$.

20. $(xD^3 + D^2)y = \dfrac{1}{x}$.

21. $(2x + 3)^2 \dfrac{d^2y}{dx^2} - (2x + 3)\dfrac{dy}{dx} - 12y = 6x$.

22. $(3x + 2)^2 \dfrac{d^2y}{dx^2} + 3(3x + 2)\dfrac{dy}{dx} - 36y = 3x^2 + 4x + 1$.

23. $(1 + x)^2 \dfrac{d^2y}{dx^2} + (1 + x)\dfrac{dy}{dx} + y = 4 \cos(\log(1 + x))$.

24. $[(x + 3)^2D^2 - 4(x + 3)D + 6]y = \log(x + 3)$.

25. $(x + a)^2 \dfrac{d^2y}{dx^2} - 4(x + a)\dfrac{dy}{dx} + 6y = x$.

5.4   LINEAR DIFFERENTIAL EQUATIONS WITH VARIABLE COEFFICIENTS

There is no general method to solve a general linear differential equation with variable coefficients. But few methods are available to solve a linear differential equation of **second order** with variable coefficients. In this section we discuss these methods. The general form of a linear differential equation of second order is

$$\frac{d^2y}{dx^2} + P.\frac{dy}{dx} + Q.y = \Phi \tag{1}$$

where P, Q and Φ are functions of x.

5.4.1 METHOD OF REDUCTION OF ORDER

In this method we transform the given differential equation by changing the dependent variable y and reducing the order of the equation by one. We start with one known solution of the corresponding homogeneous equation.

$$\frac{d^2y}{dx^2} + P.\frac{dy}{dx} + Q.y = 0 \qquad (2)$$

and construct the general solution of the given differential equation (1).
Let the known solution of equation (2) be

$$y = u(x) \qquad (3)$$

Assume that

$$y = u(x).v(x) \qquad (4)$$

is the complete solution of (1).
Differentiating (4) w.r.t x, we have

$$y' = u.v' + u'.v \qquad (5)$$

and $\qquad y'' = uv'' + 2u'v' + u''v \qquad (6)$

Substituting for y, y' and y'' in (1) we get

$$uv'' + 2u'v' + u''v + P(uv' + u'v) + Q(uv) = \Phi$$

i.e., $\qquad uv'' + (2u' + Pu)v' + (u'' + Pu' + Qu)v = \Phi \qquad (7)$

Since u is a solution of (2), $u'' + Pu' + Qu = 0$

$\therefore$ (7) becomes, $uv'' + (2u' + Pu)v' = \Phi$

i.e., $\qquad v'' + (2\frac{u'}{u} + P)v' = \dfrac{\Phi}{u}$

i.e., $\qquad v'' + P_1v' = X_1 \qquad (8)$

where $\qquad P_1 = 2\frac{u'}{u} + P, \ X_1 = \dfrac{\Phi}{u}$

Putting $\qquad v' = w$ in (8), we get $w' + P_1w = X_1$.

i.e., $\qquad \dfrac{dw}{dx} + P_1w = X_1 \qquad (9)$

Equation (9) is a linear first order differential equation in w and hence can be solved. Thus by changing the dependent variable y to w, we have reduced the order of the given differential equation by one. After computing w from (9), we can find v from $v' = w$. Thus the complete integral $y = uv$ of equation (1) can be found.

Note:
(i) In equation (9), if $X_1 = 0$, we have a first order linear differential homogeneous equation,

$$\frac{dw}{dx} + P_1w = 0$$

i.e., $v'' + (2\frac{u'}{u} + P)v' = 0$.

i.e., $\dfrac{d(v')}{v'} = -\left(2\dfrac{u'}{u} + P\right)$

Integrating both sides w.r.t. x, we get,

$\log v' = -2 \log u - \int P \, dx + \log c_1$

$\qquad = \log\left(c_1 u^{-2} e^{-\int P \, dx}\right)$

$\therefore \ v' = c_1 u^{-2} e^{-\int P \, dx}$ \hfill (10)

Solving this equation we can find v and hence the complete solution $y = uv$ of the given equation (1).

(ii)   To find a starting solution of the homogeneous equation $y'' + P.y' + Q.y = 0$, one can remember the following results.

1.   If $1 + P + Q = 0$, e^x is a solution.
2.   If $1 - P + Q = 0$, e^{-x} is a solution.
3.   If $m^2 + mP + Q = 0$, e^{mx} is a solution.
4.   If $P + Qx = 0$, $y = x$ is a solution.
5.   If $2 + 2Px + Qx^2 = 0$, $y = x^2$ is a solution.
6.   If $m(m-1) + Pmx + Qx^2 = 0$, $y = x^m$ is a solution.

5.4.2   SPECIAL TYPES OF EQUATIONS

(i) Equations in which y is absent

i.e., $\qquad \dfrac{d^2 y}{dx^2} = f\left(x, \dfrac{dy}{dx}\right)$ \hfill (1)

Put $\qquad \dfrac{dy}{dx} = p$

Then $\qquad \dfrac{d^2 y}{dx^2} = \dfrac{dp}{dx}$

$\therefore$ Equation (1) becomes, $\dfrac{dp}{dx} = f(x, p)$, which is only a first order equation.

Solving this equation, we get, $p = \varphi(x, c_1)$,

i.e., $\qquad \dfrac{dy}{dx} = \varphi(x, c_1)$ which gives $y = \chi(x, c_1, c_2)$, where c_1 and c_2 are constants.

(ii) Equations in which x is absent

i.e., $\qquad \dfrac{d^2 y}{dx^2} = f\left(y, \dfrac{dy}{dx}\right)$ \hfill (2)

Put $\qquad \dfrac{dy}{dx} = p$

Then
$$\frac{d^2y}{dx^2} = \frac{dp}{dx} = \frac{dp}{dy} \cdot \frac{dy}{dx} = \frac{dp}{dy} \cdot p$$

∴ Equation (2) reduces to the form $\frac{dp}{dy} = f(y, p)$, which is a first order equation. Solving,

we get, $\quad p = \frac{dy}{dx} = \varphi(y, c_1)$

Hence $\quad\quad \frac{dx}{dy} = \varphi_1(y, c_1)$

Solving, we get, $\quad x = \chi(y, c_1, c_2)$, where c_1 and c_2 are constants.

(iii) Equations which are homogeneous in y, y' and y" (but not in x).

We transform the dependent variable y and reduce the order of the given differential equation by one by putting $y = e^{\int z \, dx}$ and hence $y' = z . e^{\int z \, dx}$, $y'' = (z^2 + z') . e^{\int z \, dx}$.

The resulting equation in the dependent variable z is of degree one less than that of the given equation.

(iv) The given differential equation is exact.

Then the given equation can be expressed as
$$\frac{d}{dx}(\varphi(x, y, y')) = 0$$

Integrating, we get, $\quad \varphi(x, y, y') = c_1.$

This is a first order differential equation, solving which we get the complete solution of the given differential equation.

Note: The differential equation $[P_0 D^2 + P_1 D + P_2]y = \Phi$ is exact if and only if
$$P_0'' - P_1' + P_2 = 0.$$

When $P_0'' - P_1' + P_2 = 0$, we have,

$$
\begin{aligned}
(P_0 D^2 + P_1 D + P_2)\, y &= (P_0 D^2 + P_1 D + P_1' - P_0'')\, y \\
&= P_0 y'' + P_1 y' + P_1' y - P_0'' y \\
&= P_0 y'' + P_0' y' + P_1 y' + P_1' y - P_0'' y - P_0' y' \\
&= D(P_0 y' + P_1 y - P_0' y)
\end{aligned}
$$

∴ $D(P_0 y' + P_1 y - P_0' y) = \Phi$

Integrating w.r.t x, we get,

$$P_0 y' + (P_1 - P_0')\, y = \int \Phi \, dx + c_1 \tag{3}$$

Equation (3) is called the **first integral** of the given second order differential equation, which being a first order equation can be solved.

5.4.3 REDUCTION TO CANONICAL OR NORMAL FORM

Definition 5.4.1   A second order linear differential equation in y is said to be in the **canonical form or normal form** if the term containing y' is absent.

i.e., in $y'' + Py' + Qy = \Phi$, $P = 0$ i.e., $y'' + Qy = \Phi$ is the canonical form.

Assume that $y = uv$ is a solution of

$$y'' + Py' + Qy = \Phi \tag{1}$$

Differentiating $y = uv$ w.r.t x, we get,

$$y' = u'v + uv' \tag{2}$$

$$y'' = u''v + 2u'v' + uv'' \tag{3}$$

Equation (1) becomes, $u''v + 2u'v' + uv'' + P(u'v + uv') + Q.uv = \Phi$

i.e., $uv'' + (2u' + Pu)v' + (u'' + Pu' + Qu)v = \Phi \tag{4}$

If (4) is to be in the canonical form, we must have $2u' + Pu = 0$

i.e., $\dfrac{u'}{u} = -\dfrac{P}{2}$

i.e., $\log u = -\displaystyle\int \frac{P}{2} dx$

i.e., $u = e^{-\frac{1}{2}\int P dx} \tag{5}$

$\therefore y = uv = v.e^{-\frac{1}{2}\int P dx}$ transforms equation (1) into the canonical form.

(The transformation $y = uv$ changes the dependent variable from y to v)

When $u = e^{-\frac{1}{2}\int P dx}$, we have $u' = -\dfrac{P}{2}e^{-\frac{1}{2}\int P dx}$ and $u'' = \dfrac{P^2}{4}e^{-\frac{1}{2}\int P dx} - \dfrac{P'}{2}e^{-\frac{1}{2}\int P dx}$

Now the canonical form of the given differential equation (1) is

$$uv'' + (u'' + Pu' + Qu)v = \Phi$$

i.e., $v'' + \dfrac{1}{u}(u'' + Pu' + Qu)v = \dfrac{1}{u}.\Phi$

i.e., $v'' + \left(\dfrac{P^2}{4} - \dfrac{P'}{2} - \dfrac{P^2}{2} + Q\right)v = \Phi.e^{\frac{1}{2}\int P dx}$

i.e., $v'' + \left(Q - \dfrac{1}{4}P^2 - \dfrac{1}{2}P'\right)v = \Phi.e^{\frac{1}{2}\int P dx} \tag{6}$

i.e., the canonical form is $v'' + f(x).v = g(x)$

where $f(x) = Q - \dfrac{1}{4}P^2 - \dfrac{1}{2}P'$ and $g(x) = \Phi.e^{\frac{1}{2}\int P dx}$

Thus the equation $y'' + Py' + Qy = \Phi$ is reduced to the canonical form,

$$v'' + f(x).v = g(x) \tag{7}$$

where $f(x) = Q - \frac{1}{4}P^2 - \frac{1}{2}P'$ and $g(x) = \Phi.e^{\frac{1}{2}\int P \, dx}$

by the substitution, $y = v.e^{-\frac{1}{2}\int P \, dx}$

The differential equation (7) in the canonical form can be solved easily, especially when $f(x)$ is a constant.

Example 5.4.1 Solve $x.\dfrac{d^2y}{dx^2} - (2x-1)\dfrac{dy}{dx} + (x-1)y = 0$ by the method reduction of order.

Solution: The given equation is

$$\frac{d^2y}{dx^2} - (2 - \tfrac{1}{x})\frac{dy}{dx} + (1 - \tfrac{1}{x})y = 0 \tag{1}$$

$$P = -(2 - \tfrac{1}{x}), \quad Q = (1 - \tfrac{1}{x})$$

$$1 + P + Q = 1 - 2 + \tfrac{1}{x} + 1 - \tfrac{1}{x} = 0$$

$\therefore \quad u = e^x$ is a solution of (1).

Put $y = ve^x$

Then $y' = v.e^x + e^x.\dfrac{dv}{dx}$

and $y'' = e^x.\dfrac{d^2v}{dx^2} + 2e^x \dfrac{dv}{dx} + ve^x$

$\therefore$ Equation (1) becomes

$$e^x[v'' + 2v' + v] - (2 - \tfrac{1}{x}).e^x[v + v'] + (1 - \tfrac{1}{x})e^x.v = 0.$$

$\therefore \qquad v'' + 2v' + v - (2 - \tfrac{1}{x})v - (2 - \tfrac{1}{x})v' + (1 - \tfrac{1}{x})v = 0.$

i.e., $\quad v'' + 2v' + v - 2v + \tfrac{1}{x}v - 2v' + \tfrac{1}{x}v' + v - \tfrac{1}{x}v = 0.$

i.e., $\quad v'' + \tfrac{1}{x}v' = 0$

Put $v' = p.$ Then $\dfrac{dp}{dx} + \dfrac{1}{x}p = 0$

i.e., $\quad \dfrac{dp}{p} = -\dfrac{dx}{x}$

Integrating, $\log p = -\log x + \log c_1$

$\therefore \qquad p = \dfrac{c_1}{x} \qquad$ i.e., $\qquad \dfrac{dv}{dx} = \dfrac{c_1}{x}$

$$dv = c_1 \frac{1}{x}dx$$

Integrating, $v = c_1 \log x + c_2$.

$\therefore$ The general solution of the given equation is $y = uv$.

i.e., $y = e^x [c_1 \log x + c_2]$

Example 5.4.2 Solve the equation $(1 - x^2).\dfrac{d^2y}{dx^2} - 2x.\dfrac{dy}{dx} + 2y = 2$ by the method of reduction of order.

Solution: The given equation is

$$y'' - \frac{2x}{1-x^2}.y' + \frac{2}{1-x^2}.y = \frac{2}{1-x^2} \tag{1}$$

Here $P + Qx = -\dfrac{2x}{1-x^2} + \dfrac{2x}{1-x^2} = 0$

$\therefore$ $u = x$ is a solution of $y'' - \dfrac{2x}{1-x^2}.y' + \dfrac{2y}{1-x^2} = 0$

Put $y = xv$ is the given equation (1)

i.e., $y' = xv' + \dot{v},\ y'' = xv'' + 2v'$

(1) becomes $(xv'' + 2v') - \dfrac{2x}{1-x^2}.(xv' + v) + \dfrac{2vx}{1-x^2} = \dfrac{2}{1-x^2}$

i.e., $xv'' + \left(2 - \dfrac{2x^2}{1-x^2}\right).v' = \dfrac{2}{1-x^2}$

$$v'' + \left(\frac{2}{x} - \frac{2x}{1-x^2}\right).v' = \frac{2}{x(1-x^2)} \tag{2}$$

Put $w = v'$

i.e., $\dfrac{dw}{dx} = v''$

Then (2) becomes $\dfrac{dw}{dx} + \left(\dfrac{2}{x} - \dfrac{2x}{1-x^2}\right)w = \dfrac{2}{x(1-x^2)}$ (3)

Equation (3) is a linear first order equation in w. The solution is

$$w.e^{\int\left(\frac{2}{x} - \frac{2x}{1-x^2}\right)dx} = \int e^{\int\left(\frac{2}{x} - \frac{2x}{1-x^2}\right)dx}.\frac{2}{x(1-x^2)}dx + c_1 \tag{4}$$

Now $e^{\int\left(\frac{2}{x} - \frac{2x}{1-x^2}\right)dx} = e^{2\log x + \log(1-x^2)}$

$$= e^{\log\left(x^2(1-x^2)\right)} = x^2(1-x^2)$$

$\therefore$ (4) becomes $w.x^2 (1-x^2) = \int x^2(1-x^2).\dfrac{2}{x(1-x^2)}dx + c_1$

i.e., $\quad w.x^2(1-x^2) = x^2+c_1$.

i.e., $\quad \dfrac{dv}{dx} = \dfrac{1}{1-x^2} + \dfrac{c_1}{x^2(1-x^2)}$

$\therefore \quad v = \displaystyle\int \dfrac{1}{1-x^2}dx + c_1\int\dfrac{1}{x^2(1-x^2)}dx + c_2$

$\qquad = \displaystyle\int\dfrac{1}{1-x^2}dx + c_1\int\left[\dfrac{1}{x^2} + \dfrac{1}{1-x^2}\right]dx + c_2$

$\qquad = (1+c_1)\displaystyle\int\dfrac{1}{1-x^2}dx + c_1\int\dfrac{1}{x^2}dx + c_2$

i.e., $\quad v = (c_1+1)\dfrac{1}{2}\log\left(\dfrac{1+x}{1-x}\right) - \dfrac{c_1}{x} + c_2$

$\therefore$ The complete solution of equation (1) is

$\qquad y = uv$

$\qquad = (c_1+1)\dfrac{x}{2}\log\left(\dfrac{1+x}{1-x}\right) - c_1 + c_2x \quad$ (Since $u = x$).

Example 5.4.3 Solve $xy'' - 2(x+1)y' + (x+2)y = (x-2)e^x$ using the method of reduction of order.

Solution: The given equation is

$\qquad y'' - 2(1+\tfrac{1}{x})y' + (1+\tfrac{2}{x})y = (1-\tfrac{2}{x})e^x \qquad$ (1)

Here $\quad P = -2(1+\tfrac{1}{x}), \ Q = (1+\tfrac{2}{x})$

$\qquad 1 + P + Q = 1 - 2 - \tfrac{2}{x} + 1 + \tfrac{2}{x} = 0$

$\therefore u = e^x$ is a solution of the equation

$\qquad y'' - 2(1+\tfrac{1}{x})y' + (1+\tfrac{2}{x})y = 0 \qquad$ (2)

Put $\quad y = uv = e^x.v$ in (1)

i.e., $\quad y' = e^xv' + e^x.v$

and $\quad y'' = e^xv'' + 2e^xv' + e^xv$

Then $\quad (e^xv'' + 2e^xv' + e^xv) - 2(1+\tfrac{1}{x})(e^xv' + e^x.v) + (1+\tfrac{2}{x})e^xv = (1-\tfrac{2}{x})e^x$

i.e., $\quad v'' + 2v' + v - 2v' - 2v - \tfrac{2}{x}v' - \tfrac{2}{x}v + v + \tfrac{2}{x}v = (1-\tfrac{2}{x})$

i.e., $\quad v'' - \tfrac{2}{x}v' = 1 - \tfrac{2}{x}$

Put $\quad v' = \dfrac{dv}{dx} = w$

$\therefore \quad \dfrac{dw}{dx} - \tfrac{2}{x}w = 1 - \tfrac{2}{x} \qquad$ (3)

Equation (3) is a linear first order equation in w. The solution is

$$w.e^{\int -\frac{2}{x} dx} = \int e^{\int -\frac{2}{x} dx} (1 - \frac{2}{x}) dx + c_1$$

i.e., $w \dfrac{1}{x^2} = \int \dfrac{1}{x^2}(1 - \dfrac{2}{x}) dx + c_1$

i.e., $\dfrac{w}{x^2} = \dfrac{-1}{x} + \dfrac{1}{x^2} + c_1$

i.e., $\dfrac{dv}{dx} = -x + 1 + c_1 x^2$

i.e., $v = \int (-x + 1 + c_1 x^2) dx + c_2$

i.e., $v = \dfrac{-x^2}{2} + x + c_1 \dfrac{x^3}{3} + c_2$

∴ The general solution of the given differential equation is

   $y = uv$

   $= e^x \left[\dfrac{-x^2}{2} + x + c_1 \dfrac{x^3}{3} + c_2 \right]$

Example 5.4.4   Solve $y'' - x^2 y' + xy = x$ using the method of reduction of order

Solution: The given equation is

   $y'' - x^2 y' + xy = x$ (1)

   $P = -x^2, Q = x$

∴ $P + Qx = -x^2 + x^2 = 0$

∴ $u = x$ is a solution of

   $y'' - x^2 y' + xy = 0$ (2)

Let $y = uv = xv$ be a solution of (1).

Then $y' = xv' + v, y'' = xv'' + 2v'$

∴ Equation (1) becomes $(xv'' + 2v') - x^2(xv' + v) + x.xv = x$

i.e., $xv'' - x^3 v' + 2v' = x$

i.e., $v'' + (\frac{2}{x} - x^2)v' = 1$ (3)

Put $w = v' = \dfrac{dv}{dx}$

Then equation (3) becomes

   $\dfrac{dw}{dx} + (\frac{2}{x} - x^2)w = 1$

The solution of this first order differential equation is

$$w.e^{\int \left(\frac{2}{x} - x^2 \right) dx} = \int e^{\int \left(\frac{2}{x} - x^2 \right) dx} .1 dx + c_1$$

i.e., $\quad$ w.e$^{2\log x - \frac{x^3}{3}} = \int e^{2\log x - \frac{x^3}{3}} dx + c_1$

i.e., $\quad$ wx^2.e$^{-\frac{x^3}{3}} = \int e^{-\frac{x^3}{3}}.x^2 dx + c_1$

i.e., $\quad$ wx^2.e$^{-\frac{x^3}{3}} = -e^{-\frac{x^3}{3}} + c_1$

i.e., $\quad$ w$= \dfrac{-1}{x^2} + \dfrac{c_1}{x^2}.e^{\frac{x^3}{3}}$

i.e., $\quad$ $\dfrac{dv}{dx} = \dfrac{-1}{x^2} + \dfrac{c_1}{x^2}.e^{\frac{x^3}{3}}$

$\therefore \quad$ v $= \dfrac{1}{x} + c_1 \int e^{\frac{x^3}{3}}.x^{-2}dx + c_2$

Hence $y = uv = x.\left[\dfrac{1}{x} + c_1 \int e^{\frac{x^3}{3}}.x^{-2}dx + c_2 \right]$

$$= 1 + c_1 x \int e^{\frac{x^3}{3}}.x^{-2}dx + c_2 x$$

is the complete solution of the given differential equation

Example 5.4.5 $\;$ Solve $y'' + (1 - \cot x)y' - y \cot x = \sin^2 x$

Solution: $\quad y'' + (1 - \cot x)y' - y \cot x = \sin^2 x$ $\hfill$ (1)

$\qquad$ P $= 1 - \cot x$, Q $= -\cot x$

$\therefore \qquad 1 - P + Q = 1 - 1 + \cot x - \cot x = 0$

$\therefore$ u $= e^{-x}$ is a solution of the equation

$\qquad y'' + (1 - \cot x)y' - y \cot x = 0$ $\hfill$ (2)

Let $\quad$ y $= uv$

$\qquad = e^{-x}.v$ be a solution of the given equation (1)

Then $\quad y' = e^{-x}v' - e^{-x}.v$

$\qquad y'' = e^{-x}v'' - e^{-x}.v' - e^{-x}v' + e^{-x}.v$

$\qquad = e^{-x}v'' - 2e^{-x}.v' + e^{-x}.v$

Substituting the values of y, y', y'' in the given equation (1), we get,

$\qquad e^{-x}(v'' - 2v' + v) + (1 - \cot x) e^{-x}(v' - v) - \cot x.e^{-x}.v = \sin^2 x$

i.e., $\quad v'' - (1 + \cot x)v' = e^x \sin^2 x$ $\hfill$ (3)

Put $\quad$ w $= v' = \dfrac{dv}{dx}$ in (3)

Then $\dfrac{dw}{dx} - (1 + \cot x)w = e^x \sin^2 x$ (4)

$$e^{-\int(1+\cot x)\,dx} = e^{-x-\log \sin x} = \dfrac{e^{-x}}{\sin x}$$

∴ Solution of the first order equation (4) is

$$w.\dfrac{e^{-x}}{\sin x} = \int \dfrac{e^{-x}}{\sin x}.e^x.\sin^2 x\,dx + c_1$$

$$= \int \sin x\,dx + c_1$$

$$= -\cos x + c_1$$

$$\therefore w = \dfrac{dv}{dx} = -\sin x \cos x.e^x + c_1 e^x.\sin x$$

$$= -\dfrac{1}{2} e^x \sin 2x + c_1 e^x.\sin x$$

$$v = -\dfrac{1}{2}\int e^x \sin 2x\,dx + c_1 \int e^x \sin x\,dx + c_2$$

Now, $\int e^x \sin 2x\,dx = -e^x \dfrac{\cos 2x}{2} + \int e^x \dfrac{\cos 2x}{2}\,dx$

$$= \dfrac{-e^x}{2}\cos 2x + \dfrac{1}{2}\left[e^x \dfrac{\sin 2x}{2} - \int e^x \dfrac{\sin 2x}{2}\,dx\right]$$

i.e., $\int e^x \sin 2x\,dx = \dfrac{-e^x}{2}\cos 2x + \dfrac{1}{4}e^x \sin 2x - \dfrac{1}{4}\int e^x \sin 2x\,dx$

i.e., $\dfrac{5}{4}\int e^x \sin 2x\,dx = \dfrac{-e^x}{2}\cos 2x + \dfrac{1}{4}e^x \sin 2x$

$\therefore \int e^x \sin 2x\,dx = \dfrac{1}{5}\left[e^x \sin 2x - 2e^x \cos 2x\right]$

Similarly, $\int e^x \sin x\,dx = \dfrac{1}{2}\left[e^x \sin x - e^x \cos x\right]$

$\therefore v = -\dfrac{1}{2}.\dfrac{1}{5}.e^x\left[\sin 2x - 2\cos 2x\right] + \dfrac{c_1}{2}e^x\left[\sin x - \cos x\right] + c_2$

The complete solution is $y = uv$

Substititing for u and v, we get,

$$y = e^{-x}\left[-\dfrac{1}{10}.e^x\left[\sin 2x - 2\cos 2x\right] + \dfrac{c_1}{2}e^x\left[\sin x - \cos x\right] + c_2\right]$$

i.e., $y = -\dfrac{1}{10}.\left[\sin 2x - 2\cos 2x\right] + \dfrac{c_1}{2}\left[\sin x - \cos x\right] + c_2 e^{-x}$

Example 5.4.6 Solve the differential equation $\dfrac{d^2y}{dx^2} - \dfrac{x}{x-1} \cdot \dfrac{dy}{dx} + \dfrac{y}{x-1} = x-1$ using the method of reduction of order.

Solution: $P = \dfrac{-x}{x-1}, Q = \dfrac{1}{x-1}$

$\therefore \qquad P + Qx = 0$

Hence $u = x$ is a solution of the equation

$$y'' - \frac{x}{x-1} \cdot y' + \frac{y}{x-1} = 0 \qquad (1)$$

Let $y = uv = xv$ be a solution of the given equation.

Then $y' = v + xv' \qquad y'' = 2v' + xv''$

Substituting the values of y, y', y'' in the given equation, we get,

$$xv'' + 2v' - \frac{x}{x-1} \cdot (xv' + v) + \frac{xv}{x-1} = x-1$$

i.e., $\qquad xv'' + \left(2 - \dfrac{x^2}{x-1}\right)v' = x-1$

i.e., $\qquad v'' + \left(\dfrac{2}{x} - \dfrac{x}{x-1}\right)v' = 1 - \dfrac{1}{x}$

Put $\quad w = v' = \dfrac{dv}{dx}$. Then $\dfrac{dw}{dx} + \left(\dfrac{2}{x} - \dfrac{x}{x-1}\right)v' = 1 - \dfrac{1}{x}$ $\qquad (2)$

$e^{\int\left(\frac{2}{x} - \frac{x}{x-1}\right)dx} = e^{\int\left(\frac{2}{x} - 1 - \frac{1}{x-1}\right)dx}$

$\qquad = e^{2 \log x - x - \log(x-1)}$

$\qquad = e^{-x} \cdot \dfrac{x^2}{x-1}$

$\therefore$ The solution of the firsr order equation (2) is

$w.e^{-x} \cdot \dfrac{x^2}{x-1} = \displaystyle\int e^{-x} \cdot \dfrac{x^2}{x-1}\left(1 - \dfrac{1}{x}\right)dx + c_1$

$\qquad = \displaystyle\int e^{-x} \cdot x \, dx + c_1$

$\qquad = x\left(-e^{-x}\right) - \displaystyle\int -e^{-x} \, dx + c_1$

$\qquad = -x.e^{-x} - e^{-x} + c_1$

i.e., $\quad w = \dfrac{dv}{dx} = \dfrac{-(x+1)(x-1)}{x^2} + c_1 \dfrac{(x-1)}{x^2} e^x$

$\therefore \qquad v = \displaystyle\int \dfrac{-(x^2-1)}{x^2} dx + c_1 \displaystyle\int \dfrac{(x-1)}{x^2} e^x dx + c_2$

i.e., $\quad v = -\int\left(1 - \dfrac{1}{x^2}\right)dx + c_1 \int\left(\dfrac{1}{x} - \dfrac{1}{x^2}\right)e^x dx + c_2$

$\qquad = -\left(x + \dfrac{1}{x}\right) + c_1\dfrac{e^x}{x} + c_2$

∴ The complete solution is $y = uv = x\left[-x - \dfrac{1}{x} + c_1\dfrac{e^x}{x} + c_2\right] = -x^2 - 1 + c_1 e^x + c_2.x$

Example 5.4.7   Solve $x^2 y'' + xy' - 9y = 0$, given that $y = x^3$ is a solution.

Solution: The given equation is

$$y'' + \dfrac{1}{x}y' - \dfrac{9}{x^2}y = 0 \qquad\qquad (1)$$

Let $\quad u = x^3$

Assume $y = uv$ is the complete solution of (1).

i.e., $\quad y = x^3 v$

Method 1.   Now $y = uv \Rightarrow y' = u'.v + u.v'$

$\qquad\qquad\qquad y'' = u''v + 2u'v' + uv''$

∴ Equation (1) becomes

$\qquad (uv'' + 2u'v' + u''v) + \dfrac{1}{x}(u'.v + u.v') - \dfrac{9}{x^2}uv = 0$

i.e., $\quad uv'' + \left(2u' + \dfrac{u}{x}\right)v' + \left(u'' + \dfrac{u'}{x} - \dfrac{9}{x^2}u\right)v = 0$

i.e., $\quad uv'' + \left(2u' + \dfrac{u}{x}\right)v' = 0 \qquad \left(\because u \text{ is a solution of (1)}, u'' + \dfrac{u'}{x} - \dfrac{9}{x^2}u = 0\right)$

i.e., $\quad v'' + \left(2\dfrac{u'}{u} + \dfrac{1}{x}\right)v' = 0 \qquad\qquad (2)$

Put $w = v' = \dfrac{dv}{dx}$ so that $v'' = \dfrac{dw}{dx}$

∴ Equation (2) becomes $\dfrac{dw}{dx} + \left(2\dfrac{u'}{u} + \dfrac{1}{x}\right)w = 0$

∴ $\qquad w = c_1 e^{-\int\left(2\frac{u'}{u} + \frac{1}{x}\right)dx}$

$\qquad\quad = c_1 e^{-(2\log u + \log x)}$

i.e., $\quad w = c_1 u^{-2}.x^{-1} = c_1 x^{-6}.x^{-1} = c_1.x^{-7}$

i.e., $\quad \dfrac{dv}{dx} = c_1.x^{-7}$

∴ $\qquad v = c_1\int x^{-7}dx + c_2 = c_1\dfrac{x^{-6}}{-6} + c_2$

The complete solution is $y = uv = x^3 \left[-\dfrac{c_1}{6} x^{-6} + c_2 \right] = -\dfrac{c_1}{6} x^{-3} + c_2 x^3$

Method 2. Since equation (1) is homogeneous $(X = 0)$, v' is given by equation (10) in section 5.4.1.

i.e., $v' = c_1 u^{-2} e^{-\int P\,dx}$ where $u = x^3$ and $P = \dfrac{1}{x}$

i.e., $\dfrac{dv}{dx} = c_1 x^{-6} e^{-\int \frac{1}{x}\,dx} = c_1 x^{-6} . x^{-1} = c_1 . x^{-7}$

$\therefore \qquad v = c_1 \dfrac{x^{-6}}{-6} + c_2$

and $y = uv = x^3 \left[-\dfrac{c_1}{6} x^{-6} + c_2 \right]$

i.e., $y = -\dfrac{c_1}{6} x^{-3} + c_2 x^3$ is the complete solution of the given differential equation.

Example 5.4.8 Solve $x^2 y'' + xy' - (x^2 + \dfrac{1}{4})y = 0$ by reducing it to the canonical form.

Solution: The given equation is

$$y'' + \dfrac{1}{x} y' - (1 + \dfrac{1}{4x^2})y = 0 \qquad\qquad (1)$$

Comparing (1) with $y'' + Py' + Qy = \Phi$, we have

$P = \dfrac{1}{x},\ Q = -1 - \dfrac{1}{4x^2}$ and $\Phi = 0$

$u = e^{-\frac{1}{2}\int P\,dx} = e^{-\frac{1}{2}\int \frac{1}{x}\,dx} = e^{-\frac{1}{2}\log x} = x^{-\frac{1}{2}}$

Substituting $y = uv$ in (1), it reduces to the canonical form.

$$v'' + (Q - \dfrac{1}{4} P^2 - \dfrac{1}{2} P')v = \Phi\, e^{\frac{1}{2}\int P\,dx} \qquad \text{(Refer section 5.4.3)}$$

i.e., $v'' + (-1 - \dfrac{1}{4x^2} - \dfrac{1}{4x^2} + \dfrac{1}{2x^2})v = 0$

i.e., $v'' - v = 0$

i.e., $(D^2 - 1)v = 0$

A.E. is $m^2 - 1 = 0$

$\therefore \qquad m = 1, -1$ and $v = c_1 e^x + c_2 e^{-x}$

Hence the complete solution of (1) is $y = uv = x^{-\frac{1}{2}} (c_1 e^x + c_2 e^{-x})$

Example 5.4.9 Solve $x^2 y'' - 2(x^2 + x)y' + (x^2 + 2x + 2)y = 0$

by reducing it to the normal form.

Solution: The given equation is

$$y'' - \frac{2}{x^2}(x^2 + x)y' + \frac{1}{x^2}(x^2 + 2x + 2)y = 0 \qquad (1)$$

Comparing equation (1) with $y'' + Py' + Qy = \Phi$,

We have $P = -\frac{2}{x^2}(x^2 + x) = -2\left(1 + \frac{1}{x}\right)$

$Q = \frac{1}{x^2}(x^2 + 2x + 2) = 1 + \frac{2}{x} + \frac{2}{x^2}$ and $\Phi = 0$

$$u = e^{-\frac{1}{2}\int P\,dx} = e^{-\frac{1}{2}\int(-2)\left(1 + \frac{1}{x}\right)dx} = e^{x + \log x} = xe^x$$

Substituting $y = uv$ in (1), it reduces to the normal form

$$v'' + \left(Q - \frac{1}{4}P^2 - \frac{1}{2}P'\right)v = \Phi\, e^{\frac{1}{2}\int P\,dx}$$

i.e., $v'' + \left[1 + \frac{2}{x} + \frac{2}{x^2} - \left(1 + \frac{1}{x}\right)^2 - \frac{1}{2}(-2)\left(-\frac{1}{x^2}\right)\right]v = 0$

i.e., $v'' + \left[1 + \frac{2}{x} + \frac{2}{x^2} - 1 - \frac{2}{x} - \frac{1}{x^2} - \frac{1}{x^2}\right]v = 0$

i.e., $v'' = 0$

$\therefore \qquad v = c_1 + c_2 x$

Hence the general solution of the given differential equation is $y = uv = xe^x(c_1 + c_2 x)$.

Example 5.4.10   Solve $y'' - 4xy' + (4x^2 - 3)y = e^{x^2}$ by removing the first order term.

Solution: Comparing the given equation with $y'' + Py' + Qy = \Phi$,

we have $P = -4x$, $Q = 4x^2 - 3$ and $\Phi = e^{x^2}$

$$u = e^{-\frac{1}{2}\int P\,dx} = e^{-\frac{1}{2}\int(-4x)\,dx} = e^{x^2}$$

Substituting $y = uv$ in the given equation, it reduces to the following equation in which the first order term is absent.

$$v'' + \left(Q - \frac{1}{4}P^2 - \frac{1}{2}P'\right)v = \Phi\, e^{\frac{1}{2}\int P\,dx}$$

i.e., $v'' + \left[4x^2 - 3\frac{16x^2}{4} - \frac{1}{2}(-4)\right]v = e^{x^2}.e^{\frac{1}{2}\int(-4x)\,dx}$

i.e., $v'' + (4x^2 - 3 - 4x^2 + 2)v = e^{x^2}.e^{-x^2}$

i.e., $v'' + (-1)v = 1$

i.e., $v'' - v = 1$

i.e., $(D^2 - 1)v = 1$

A.E is $m^2 - 1 = 0$

$\therefore \qquad m = 1, -1$

C.F $= c_1 e^x + c_2 e^{-x}$

P.I $= \dfrac{1}{D^2 - 1}(1) = -(1 - D^2)^{-1}.(1)$

$\qquad\qquad = -(1 + D^2)(1)$

$\qquad\qquad = -1$

$\therefore \qquad v = c_1 e^x + c_2 e^{-x} - 1$

The complete solution of the given equation is $y = uv = e^{x^2}(c_1 e^x + c_2 e^{-x} - 1)$

Example 5.4.11 Solve $y'' - 2\tan x.y' + 5y = \sec x.e^x$ by removing the first order term.

Solution: Comparing the given equation with $y'' + Py' + Qy = \Phi$,
we have $P = -2\tan x$, $Q = 5$ and $\Phi = \sec x.e^x$

$$u = e^{-\frac{1}{2}\int P\,dx} = e^{\int \tan x\,dx} = e^{\log \sec x} = \sec x$$

Substituting $y = uv$ in the given equation, it reduces to one in which there is no first order term. The reduced form is

$$v'' + \left(Q - \frac{1}{4}P^2 - \frac{1}{2}P'\right)v = \Phi\, e^{\frac{1}{2}\int P\,dx}$$

i.e., $\quad v'' + \left[5 - \dfrac{1}{4}4\tan^2 x - \dfrac{1}{2}(-2)\sec^2 x\right]v = \sec x\, e^x.e^{\frac{1}{2}\int(-2)\tan x\,dx}$

i.e., $\quad v'' + [5 - \tan^2 x + \sec^2 x]v = \sec x\, e^x.\dfrac{1}{\sec x}$

i.e., $\quad v'' + [5 - \tan^2 x + 1 + \tan^2 x]v = e^x$

i.e., $\quad v'' + 6v = e^x$

i.e., $\quad (D^2 + 6)v = e^x$

A.E is $m^2 + 6 = 0 \qquad \therefore m = \sqrt{6}i,\ -\sqrt{6}i$

C.F. $= c_1\cos\sqrt{6}\,x + c_2\sin\sqrt{6}\,x$

P.I $= \dfrac{1}{D^2 + 6}e^x = \dfrac{e^x}{7}$

$\therefore \qquad v = c_1\cos\sqrt{6}\,x + c_2\sin\sqrt{6}\,x + \dfrac{e^x}{7}$

The complete solution of the given equation is

$$y = uv = \sec x.\left[\, c_1\cos\sqrt{6}\,x + c_2\sin\sqrt{6}\,x + \dfrac{e^x}{7}\,\right]$$

Example 5.4.12 Solve $\dfrac{d^2y}{dx^2} + 2x.\dfrac{dy}{dx} + (x^2 + 5)y = xe^{-\frac{x^2}{2}}$ by removing the first derivative.

Solution: Comparing the given equation with the equation $y'' + Py' + Qy = \Phi$,

we have $P = 2x$, $Q = x^2 + 5$ and $\Phi = xe^{-\frac{x^2}{2}}$

$$u = e^{-\frac{1}{2}\int P\,dx} = e^{-\frac{1}{2}\int 2x\,dx}$$

$$= e^{-\frac{x^2}{2}}$$

Substituting $y = uv$ in the given equation, it reduces to

$$v'' + (Q - \frac{1}{4}P^2 - \frac{1}{2}P')v = \Phi\, e^{\frac{1}{2}\int P\,dx}$$

i.e., $v'' + [x^2 + 5 - \frac{1}{4}4x^2 - \frac{1}{2}(2)]\,v = xe^{-\frac{x^2}{2}}\,e^{\int x\,dx}$

i.e., $v'' + (4)\,v = xe^{-\frac{x^2}{2}}\,e^{\frac{x^2}{2}}$

i.e., $v'' + 4v = x$

i.e., $(D^2 + 4)v = x$

A.E is $m^2 + 4 = 0$

i.e., $m = 2i, -2i$

C.F $= c_1\cos 2x + c_2\sin 2x$

$$P.I = \frac{1}{D^2+4}x = \frac{1}{4}\left(1 + \frac{D^2}{4}\right)^{-1}x$$

$$= \frac{1}{4}\left(1 - \frac{D^2}{4}\right)x$$

$$= \frac{1}{4}.x$$

$\therefore$ $v = c_1\cos 2x + c_2\sin 2x + \frac{x}{4}$

The complete solution of the given equation is

$$y = uv = e^{-\frac{x^2}{2}}\left[c_1\cos 2x + c_2\sin 2x + \frac{x}{4} \right]$$

Example 5.4.13 Show that the equation $(1 + x^2)\,y'' + 3xy' + y = 0$ is exact and hence solve it.

Solution: Comparing the given equation with $P_0 y'' + P_1 y' + P_2 y = \Phi$, we have,

$P_0 = 1 + x^2$, $P_1 = 3x$, $P_2 = 1$ and $\Phi = 0$.

Now $P_0'' - P_1' + P_2 = 2 - 3 + 1 = 0$

$\therefore$ The given equation is exact.

The first order differential equation is $P_0 y' + (P_1 - P_0')y = c_1 + \int \Phi\,dx$

$(1 + x^2)y' + (3x - 2x)y = c_1$

$$y' + \frac{x}{1+x^2}y = \frac{c_1}{1+x^2}$$

General solution of this first order differential equation is

$$y.e^{\int \frac{x}{1+x^2}dx} = \int e^{\int \frac{x}{1+x^2}dx}\left(\frac{c_1}{1+x^2}\right)dx + c_2$$

$$y.e^{\frac{1}{2}\log(1+x^2)} = \int e^{\frac{1}{2}\log(1+x^2)}.\frac{c_1}{1+x^2}dx + c_2$$

$$y\sqrt{1+x^2} = \int \sqrt{1+x^2}.\frac{c_1}{1+x^2}dx + c_2$$

$$y\sqrt{1+x^2} = c_1\int \frac{dx}{\sqrt{1+x^2}} + c_2$$

$$y\sqrt{1+x^2} = c_1.\log(x+\sqrt{1+x^2}) + c_2$$

$$y = \frac{c_1}{\sqrt{1+x^2}}.\log(x+\sqrt{1+x^2}) + \frac{c_2}{\sqrt{1+x^2}}$$

Example 5.4.14 Solve $x^2y'' + 3xy' + y = \frac{1}{(1-x)^2}$, after testing whether it is exact.

Solution: Comparing the given differential equation with $P_0y'' + P_1y' + P_2y = \Phi$, we have, $P_0 = x^2$, $P_1 = 3x$, $P_2 = 1$ and $\Phi = \frac{1}{(1-x)^2}$.

Now $P_0'' - P_1' + P_2 = 2 - 3 + 1 = 0$

Hence the given equation is exact.

The first order differential equation is $P_0y' + (P_1 - P_0')y = \int \Phi dx + c_1$

$$x^2y' + (3x - 2x)y = \int \frac{1}{(1-x)^2}dx + c_1$$

$$x^2y' + x.y = \frac{1}{1-x} + c_1$$

$$y' + \frac{1}{x}y = \frac{1}{x^2(1-x)} + \frac{c_1}{x^2}$$

Solving,

$$y.e^{\int \frac{1}{x}dx} = \int e^{\int \frac{1}{x}dx}\left[\frac{1}{x^2(1-x)} + \frac{c_1}{x^2}\right]dx$$

$$y.x = \int x \cdot \left[\frac{1}{x^2(1-x)} + \frac{c_1}{x^2} \right] dx$$

$$= \int \left[\frac{x}{x^2(1-x)} + \frac{c_1}{x} \right] dx$$

$$= \int \left[\frac{1}{x} + \frac{1}{1-x} + \frac{c_1}{x} \right] dx$$

$$= \log x - \log(1-x) + c_1 \log x + c_2$$

$$\therefore \qquad y = \frac{1}{x} \left[\log \frac{x}{1-x} + c_1 \log x + c_2 \right]$$

EXERCISE 5.4

PART – A

1. If $y = u$ and $y = uv$ are solutions of $y'' + Py' + Qy = 0$, write down the second order equation satisfied by v.
2. If $y = u$ and $y = uv$ are solutions of $y'' + Py' + Qy = 0$, write down the first order equation satisfied by v.
3. When is a second order linear differential equation said to be in normal form or canonical form?
4. Write down the transformation that transforms $y'' + Py' + Qy = \Phi$ into the canonical form.
5. When $y'' + Py' + Qy = \Phi$ is transformed into $v'' + Q_1.v = X_1$ by the transformation $y = v.e^{-\frac{1}{2}\int P\, dx}$, what are the values of Q_1 and X_1?
6. Solve $v'' - v = 0$.
7. Solve $v'' + v = 0$.
8. What is the condition for the equation $y'' + Py' + Qy = \Phi$ to be exact?
9. If $y'' + Py' + Qy = \Phi$ is exact, what is the first integral?
10. What is the condition on $y'' + Py' + Qy = 0$ to have $y = e^x$ as a solution?
11. What is the condition on $y'' + Py' + Qy = 0$ to have $y = x$ as a solution?
12. What is the condition on $y'' + Py' + Qy = 0$ to have $y = e^{mx}$ as a solution?
13. What is the condition on $y'' + Py' + Qy = 0$ to have $y = x^m$ as a solution?
14. Is $(ax - bx)^2 \frac{d^2y}{dx^2} + 2a \frac{dy}{dx} + 2by = x$ an exact differential equation?
15. Show that $x^2y'' + 3y' - 2y = x$ is exact and find the first integral.

PART-B

Solve the following differential equations by the method of reduction of order.

16. $x^2y'' - (x^2 + 2x)y' + (x + 2)y = x^3e^x$

17. $xy'' - (2x - 1)y' + (x - 1)y = 0.$

18. $xy'' - (2x + 1)y' + (x + 1)y = (x^2 + x - 1).e^{2x}.$

19. $(x + 2)y'' - (2x + 5)y' + 2y = (x + 1)e^x.$

20. $(x \sin x + \cos x)y'' - x \cos x.y' + y \cos x = 0.$

21. $x^2(x + 1).y'' - x(2 + 4x + x^2)y' + (2 + 4x + x^2)y = -x^4 - 2x^3.$

22. $(x - 2)y'' - (4x - 7)y' + (4x - 6)y = 0.$

23. $y'' - 2 \tan x.y' + 3y = 2 \sec x.$

24. $[(x + 3)D^2 - (2x + 7)D + 2]y = (x + 3)^2.e^x.$

25. $(D^2 - 4xD + 4x^2)y = x.e^{x^2}$

26. $[(x^2 - 2x)D^2 - (x^2 - 2)D + 2(x - 1)]y = 0.$

27. $(x - 1)y'' - xy' + y = 4x - 4 - 2x^2.$

28. $(x + 1)y'' - 2(x + 3)y' + (x + 5)y = e^x.$

29. $y'' - \cot x.y' - (1 - \cot x)y = e^x \sin x.$

30. $x^2y'' - 2x(1 + x)y' + 2(1 + x)y = x^3$

31. $(x^2 \log x - x^2)y'' - xy' + y = 0.$

32. $xy'' - y' - 4x^3y = -4x^5$ given $y = e^{x^2}$ is a solution of the corresponding homogeneous equation.

33. $(\sin x - x \cos x) y'' - x \sin x.y' + y \sin x = 0$ given $y = \sin x$ is a solution.

34. $y'' + y = \sec x$, given $y = \cos x$ is a solution of the corresponding homogeneous equation.

35. $x(1 + 3x^2)y'' + 2y' - 6xy = 0$, given $y = 1/x$ is a solution.

36. $(x^2 - 1)y'' - 6y = 1$, given that $y = x(1 - x^2)$ is a solution of the corresponding homogeneous equation.

Reduce the following equations to the canonical form and hence solve them.

37. $4x^2y'' + 4xy' + (x^2 - 1)y = 0$

38. $x^2y'' + xy' + \left(x^2 - \frac{1}{4}\right)y = 0$

39. $4x^2y'' + 4xy' + (16x^2 - 1)y = 4x^{3/2} \sin x$

40. $y'' - 4xy' + (4x^2 - 1)y = -3 e^{x^2} \sin 2x.$

41. $y'' - \frac{2}{x}y' + \left(1 + \frac{2}{x^2}\right)y = 0$

42. $x^2y'' - 2x(3x - 2)y' + 3x(3x - 4)y = e^{3x.}$

43. $y'' + 2n \cot nx.y' + (a^2 - n^2)y = 0.$

44. $xy'' + 2y' + xy = \sin 2x.$

45. $y'' + 4xy' + 4x^2y = 0.$

Show that the following equations are exact and hence solve them.

46. $(1 - x^2)y'' - 3xy' - y = 1$.
47. $xy'' + (x + 2)y' + y = e^{-x}$.
48. $x(2x + 3)y'' + (6x + 3)y' + 2y = (x + 1)e^x$.
49. $\sin x\, y'' - \cos x\, y' + 2y \sin x = 0$.
50. $x(a - bx)y'' + 2ay' + 2by = x$.

5.5 METHOD OF VARIATION OF PARAMETERS

The method of variation of parameters is used to find the general solution of the second order differential equation

$$\frac{d^2y}{dx^2} + P.\frac{dy}{dx} + Q.y = \Phi \qquad (1)$$

where P, Q, Φ are functions of x.

Let the C.F $= c_1y_1 + c_2y_2$ where y_1 and y_2 are two independent solutions of the corresponding homogeneous equation

$$\frac{d^2y}{dx^2} + P.\frac{dy}{dx} + Q.y = 0 \qquad (2)$$

i.e., $y_1'' + P y_1' + Q y_1 = 0, \qquad y_2'' + P y_2' + Q y_2 = 0 \qquad (3)$

In order to find the general solutions of (1), replace c_1 and c_2 by unknown functions u(x) and v(x). Let a particular integral of (1) be

$$y = u\, y_1 + v\, y_2 \qquad (4)$$

Then $y' = u'y_1 + u\, y_1' + v'y_2 + v\, y_2'$

Assuming

$$u'y_1 + v'y_2 = 0 \qquad (5)$$

we have, $y' = u\, y_1' + v\, y_2' \qquad (6)$

Differentiating (6) w.r.t x, we get,

$$y'' = u\, y_1'' + u'\, y_1' + v\, y_2'' + v'\, y_2'$$

Substituting for y, y', y'' in (1), we get,

$$u\, y_1'' + u'\, y_1' + v\, y_2'' + v'\, y_2' + P(u\, y_1' + v\, y_2') + Q(u\, y_1 + v\, y_2) = \Phi$$

i.e., $u\, (y_1'' + P y_1' + Q y_1) + v\, (y_2'' + P y_2' + Q y_2) + u'\, y_1' + v'\, y_2' = \Phi$

i.e., $u'\, y_1' + v'\, y_2' = \Phi \qquad (7)$

Solving for u' and v' from (5) and (7), we get,

$$u' = \frac{-y_2\Phi}{y_1y_2' - y_1'y_2}, \quad v' = \frac{y_1\Phi}{y_1y_2' - y_1'y_2}$$

Integrating $u = -\int \frac{y_2\Phi}{y_1y_2' - y_1'y_2}dx, \quad v = \int \frac{y_1\Phi}{y_1y_2' - y_1'y_2}dx$

i.e., $\quad u = -\int \dfrac{y_2 \Phi}{W} dx, \ v = \int \dfrac{y_1 \Phi}{W} dx \quad$ where $W = y_1 y_2' - y_1' y_2$ $\qquad$ (8)

Substituting these values of u and v in (4), we get, the $P.I = uy_1 + vy_2$.
Hence the general solution of the given differential equation (1) is
$y = c_1 y_1 + c_2 y_2 + uy_1 + vy_2$ where c_1 and c_2 are arbitrary constants.

Note:
(i) Since the solutions y_1 and y_2 are independent, the Wronskian of y_1 and y_2 does

not vanish. i.e., $W = \begin{vmatrix} y_1 & y_2 \\ y_1' & y_2' \end{vmatrix} = y_1 y_2' - y_1' y_2 \neq 0$

(ii) Since the complete solution is obtained from the C.F. = $c_1 y_1 + c_2 y_2$, by treating
c_1 and c_2 as parameters or functions of x, this method is known as the **method of
variation of parameters.**

Example 5.5.1 Solve by the method of variation of parameters $\dfrac{d^2 y}{dx^2} + y = \text{cosec } x.$

Solution: The given equation is
$\qquad (D^2 + 1)y = \text{cosec } x$ $\qquad\qquad$ (1)
A.E is $m^2 + 1 = 0.$
$\therefore \qquad m = \pm i.$
$\qquad$ C.F. = $c_1 \cos x + c_2 \sin x$ $\qquad\qquad$ (2)
Let $\quad y = u \cos x + v \sin x$ be a P.I of (1)

Method 1. Then $y' = u (-\sin x) + u' \cos x + v \cos x + v' \sin x$
$\qquad\qquad\qquad = -u \sin x + v \cos x$ where $u' \cos x + v' \sin x = 0$ $\qquad$ (3)
Then $\quad y'' = -u'\sin x - u \cos x + v' \cos x - v \sin x$
Substituting for y, y', y" in (1), we get $y'' + y = \text{cosec } x.$
i.e., $\quad -u'\sin x + v' \cos x - u \cos x - v \sin x + u \cos x + v \sin x = \text{cosec } x$
i.e., $\quad -u'\sin x + v' \cos x = \text{cosec } x.$ $\qquad\qquad$ (4)
Solving (3) and (4), we get $u' = -1, v' = \cot x.$
Integrating $u = \int (-1) dx = -x$ and $v = \int \cot x \, dx = \log \sin x$
$\therefore \qquad\qquad P.I = -x \cos x + (\log \sin x) \sin x$
The complete solution is $y = c_1 \cos x + c_2 \sin x - x \cos x + (\log \sin x) \sin x$

Method 2. Let $y_1 = \cos x, y_2 = \sin x$ and $\Phi = \text{cosec } x$
Then $\quad W = y_1 y_2' - y_1' y_2 = \cos x(\cos x) - (-\sin x) \sin x$
$\qquad\qquad\qquad = 1$
$\therefore \qquad u = -\int \dfrac{y_2 \Phi}{W} dx = -\int \dfrac{\sin x}{1} \cdot \text{cosec } x \, dx = -\int 1.dx = -x$ (By using (8))

Also, $v = \int \dfrac{y_1 \Phi}{W} dx = \int \dfrac{\cos x}{1}.\csc x\, dx = \int \cot x\, dx = \log \sin x$

$\therefore$ P.I $= uy_1 + vy_2 = -x \cos x + (\log \sin x) \sin x.$

The general solution is $y = c_1 \cos x + c_2 \sin x - x \cos x + (\log \sin x) \sin x$

Example 5.5.2 Solve by the method of variation of parameters, $\dfrac{d^2 y}{dx^2} - y = \dfrac{2}{1 + e^x}$.

Solution: The given equation is $y'' - y = \dfrac{2}{1 + e^x}$

A.E. is $m^2 - 1 = 0$

$\qquad m = 1, -1.$

$\therefore \qquad$ C.F. $= c_1 e^x + c_2 e^{-x}$

Let $y_1 = e^x,\ y_2 = e^{-x},\ \Phi = \dfrac{2}{1 + e^x}$

Let the P.I $= ue^x + ve^{-x}$

$\qquad W = y_1 y_2' - y_1' y_2 = e^x(-e^{-x}) - e^x e^{-x} = -2.$

$\therefore\ u = -\int \dfrac{y_2 \Phi}{W} dx = -\int \dfrac{e^{-x}.2}{(-2)(1 + e^x)}.dx$

$\qquad\qquad = \int \dfrac{e^{-x}}{1 + e^x}.dx$

$\qquad\qquad = \int \dfrac{1}{z^2(1 + z)}.dz \qquad\qquad \left(\text{Put } e^x = z,\ dx = \dfrac{dz}{z}\right)$

$\qquad\qquad = \int \left(\dfrac{1}{z^2} - \dfrac{1}{z} + \dfrac{1}{1 + z}\right).dz$

$\qquad\qquad = -\dfrac{1}{z} - \log z + \log(1 + z)$

$\qquad\qquad = -\dfrac{1}{z} + \log\left(\dfrac{1 + z}{z}\right)$

$\qquad\qquad = -e^{-x} + \log\left(\dfrac{1 + e^x}{e^x}\right)$

$v = \int \dfrac{y_1 \Phi}{W} dx = \int \dfrac{e^x.2}{(-2)(1 + e^x)}.dx$

$\qquad\qquad = -\log(1 + e^x)$

$\therefore\ \text{P.I} = \left[-e^{-x} + \log\left(\dfrac{1 + e^x}{e^x}\right)\right] e^x - \log(1 + e^x).e^{-x}$

Complete solution is $y = c_1 e^x + c_2 e^{-x} - 1 + e^x \log\left(\dfrac{1+e^x}{e^x}\right) - e^{-x} \log(1+e^x)$

Example 5.5.3 Solve by the method of variation of parameters, $x^2 y'' + x.y' - y = x^2 e^x$.

Solution: The given differential equation is

$$y'' + \frac{1}{x} y' - \frac{1}{x^2} y = e^x \tag{1}$$

$$P = \frac{1}{x}, \quad Q = -\frac{1}{x^2}$$

$\therefore \quad P + Qx = \dfrac{1}{x} - \dfrac{1}{x} = 0.$

$\therefore \quad y = x$ is a solution of the corresponding homogeneous equation.

$$y'' + \frac{1}{x} y' - \frac{1}{x^2} y = 0 \tag{2}$$

Let the C.F be $y = vx$.

Then, $y' = v + x.v'$

$\qquad y'' = 2v' + xv''$

Substituting for y, y', y'' in equation (2), $2v' + xv'' + \dfrac{1}{x}(v + x.v') - \dfrac{1}{x^2} vx = 0$

i.e., $\quad xv'' + 3v' = 0$

i.e., $\quad \dfrac{v''}{v'} = -\dfrac{3}{x}$

Integrating $\log v' = -3\log x + \log c_1$

$\therefore \qquad v' = c_1 x^{-3}$

i.e., $\quad \dfrac{dv}{dx} = c_1 x^{-3}$

Again integrating, $v = \int c_1 x^{-3} dx = c_1\left(-\dfrac{1}{2x^2}\right) + c_2$

i.e., $\quad v = -\dfrac{c_1}{2x^2} + c_2$

$\therefore \qquad$ C.F $= vx = -\dfrac{c_1}{2x} + c_2 x \tag{3}$

Let $\quad y_1 = \dfrac{-1}{2x}, \; y_2 = x, \; \Phi = e^x$

Then $W = y_1 y_2' - y_1' y_2 = \dfrac{-1}{2x}.1 - \dfrac{1}{2x^2}.x = -\dfrac{1}{2x} - \dfrac{1}{2x} = -\dfrac{1}{x}$

If the P.E $= Uy_1 + Vy_2$

$$U = -\int \frac{y_2 \Phi}{W} dx = -\int \frac{x.e^x}{(-\frac{1}{x})}.dx = \int x^2.e^x dx$$

$$= x^2 e^x - 2xe^x + 2e^x$$

$$V = \int \frac{y_1 \Phi}{W} dx = \int \frac{-\frac{1}{2x}.(e^x)}{(-\frac{1}{x})}.dx = \frac{1}{2}e^x$$

$$\therefore \text{P.I} = e^x(x^2 - 2x + 2)\left(-\frac{1}{2x}\right) + \frac{1}{2}e^x.x = -e^x.\frac{x}{2} + e^x - \frac{e^x}{x} + e^x\frac{x}{2} \qquad (4)$$

From (3) and (4), we get the complete solution of the given equation (1) as

$$y = -\frac{c_1}{2x} + c_2 x - e^x.\frac{x}{2} + e^x - \frac{e^x}{x} + e^x\frac{x}{2}$$

$$= -\frac{c_1}{2x} + c_2 x + e^x - \frac{e^x}{x}.$$

Example 5.5.4 Solve $y'' + y = \tan x$ by the method of variation of parameters.

Solution: The given equation is
$$y'' + y = \tan x \qquad (1)$$
The A.E. is $m^2 + 1 = 0$
i.e., $\quad m = i, -i.$
The C.F. $= c_1 \cos x + c_2 \sin x$
Let $\quad y_1 = \cos x, y_2 = \sin x$ and $\Phi = \tan x$
If the P.E $= uy_1 + vy_2$

$$u = -\int \frac{y_2 \Phi}{W} dx, \quad v = \int \frac{y_1 \Phi}{W} dx$$

where $W = y_1 y_2' - y_1' y_2 = \cos x(\cos x) - (-\sin x) \sin x = \cos^2 x + \sin^2 x = 1$

i.e., $u = -\int \frac{\sin x.\tan x}{1} dx = -\int \frac{\sin^2 x}{\cos x} dx = -\int (\sec x - \cos x) dx$

$$= -\log (\sec x + \tan x) + \sin x$$

$$v = \int \cos x.\tan x \, dx = \int \sin x \, dx = -\cos x$$

$\therefore$ P.I $= u \cos x + v \sin x$

$$= -\cos x \log (\sec x + \tan x) + \cos x.\sin x - \cos x. \sin x$$

$$= -\cos x \log (\sec x + \tan x)$$

The complete solution is $y = \text{C.F} + \text{P.I} = c_1 \cos x + c_2 \sin x - \cos x \log (\sec x + \tan x)$

Example 5.5.5 Solve using the method of variation of parameters, $y'' + 7y' - 8y = e^{2x}$.

Solution: A.E is $m^2 + 7m - 8 = 0$

i.e., $(m + 8)(m - 1) = 0$

i.e., $m = 1, -8$.

$C.F = c_1 e^x + c_2 e^{-8x}$ ... (1)

Let a particular integral of the given equation be $y = u.e^x + v.e^{-8x}$ ... (2)

Let $y_1 = e^x$, $y_2 = e^{-8x}$ and $\Phi = e^{2x}$

Then $W = y_1 y_2' - y_1' y_2 = e^x e^{-8x}(-8) - e^x e^{-8x} = -9e^{-7x}$

$$\therefore \ u = -\int \frac{y_2 \Phi}{W} dx = -\int \frac{e^{-8x}.e^{2x}}{-9e^{-7x}}.dx = \frac{1}{9}\int e^x.dx = \frac{e^x}{9}$$

$$\therefore \ v = \int \frac{y_1 \Phi}{W} dx = \int \frac{e^x.e^{2x}}{-9e^{-7x}}.dx = -\frac{1}{9}\int e^{10x}.dx = -\frac{e^{10x}}{90}$$

$$\therefore \ \text{From (2) P.I} = \frac{e^x}{9}e^x - \frac{e^{10x}}{90}e^{-8x} = \frac{e^{2x}}{9} - \frac{e^{2x}}{90} = \frac{e^{2x}}{10}$$

The complete solution is $y = C.F + P.I$

i.e., $\quad y = c_1 e^x + c_2 e^{-8x} + \dfrac{e^{2x}}{10}$

Example 5.5.6 Solve by the method of variation of parameters, $x^2 y'' - 2x(1 + x)y' + 2(1 + x)y = x^3$.

Solution: The given equation is $y'' - \dfrac{2}{x}(1 + x)y' + \dfrac{2(1 + x)}{x^2}y = x$

$$P + Qx = -\frac{2}{x}(1 + x) + \frac{2(1 + x)}{x^2}.x = 0$$

$\therefore \quad y = x$ is a solution of corresponding homogeneous equation.

Let the C.F be $y = vx$

Then $\quad y' = v'.x + v$

$\qquad\quad y'' = v''.x + 2v'$

Substituting for y, y', y'' in the homogeneous equation,

$y'' - \dfrac{2}{x}(1 + x)y' + \dfrac{2(1 + x)}{x^2}y = 0$, we get,

$v''x + 2v' - \dfrac{2}{x}(1 + x)(v'x + v) + \dfrac{2(1 + x)}{x^2}vx = 0$

i.e., $\quad v''x + 2v' - 2(1 + x).v' = 0$

i.e., $\quad v'' - 2v' = 0$

i.e., $\quad (D^2 - 2D)v = 0$

A.E. is $m^2 - 2m = 0$

$\qquad\quad m = 0, 2$.

$\therefore \qquad v = c_1 + c_2 e^{2x}$.

$\therefore \qquad C.F = vx = c_1 x + c_2 xe^{2x}$.

Then, $y_1 = x$, $y_2 = xe^{2x}$ and $\Phi = x$

$W = y_1 y_2' - y_1' y_2$

$\quad = x[x.e^{2x} 2 + e^{2x}] - 1.x.e^{2x} = 2x^2 e^{2x}$

If a P.E $= ux + v.xe^{2x}$, then,

$$u = -\int \frac{y_2 \Phi}{W} dx = -\int \frac{x.e^{2x}.x}{2x^2 e^{2x}}.dx = -\frac{1}{2}\int 1.dx = -\frac{1}{2}x$$

$$v = \int \frac{y_1 \Phi}{W} dx = \int \frac{x..x}{2x^2 e^{2x}}.dx = \frac{1}{2}\int e^{-2x}.dx = -\frac{e^{-2x}}{4}$$

$$\therefore \ P.E = \left(-\frac{1}{2}x\right).x + \left(\frac{e^{-2x}}{-4}\right).x.e^{2x} = -\frac{x^2}{2} - \frac{x}{4}$$

The complete integral is $y = C.F + P.I$

i.e., $y = c_1 .x + c_2 xe^{-8x} - \dfrac{x^2}{2} - \dfrac{x}{4}$

Example 5.5.7 Solve by the method of variation of parameters,
$x^2 y'' - xy' + y = x \log x$.

Solution: The given equation is

$$y'' - \frac{1}{x}y' + \frac{1}{x^2}y = \frac{1}{x}\log x \qquad\qquad (1)$$

The corresponding homogeneous equation is $y'' - \dfrac{1}{x}y' + \dfrac{1}{x^2}y = 0$ \qquad (2)

$$P + Qx = -\frac{1}{x} + \frac{1}{x^2}.x = 0$$

$\therefore$ \qquad $y = x$ is a solution of (2)

If the   C.F is $y = vx$, then,

$\qquad y' = v'.x + v$ and $y'' = v''.x + 2v'$

Substituting for y, y', y'' in the homogeneous equation,

$$v''x + 2v' - \frac{1}{x}(v'x + v) + \frac{1}{x^2}vx = 0$$

i.e., $v''x + v' = 0$

i.e., $x^2 v'' + xv' = 0$

Put $x = e^t$ or $t = \log x$

Then $(\theta(\theta - 1) + \theta) v = 0$ where $\theta \equiv \dfrac{d}{dt}$

i.e., $\theta^2 .v = 0$

$\therefore$ \qquad $v = c_1 t + c_2$

i.e., $v = c_1 \log x + c_2$.

C.F = vx

$\qquad$ = c_1.x log x + c_2.x $\qquad\qquad\qquad\qquad$ (3)

Let $\quad$ y_1 = x log x, y_2 = x and Φ = $\dfrac{1}{x}$ log x.

Then $\quad$ W = $y_1 y_2' - y_1' y_2$

$\qquad\qquad$ = x log x.1 - (x.$\dfrac{1}{x}$ + log x).x = -x.

If a P.I is y = u. y_1 + v. y_2

$\qquad\qquad\qquad$ = u.x log x + v.x, then,

$$u = -\int \frac{y_2 \Phi}{W} dx = -\int \frac{x.\log x}{(-x).x}.dx = \int \frac{\log x}{x}.dx = \frac{(\log x)^2}{2}$$

$$v = \int \frac{y_1 \Phi}{W} dx = \int \frac{x.\log x}{(-x)}.\frac{\log x}{x} dx = -\int \frac{(\log x)^2}{x} dx = -\frac{(\log x)^3}{3}$$

$\therefore$ P.E $= \dfrac{(\log x)^2}{2}$.x log x $- \dfrac{x}{3}(\log x)^3 = \dfrac{1}{6}$x.$(\log x)^3$

The complete solution is y = C.F + P.I

i.e., $\quad$ y = c_1 .x log x + c_2 x $+\dfrac{1}{6}$x.$(\log x)^3$

EXERCISE 5.5

PART-A
1. Explain the method of 'variation of parameters'
2. If y = u.y_1 + v.y_2 is a solution of y" + Py' + Qy = Φ, where y_1, y_2 are two independent solutions of y" + Py' + Qy = 0, write down the integrals giving u and v.

PART-B
Solve the following differential equations by the method of variation of parameters.

3. y" + 4y = 4 tan 2x.
4. y" + (1 - cot x)y' - y cot x = $\sin^2$ x.
5. y"- 2y' + y = e^x log x.
6. $(D^2 + 2D + 5)$y = e^{-x} tan x.
7. $(1 - x)D^2 y + xDy - y = (1 - x)^2$
8. $(D^2 -1)$y = e^{-2x} sin (e^{-x}).
9. $(D^2 +1)$y = x sin x.
10. $(D^2 - 6D + 9)$y = $\dfrac{e^{3x}}{x^2}$.

11. $(x^2D^2 + 4xD + 2)y = x^2 + \dfrac{1}{x^2}$.

12. $(x^2D^2 - 2xD - 4)y = 32 (\log x)^2$

13. $(D^2 + a^2)y = \sec ax$.

14. $(x + 2)\dfrac{d^2y}{dx^2} - (2x + 5)\dfrac{dy}{dx} + 2y = (1 + x).e^x$.

15. Verify that e^x and x are linearly independent solutions of the homogeneous equation corresponding to $(1 - x) y_2 + xy_1 - x = 2(x - 1)^2.e^{-x}$, $0 < x < 1$. Hence find the general solution by the method of variation of parameters.

16. Solve $x^2y'' - x(x + 2)y' + (x + 2)y = x^3$, given that $y = x$ and $y = xe^x$ are two linearly independent function of C.F.

17. Use the method of variation of parameters to solve $(x^2 + 1) y'' - 2xy' + 2y = 6(x^2 + 1)^2$.

ANSWERS

EXERCISE 5.1

(2) Order = 2, Degree = 2

(5) $y = c.e^{ax}$

(6) $y = e^{ax} \int Q.e^{-ax}dx$

(7) $(c_1 + c_2x) e^{2x}$

(8) $(c_1 + c_2x)\cos x + (c_3 + c_4x)\sin x$.

(9) $c_1e^{-2x} + e^x(c_2 \cos \sqrt{3}x + c_3 \sin \sqrt{3}x)$

(10) $y = c_1 + c_2x + (c_3 + c_4x)e^x$

(11) $y = (c_1 + c_2x)\cos x + (c_3 + c_4x)\sin x$.

(12) $y = c_1e^x + c_2e^{-x} + c_3\cos x + c_4\sin x$.

(13) $x^2 - 2x$

(14) $\dfrac{-1}{8}x(2x^2 + 3)$

(15) $c_1\sin 2x + c_2 \cos 2x - \dfrac{x}{2}\cos 2x$

(16) $y = c_1 + c_2e^x + c_3e^{-x} + \dfrac{x}{2}(e^x + e^{-x})$.

(17) $y = (c_1 + c_2x)e^x + (c_3 + c_4x)e^{-x} + 10x^2\cosh x$.

(18) $y = c_1e^x + c_2e^{2x} - \dfrac{1}{130}(-9 \cos 3x + 7 \sin 3x)$

(19) $y = c_1e^x + c_2e^{3x} + \dfrac{1}{884}(10 \cos 5x - 11 \sin 5x) + \dfrac{1}{20}(\sin x + 2 \cos x)$.

(20) $y = c_1 \cos 4x + c_2 \sin 4x + \dfrac{2}{7}e^{-3x} + \dfrac{x}{8}\sin 4x$.

(21) $y = c_1e^{-x} + c_2e^{-2x} + c_3e^{-3x} - \dfrac{1}{5}\cos x$.

(22) $y = c_1e^x + c_2e^{-x} + c_3\cos x + c_4\sin x + x(e^x + \cos x)$

(23) $y = e^{-x/2}\left(c_1 \cos \frac{\sqrt{3}}{2}x + c_2 \sin \frac{\sqrt{3}}{2}x\right) + x^2 - 2x$.

(24) $y = c_1 e^x + c_2 e^{-2x} - \frac{1}{32}(8x^2 - 4x + 1)$.

(25) $y = c_1 + c_2 e^{-2x} + c_3 e^{3x} - \frac{1}{18}x^3 + \frac{1}{36}x^2 - \frac{25}{108}x$

(26) $y = c_1 + c_2 x + c_3 \cos 2x + c_4 \sin 2x + 2x^4 - 6x^2 + 3$.

(27) $y = c_1 e^{-x} + e^{x/2}\left(c_2 \cos \frac{\sqrt{3}}{2}x + c_4 \sin \frac{\sqrt{3}}{2}x\right) + \frac{e^{2x}}{130}(3 \sin x - 11 \cos x)$.

(28) $y = c_1 e^{-x} + e^x(c_2 \cos 2x + c_3 \sin 2x) - \frac{x}{16}e^x(\cos 2x - \sin 2x)$.

(29) $y = (c_1 + c_2 x)e^x + \frac{1}{2}\cos x + \frac{e^x x^4}{12}$.

(30) $y = c_1 + c_2 x + (c_3 + c_4 x)e^x + \frac{x^5}{20} + \frac{x^4}{2} + 3x^3 + 12x^2$.

(31) $y = (c_1 + c_2 x)\cos 2x + (c_3 + c_4 x)\sin 2x + \frac{1}{16}(16x + 10)$.

(32) $y = (c_1 + c_2 x)e^{-2x} - \frac{1}{25}e^{-x}(4 \cos 2x + 3 \sin 2x)$.

(33) $y = c_1 \cos 3x + c_2 \sin 3x + \frac{e^{3x}}{18}\left(x^2 - \frac{2}{3}x + \frac{10}{9}\right)$

(34) $y = c_1 e^{-x} + c_2 e^{-3x} + \frac{e^x}{40}(3 \sin 2x + \cos 2x) - \frac{1}{30}(2 \sin 3x - \cos 3x)$

(35) $y = (c_1 + c_2 x)e^x + \frac{e^{3x}}{4}\left(x^2 - 2x + \frac{3}{2}\right)$.

(36) $y = c_1 e^x + c_2 e^{-x} + c_3 \cos x + c_4 \sin x - \frac{e^x}{5}\cos x$.

(37) $y = (c_1 + c_2 x)e^x + \frac{1}{2}x \cos x + \frac{1}{2}\cos x - \frac{1}{2}\sin x$

(38) $y = c_1 e^x + c_2 e^{-x} + x \sin x + \frac{1}{2}(1 - x^2)\cos x$.

(39) $y = e^x(c_1 + c_2 x - x \sin x - 2 \cos x)$.

(40) $y = c_1 + (c_2 + c_3 x)e^{-x} + \frac{e^{2x}}{18}\left(x^2 - \frac{7}{8}x + \frac{11}{6}\right) + \frac{1}{100}(3 \sin 2x + 4 \cos 2x)$.

(41) $y = c_1 \cos x + c_2 \sin x - \frac{1}{27}(24x \cos 2x + (9x^2 - 26)\sin 2x)$.

(42) $y = \dfrac{1}{5}e^x + \dfrac{1}{20}e^{6x} - \dfrac{1}{4}e^{2x}.$

(43) $y = 2e^{-x} - 8e^{-5x} + \dfrac{1}{2}e^{3x} - \dfrac{5}{3}e^{-2x}$

(44) $\dot{x} = c_1 \cos wt + c_2 \sin wt + \dfrac{at}{2w}\sin wt.$

(45) $x = c_1 \cos nt + c_2 \sin nt + \dfrac{E}{n^2 - p^2}\cos pt,$

   $x = c_1 \cos nt + c_2 \sin nt + \dfrac{E.t}{2n}\sin nt.$

(46) $y = c_1 \cos ax + c_2 \sin ax - \dfrac{x}{a}\sin ax - \dfrac{1}{a^2}\cos ax.\ \log \sec ax.$

EXERCISE 5.2

1) 2   (2) 2   (3) 3   (4) $x = A \cos t + B \sin t + 2$

(5) $x = (c_1 + c_2 t)e^{-t} + (c_3 + c_4 t)e^{\,t}$ (6) 1

(7) $x = c_1 e^t - c_2 e^{-5t},\ y = c_1 e^t + c_2 e^{-5t}$

(8) $x = c_2 \cos 2t - c_1 \sin 2t - \cos t,\ y = c_1 \cos 2t + c_2 \sin 2t - \sin t$

(9) $x = -c_1 e^{-5t} + c_2 e^t + \dfrac{3}{7}e^{2t} - \dfrac{2}{5}t - \dfrac{13}{25},$

   $y = c_1 e^{-5t} + c_2 e^t + \dfrac{4}{7}e^{2t} - \dfrac{3}{5}\left(t + \dfrac{4}{5}\right)$

(10) $x = \dfrac{-1}{27}(1 + 6t)e^{-3t} + \dfrac{1}{27}(1 + 3t),$

   $y = \dfrac{-2}{27}(2 + 3t)e^{-3t} + \dfrac{9}{27}(2 - 3t)$

(11) $x = e^t(c_1 \cos t + c_2 \sin t) - \dfrac{1}{2}\cos 2t,$

   $y = e^t(c_1 \sin t - c_2 \cos t) - \dfrac{1}{2}\sin 2t$

(12) $x = \dfrac{1}{2}\Big[(5+\alpha)c_1 e^{\alpha t} + (5+\beta)c_2 e^{\beta t}\Big] + \dfrac{1}{89}(41 \sin 2t - 30 \cos 2t)$

   $y = c_1 e^{\alpha t} + c_2 e^{\beta t} - \dfrac{1}{89}(8 \cos 2t - 5 \sin 2t)$ where $\alpha = -4 + \sqrt{2},\ \beta = -4 - \sqrt{2}$

(13) $x = c_1 e^{-t} + c_2 e^{-3t},\ y = c_1 e^{-t} + 3c_2 e^{-3t} + \cos t.$

(14) $x = c_1 e^{\,t} + c_2 e^{-t},\ y = -c_1 e^{\,t} + c_2 e^{-t} + 1 + \sin t.$

(15) $x = c_1 e^{\sqrt{2}t} + c_2 e^{-\sqrt{2}t} + 3\cos t$,

$\quad y = c_1(\sqrt{2}+1)e^{\sqrt{2}t} + c_2(1-\sqrt{2})e^{-\sqrt{2}t} + 2\sin t$

(16) $x = (c_1 + c_2 t)e^{3t}$, $\quad y = (1-2t)(c_2 - 2c_1)e^{3t}$

(17) $x = e^{6t}(c_1\cos t + c_2\sin t)$,

$\quad y = e^{6t}((c_1 - c_2)\cos t + (c_1 + c_2)\sin t)$

(18) $x = -t - \dfrac{2}{3}$, $\quad y = \dfrac{t^2}{2} + \dfrac{4}{3}t + c$

(19) $x = c_1 e^t + c_2 e^{-2t} + 2e^{-t}$, $\quad y = 3c_1 e^t + 2c_2 e^{-2t} + 3e^{-t}$

(20) $x = c_1\cos t + c_2\sin t + t^2 - 1$, $\quad y = -c_1\sin t + c_2\cos t + t$.

(21) $x = c_1 e^{-5t} + c_2 e^{-t/3} + \dfrac{1}{65}(8\sin t + \cos t)$,

$\quad y = \dfrac{-4}{3}c_1 e^{-5t} + c_2 e^{-t/3} + \dfrac{1}{130}(61\sin t - 33\cos t)$

(22) $x = e^{-4t}(c_1\cos t + c_2\sin t) + \dfrac{31}{26}e^t - \dfrac{93}{17}$,

$\quad y = e^{-4t}((c_1 - c_2)\sin t - (c_1 + c_2)\cos t) - \dfrac{2}{13}e^t + \dfrac{6}{17}$

(23) $x = c_1\cos 2t + c_2\sin 2t - t\sin 2t$, $\quad y = c_1\sin 2t - c_2\cos 2t + t\cos 2t$

(24) $x = c_1 e^{-2t} + c_2 e^t$, $\quad y = -\dfrac{3}{5}c_1 e^{-2t} + \left(-\dfrac{3}{4}c_2 + \dfrac{1}{4}\right)e^t$

(25) $y = e^x + c_1 + c_2 e^{-2x}$, $\quad z = e^x + c_1 - c_2 e^{-2x}$

(26) $x = \dfrac{a}{w+p}(\sin wt + \sin pt)$, $\quad y = \dfrac{a}{w+p}(\cos wt - \cos pt)$

(27) $x = \dfrac{1}{5}\left(e^{-4t} - e^t\right)$, $\quad y = \dfrac{-1}{10}\left(3e^t + 2e^{-4t}\right)$

(28) $x = (c_1 + c_3 t)\cos t + (c_2 + c_4 t)\sin t + \dfrac{e^t}{25}(4\sin t - 3\cos t)$

$\quad y = -(c_1 + c_3 t)\sin t + (c_2 + c_4 t)\cos t - \dfrac{e^t}{25}(3\sin t + 4\cos t)$

(29) $x = (c_1 + c_2 t)e^t + c_3 e^{-3t/2} - \dfrac{t}{2}$, $\quad y = (-2c_1 + 6c_2 - 2c_2 t)e^t - \dfrac{c_3}{3}e^{-3t/2} - \dfrac{1}{3}$

(30) $x = c_1 e^{2t} + c_2 e^{-2t} + c_3\cos 2t + c_4\sin 2t - \dfrac{4}{15}\sin t + \dfrac{3}{16}t$

$\quad y = \dfrac{c_1}{3}e^{2t} + \dfrac{c_2}{3}e^{-2t} + 3c_3\cos 2t + 3c_4\sin 2t - \dfrac{1}{5}\sin t + \dfrac{5}{16}t$

EXERCISE 5.3

(6) $y = x(A \log x + B)$

(7) $y = c_1 x^{-3} \cos(\log x^2 + c_2)$

(8) $y = x(c_1 + c_2 \log x) + c_3/x^2$

(9) $y = (c_1 + c_2 \log x)x^2 + x^2 (\log x)^2$

(10) $y = \dfrac{1}{4}x^2 - x^4 - 2x^3$

(11) $y = c_1 x^2 + \dfrac{c_2}{x} + \dfrac{1}{3}x^2 \log x - \dfrac{1}{3x} \log x.$

(12) $y = c_1 x^3 + \dfrac{c_2}{x} - \dfrac{1}{3}x^2 (\log x + \dfrac{2}{3}).$

(13) $y = x^2(c_1 \cos (\log x) + c_2 \sin (\log x) - \dfrac{x^2}{2} \log x \cos (\log x))$

(14) $y = x^m [c_1 \cos (n \log x) + c_2 \sin (n \log x) + x^m \log x]$

(15) $y = c_1 x + c_2 x^2 + x^3 - \log x(x +1) + \dfrac{x^2}{2}.$

(16) $y = (c_1 + c_2 \log x)x + c_3 x^2 + \dfrac{x^3}{4} - \dfrac{3}{2}x(\log x)^2$

(17) $y = x^{-3}(c_1 \cos (2 \log x) + c_2 \sin (2 \log x)) + \dfrac{1}{13} (\log x - \dfrac{6}{13})$

(18) $y = c_1 \cos (\log x) + c_2 \sin (\log x) - 2 \log x \cos (\log x)$

(19) $y = c_1 x^2 + c_2 x^3 + \dfrac{1}{x^4}$

(20) $y = c_1 + (c_2 + c_3 \log x)x + \dfrac{x}{2}(\log x)^2$

(21) $y = c_1 e^{m_1 t} + c_2 e^{m_2 t} - \dfrac{3}{14}e^t + \dfrac{3}{4}$, where $t = \log (1 + x)$.

(22) $y = c_1 t^2 + c_2 t^{-2} + \dfrac{1}{108}(t^2 \log t)$ where $t = \log (3x + 2)$.

(23) $y = c_1 \cos t + c_2 \sin t + 2t \sin t$ where $t = \log (1 + x)$.

(24) $y = c_1(x + 3)^2 + c_2(x + 3)^3 + \dfrac{1}{6}(\log (x + 3) + \dfrac{5}{6})$

(25) $y = c_1(x + a)^2 + c_2(x + a)^3 + \dfrac{1}{2}(x + a) - \dfrac{1}{6}a.$

EXERCISE 5.4

(1) $v'' + (2\dfrac{u'}{u} + P) v' = 0$

(2) $v' = c_1 u^{-2} e^{-\int P\,dx}$

(4) $y = v \cdot e^{-\frac{1}{2}\int P\,dx}$

(5) $Q_1 = Q - \dfrac{1}{4}P^2 - \dfrac{1}{2}P', X_1 = \Phi\, e^{\frac{1}{2}\int P\,dx}$

(6) $v = c_1 e^x + c_2 e^{-x}$

(7) $v = c_1\cos x + c_2 \sin x$

(8) $P_0'' - P_1' + P_2 = 0$

(9) $P_0 y' + (P_1 - P_0')y = \int \Phi \, dx + c_1$

(10) $1 + P + Q = 0$

(11) $P + Qx = 0$

(12) $m^2 + mP + Q = 0$

(13) $m(m-1) + Pmx + Qx^2 = 0$

(15) $x^2 y' + (3 - 2x)y = \dfrac{x^2}{2} + c_1$

(16) $y = (x^2 - x + c_1 x)e^x + c_2 x$

(17) $y = e^x(c_1 \log x + c_2)$

(18) $y = (xe^x + c_1 \cdot \dfrac{x^2}{2} + c_2)x$

(19) $y = \left[-e^{-x} - \dfrac{c_1}{4}(2x + 5) e^{-2x} + c_2 \right].e^{2x}$

(20) $y = -c_1 \cos x + c_2 x$

(21) $y = c_1 x^2 e^x + c_2 x + x^2$

(22) $y = c_1 e^{2x}(x - 2)^2 + c_2 e^{2x}$

(23) $y = c_1 \sin x.\cot 2x + c_2 \sin x + \sin x.\operatorname{cosec} 2x$

(24) $y = c_1 e^{2x} + c_2(2x + 7) - (x + 4)e^x$

(25) $y = e^{x^2}\left[c_1 \cos \sqrt{2}x + c_2 \sin \sqrt{2}x + \dfrac{x}{2} \right]$

(26) $y = c_1 x^2 + c_2 e^x$

(27) $y = c_1 e^x + c_2 x + 2x^2$

(28) $y = \left[-\dfrac{x}{4} + \dfrac{c_1}{5}(x + 1)^5 + c_2 \right].e^x$

(29) $y = -\dfrac{1}{2}\cos x.e^x - \dfrac{c_1}{5}(\cos x + 2\sin x).e^{-x} + c_2 e^x$

(30) $y = -\dfrac{1}{2}x^2 + \dfrac{c_1}{2} xe^{3x} + c_2 x$

(31) $y = c_1 x + c_2 \log x$

(32) $y = x^2 - \dfrac{c_1}{4} e^{-x^2} + c_2 e^{x^2}$

(33) $y = c_1 x + c_2 \sin x$

(34) $y = c_1 \cos x + c_2 \sin x + x \sin x + \cos x \log \cos x$

(35) $y = c_1(1 + x^2) + c_2\dfrac{1}{x}$

(36) $y = c_1(x - x^3) + c_2(4 - 3x - 6x^2 + 3x^3).\log\left(\dfrac{1+x}{1-x}\right) - \dfrac{1}{6}$

(37) $y = x^{-\frac{1}{2}}\left(c_1 \cos \tfrac{x}{2} + c_2 \sin \tfrac{x}{2}\right)$

(38) $y = x^{-\frac{1}{2}}\left(c_1 \cos x + c_2 \sin x\right)$

(39) $y = x^{-\frac{1}{2}}\left(c_1 \cos 2x + c_2 \sin 2x + \dfrac{1}{3}\sin x \right)$

(40) $y = (c_1\cos x + c_2\sin x + \sin 2x) e^{x^2}$ (41) $y = x(c_1\cos x + c_2\sin x)$

(42) $y = \left(c_1 x^2 + \dfrac{c_2}{x} + \dfrac{x^2}{3}\log x \right)\dfrac{e^{3x}}{x^2}$ (43) $y = (c_1 \cos ax + c_2 \sin ax)\,\text{cosec } nx$

(44) $y = (c_1\cos x + c_2\sin x - \dfrac{1}{3}\sin 2x)\dfrac{1}{x}$

(45) $y = e^{-x^2}\left(c_1 e^{\sqrt{2}x} + c_2 e^{-\sqrt{2}x} \right)$ (46) $y = c_1 \sin^{-1}\left(\dfrac{x}{\sqrt{1-x^2}} \right) + c_2 \dfrac{1}{\sqrt{1-x^2}} + 1$

(47) $y = (c_1 e^x + c_2 - x)x^{-1}e^{-x}$ (48) $y(3+2x) = e^x + c_1\log x + c_2$

(49) $y = \dfrac{c_1}{2}\left[\sin^2 x.\log \tan \dfrac{x}{2} - \cos x \right] + c_2 \sin^2 x$

(50) $y = \dfrac{x^2}{6a} + \dfrac{c_1}{3bx} + \dfrac{c_2}{x}(a - bx)^3$

EXERCISE 5.5

(3) $y = c_1 \cos 2x + c_2 \sin 2x + (\sin 2x - \log(\sec 2x + \tan 2x))\cos 2x - \cos 2x.\sin 2x.$

(4) $y = c_1(\cos x - \sin x) + c_2 e^{-x} - \dfrac{1}{10}(\sin 2x - 2\cos 2x).$

(5) $y = (c_1 x + c_2)e^x + \dfrac{x^2 e^x}{4}(2\log x - 3)$

(6) $y = (c_1\cos 2x + c_2\sin 2x)e^{-x} - \dfrac{x}{2} + \dfrac{\sin 2x}{4} - \dfrac{\cos 2x}{2} + \log(\cos x)$

(7) $y = c_1 e^x + c_2 x + 1 + x + x^2.$

(8) $y = c_1 e^x + c_2 e^{-x} - \sin(e^{-x}) - e^x\cos(e^{-x}).$

(9) $y = c_1\cos x + c_2\sin x + \dfrac{x}{2}\sin x - \dfrac{x^2}{4}\cos x$

(10) $y = (c_1 + c_2 x)e^{4x} - e^{3x}.\log x.$ (11) $y = \dfrac{c_1}{x} + \dfrac{c_2}{x^2} + \dfrac{x^2}{12} - \dfrac{\log x}{x^2}$

(12) $y = c_1.x^4 + c_2.\dfrac{1}{x} - 8(\log x)^2 + 12\log x + 13$

(13) $y = c_1\cos ax + c_2\sin ax + \dfrac{x}{a}\sin ax + \dfrac{1}{a^2}\cos ax.\log(\cos ax)$

(14) $y = c_1(2x + 5) + c_2 e^{2x} - e^x$ (15) $y = c_1 x + c_2 e^x + \left(\dfrac{1}{2} - x \right)e^{-x}$

(16) $y = c_1 x + c_2 x.e^x - x^2 - x$ (17) $y = c_1 x + c_2(x^2 - 1) + x^4 + 3x^2.$

INDEX

Adjoint of a matrix 1.5
Augmented matrix 1.10
Auxiliary equation 5.9

Canonical form 1.48, 1.40, 1.50
    Normal form 5.50
Cayley-Hamilton theorem 1.33
Characteristic equation 1.18
    Matrix 1.18
    Polynomial 1.18
    Roots 1.18
    Vectors 1.18
Coefficient matrix 1.10
Cofactors 1.5
Column vector 1.1
    Column matrix 1.1
Complementary function 5.3, 5.6
Conformable matrices 1.3
Consistent system 1.10
    Inconsistent system 1.10
Curvature 3.7
    Average curvature 3.7
    Center of curvature 3.12, 3.13
    Circle of curvature 3.12
    Curvature of circle 3.8
    Radius of curvature 3.9, 3.10, 3.14

Derivative 3.1
    Length of arc 3.3, 3.6
    Rate of change 3.2
Determinant of a matrix 1.2
Diagonalisation 1.34
Differential equations 5.1
    Degree 5.2

Order 5.2
Ordinary Differential Equations 5.1
Partial Diff. Equations 5.2
Variable Coefficients 5.46
Direction angles 2.4
Direction cosines 2.4
Direction ratios 2.6

Eigen values 1.18
    Vectors 1.18
Elementary transformations 1.6
    Column operations 1.6
    Row operations 1.6
Envelopes 3.33, 3.34
    One parameter family 3.33, 3.37
    Two parameter family 3.41
Equivalent matrices 1.6
Error relation 4.24
    Absolute error 4.25
    Percentage error 4.25
    Relative error 4.25
Euler's differential equations 5.35, 5.36
Euler's theorem 4.6
Evolute 3.23
Extreme values 4.32
    Constrained extreme values 4.34

First integral 5.49
Functions 4.1
    Continuous functions 4.1

Harmonic functions 4.4
Homogeneous functions 4.6
    Polynomials 4.6

Inner product 1.1
Integral 5.2
    Complete solution 5.2
    General solution 5.2
    Particular solution 5.2
    Solution 5.2
Integrating factor 5.8
Inverse of a matrix 1.4
    Invertible matrix 1.4
Involute 3.23
Involutory matrix 1.4

Jacobian 4.52
    Composite functions 4.53

Lagrange's function 4.35
    Auxiliary function 4.35
    Multiplier 4.35
Laplace equation 4.4
Latent roots 1.18
    Vectors 1.18
Legendre's differential equations 5.37
Leibnitz's rule 4.61, 4.62
Limit 4.1
Line 2.39
    Coplanar lines 2.42
    Equation 2.39
    Symmetric form 2.39
Linear differential equations 5.3
    Nonlinear Dff.Eqns 5.3
Linear homogeneous Diff.Eqns 5.3
Linear systems of equations 1.9
Linear transformations 1.48
Linearly dependent vectors 1.9
Linearly independent vectors 1.9

Maclaurin's expansion 4.23
Matrix 1.1
    Identity 1.2
    Idempotent 1.2
    Lower triangular 1.2
    Non-singular 1.2
    Order 1.1

Product 1.3
Scalar 1.2
Scalar multiplication 1.3
Singular 1.2
Skew-symmetric 1.2
Symmetric 1.2
Upper triangular 1.2
Minors 1.2, 1.5
Modal matrix 1.34

Neighbourhood 4.1
Nilpotent matrix 1.4
Normalized model matrix 1.42

Orthogonal matrices 1.4, 1.5
Orthogonal reduction 1.43
Orthogonal spheres 2.66
    Condition 2.67
Orthogonal transformations 1.49

Partial derivative 4.2, 4.5
Particular integral 5.3, 5.6
Particular solution 5.2
Plane 2.20
    Equation 2.20
    Intercept form 2.21
    Normal form 2.21
Polar coordinates 3.4
    Length of arc 3.6
    Radius vector 3.4
    Vectorial angle 3.4
Powers of a matrix 1.36
Principal minors 1.50
Projection of a point 2.3
    Line 2.3

Quadratic form 1.47
    Determinant 1.48
    Index 1.49
    Modulus 1.48
    Non-singular 1.48
    Signature 1.49
    Singular 1.59

Radical plane 2.67
   Line 2.62, 2.78
   Point 2.62
Rank of a matrix 1.5
Rank of a quadratic form 1.48
Reduction of order 5.47
Relative maximum 4.32
   Local maximum 4.32
Relative minimum 4.32
   Local minimum 4.32
Rouche's theorem 1.10
Row vector 1.1
   Row matrix 1.1

Saddle point 4.33
Similar matrices 1.5
Similarity transformation 1.5
Skew lines 2.3
   Equation of S.D line 2.44

Shortest distance 2.43
Spectral matrix 1.34
Sphere 2.62
   Equation 2.62
Stationary value 4.33
   Point 4.33

Tangent plane 2.64
   Equation 2.65
   Length of tangent 2.64
Taylor's theorem 4.22
Total derivative 4.11
   Total differential 4.11
Trace of a matrix 1.1
Transpose of a matrix 4.4

Variation of parameters 5.66
   Method 5.66

Wronskian 5.4

ABOUT THE AUTHORS

Dr. A Chandra Babu is the head of the post graduate department of Mathematics of The American College, Madurai. He has recvied Ph.D in Applied Mathematics from Madurai Kamaraj University. He has studied at the Indian Institute of Science, Bangalore and has worked in the department of Mathematics and Computer Science, at the Beloit College, Wisconsin, USA. He has thirty-five years of experience in teaching Mathematics and Computer Science at both undergraduate and postgraduate levels. His area of specialization is applications of Stochastic Processes in Stock Market Analysis and he has published many papers.

Dr. C R Seshan is the former head of the post graduate department of Mathematics of The American College, Madurai. He has received Ph.D in Operations Research from the Indian Institute of Science, Bangalore. He has thirty-five years of experience in teaching Pure and Applied Mathematics at both undergraduate and postgraduate levels. He is a member of the board of studies and Chairman of the board of examiners of several Universities and Autonomous Colleges. His area of specialization is Optimization Techniques and he has published several papers in national and international journals.